BOTANICAL RESEARCH AND PRACTICES

EMERGING GEMINIVIRAL DISEASES AND THEIR MANAGEMENT

BOTANICAL RESEARCH AND PRACTICES

Additional books in this series can be found on Nova's website under the Series tab.

Additional E-books in this series can be found on Nova's website under the E-books tab.

BOTANICAL RESEARCH AND PRACTICES

EMERGING GEMINIVIRAL DISEASES AND THEIR MANAGEMENT

PRADEEP SHARMA
RAJARSHI K. GAUR
MASATO IKEGAMI
EDITORS

Nova Science Publishers, Inc.
New York

LIBRARY OF CONGRESS CATALOGING-IN-PUBLICATION DATA

Emerging geminiviral diseases and their management / editors: Pradeep Sharma, Rajarshri K. Gaur, Masato Ikegami.
p. cm.
Includes index.
ISBN 978-1-61668-620-8 (hardcover)
1. DNA viruses. 2. Plant viruses--Genetics. 3. Geminiviridae. I. Sharma, Pradeep K., 1953- II. Gaur, Rajarshi Kumar. III. Ikegami, Masato, 1947-
QR394.5.E44 2010
579.2'4--dc22
2010012167

Published by Nova Science Publishers, Inc. ✣ New York

CONTENTS

PREFACE

Plant pathogenic geminiviruses have been proliferating worldwide and have, therefore, attracted considerable scientific interest during the past three decades. A widespread occurrence and vast spreading of geminivirus, resulting in yield losses of economically important food and fibre crops, poses a serious threat to sustainable agriculture. The major concerns about the frequent evolution and recombination of new viruses or strains lead to failure of resistance over time and space. This new book reviews recent developments in the study of geminiviruses and the progress made in this field.

Chapter 1- Phytopathogenic viruses are a major biotic constraint to agricultural productivity and, throughout the warmer parts of the world, many of the most important diseases of crops are caused by arthropod-borne viruses of the family *Geminiviridae*. With the increasing concerns over the use of insecticides as well as the adoption of agricultural practices that favour the build-up of both vector and virus populations including, for example, widespread monoculture and overlapping cropping seasons, as well as the lack of suitable sources of natural resistance (so called host-plant resistance) in many crop species the problems due to these viruses are increasing. The advent of transgensis for plants in the 1980s paved the way to a virtually unlimited source of virus resistance for application in agriculturally important plant species. This chapter will review the efforts aimed at developing transgenic resistance to geminiviruses, the drawbacks of some of these approaches and possible strategies to overcome these problems are also discussed.

Chapter 2- Although not the first virus geminivirus discovered, *Tomato yellow leaf curl virus* (TYLCV) has been one of the most studied. TYLCV impairs tomato cultures worldwide. Within less than 20 years TYLCV has spread from the Middle East to North America, Africa, Europe and Far East Asia. Molecular analyses have shown that the Tomato leaf curl disease is caused by several virus species and many strains related but different from the Middle Eastern TYLCV. This virus is an ideal object to study such topics in molecular virology as diverse as replication, expression, evolution, recombination and interaction with host factors. TYLCV is transmitted by the whitefly vector to plants in a circulative manner. The interactions between the virus, the vector and the target plant are the object of intense scrutiny. TYLCV causes a serious disease of tomato that needs to be avoided by all means. Hence ways to escape inoculation by whiteflies have been implemented, which include nets, chemical treatment to curb down whitefly populations, traps and natural enemies. Breeding for resistance has started already 30 years ago by introgressing resistance traits found in wild tomato species into the susceptible domesticated tomato. Breeding has been backed by genetic studies of resistance, discovery of QTLs and polymorphic DNA markers linked to resistance, and screening for resistance genes. Commercial lines with good tolerance to TYLCV are available; some are effective against several different TYLCV species, others are not. Altogether the study of TYLCV and the results achieved in molecular virology, genetics, plant-virus-insect interactions, IPM and breeding for resistance, can serve as model for similar studies involving other deleterious geminiviruses infecting important crops, both for basic and applied interests.

Chapter 3- Although not the first virus geminivirus discovered, *Tomato yellow leaf curl virus* (TYLCV) has been one of the most studied. TYLCV impairs tomato cultures worldwide. Within less than 20 years TYLCV has spread from the Middle East to North America, Africa, Europe and Far East Asia. Molecular analyses have shown that the Tomato leaf curl disease is caused by several virus species and many strains related but different from the Middle Eastern TYLCV. This virus is an ideal object to study such topics in molecular virology as diverse as replication, expression, evolution, recombination and interaction with host factors. TYLCV is transmitted by the whitefly vector to plants in a circulative manner. The interactions between the virus, the vector and the target plant are the object of intense scrutiny. TYLCV causes a serious disease of tomato that needs to be avoided by all means. Hence ways to escape inoculation by whiteflies have been implemented, which include nets, chemical treatment to curb down whitefly populations, traps and natural enemies. Breeding for resistance has started already 30 years ago by introgressing resistance traits found in wild tomato species into the susceptible domesticated tomato. Breeding has been backed by genetic studies of resistance, discovery of QTLs and polymorphic DNA markers linked to resistance, and screening for resistance genes. Commercial lines with good tolerance to TYLCV are available; some are effective against several different TYLCV species, others are not. Altogether the study of TYLCV and the results achieved in molecular virology, genetics, plant-virus-insect interactions, IPM and breeding for resistance, can serve as model for similar studies involving other deleterious geminiviruses infecting important crops, both for basic and applied interests.

Chapter 4- Chilli crop is susceptible to a number of plant viruses belonging to various groups viz. *Cucumo, Poty, Tospo, Tobamo, Tobra, Tombus* and Geminiviruses etc. Among them the whitefly-transmitted geminiviruses (WTGs) pose a constraint to chilli cultivation and are highly prevalent in tropical climatic conditions due to the *Bemisia tabaci* complex. Apart from the tropical countries, the whiteflies have also gained importance in temperate regions in Europe because of their adaptability in greenhouse. Numerous WTGs have been reported on chilli from India, Pakistan, Sri Lanka, Bangladesh, China, Malaysia, Mexico, U.S.A., Thailand, Jamaica and Barbados. The most prevalent viral disease of chilli is the leaf curl and yellow mosaic.

Chilli plants often show symptoms similar to tomato leaf curl, such as yellowing, leaf curling, reduction in leaf size and stunting. Since chilli and tomato crops overlap in the field, hence it is likely that chilli may become infected with tomato begomoviruses and tomato might be infected by chilli infecting begomoviruses as the severe epidemics caused by *Pepper golden mosaic virus* in Mexico and *Pepper hausteco yellow vein virus* in Western Hemisphere has been reported. Recently infection of *Pepper leaf curl Bangladesh virus* in tomato (PepLCBV) in West Bengal, India has been reported. The association of DNA-β

molecules has also been reported with bipartite begomoviruses for example: *Tomato leaf curl New Delhi virus* (ToLCNDV) causing leaf curl disease in chilli. The round year cultivation of the crops must have resulted to be susceptible for numerous economically important plant viruses and have become the challenge for the plant pathologist. The leaf curl disease is of moderate to high incidence and widely distributed wherever chilli crop is grown. Recent reports indicate that there has been continuous increase in the incidence of chilli leaf curl disease resulting in the severe economic losses. This chapter deals with the detailed work done at Molecular Plant Virology Laboratory, NBRI, Lucknow, India and by various workers from India and abroad on biological, molecular and management studies of geminivirus infecting chilli crops.

Chapter 5- Begomoviruses are one of the most important plant pathogens all over the world, especially in tropical and sub-tropical parts. The viruses from this group infect a range of crops viz. vegetables, cash crops, ornamentals, tropical fruits and even weeds which sometimes can act as latent hosts of the virus. Among these, solanaceous plants are important as food crops and are grown in a large part of the world and also cultivated throughout India. Hot and humid climate in some parts of India is suitable for spread of begomoviruses, especially affecting communities dependent on produce from these crops as sole source of very valuable protein component of diet. Many begomoviruses have been detected and characterized from important solanceous crops like tomato, chili, potato and tobacco. Due to global warming, green house cultivation, appearance and increase in population of whitefly biotype 'B' and its intrusion into new areas, begomovirus diseases are emerging as serious problem in more temperate areas of India. Human activities like introduction of new crops, exchange of germplasm and movement of infected plant materials from one place to another have also contributed significantly in the spread of these diseases. Moreover, begomoviruses are evolving very fast due to mutations, pseudo-recombination and interspecific recombination in mixed infections and new species are emerging frequently. Association of satellite DNAs with many begomoviruses has further increased the severity of the problem. One of the primary steps in managing these diseases is to create a clear picture of diversity and geographical distribution of the causal pathogens. This chapter summarizes symptomatology, virus transmission and molecular characterization of various emerging begomoviruses in the last ten years, in different parts of India.

Chapter 6- Begomoviruses are one of the most important plant pathogens all over the world, especially in tropical and sub-tropical parts. The viruses from this group infect a range of crops viz. vegetables, cash crops, ornamentals, tropical fruits and even weeds which sometimes can act as latent hosts of the virus. Among these, solanaceous plants are important as food crops and are grown in a large part of the world and also cultivated throughout India. Hot and humid climate in some parts of India is suitable for spread of begomoviruses, especially affecting communities dependent on produce from these crops as sole source of very valuable protein component of diet. Many begomoviruses have been detected and characterized from important solanceous crops like tomato, chili, potato and tobacco. Due to global warming, green house cultivation, appearance and increase in population of whitefly biotype 'B' and its intrusion into new areas, begomovirus diseases are emerging as serious problem in more temperate areas of India. Human activities like introduction of new crops, exchange of germplasm and movement of infected plant materials from one place to another have also contributed significantly in the spread of these diseases. Moreover, begomoviruses are evolving very fast due to mutations, pseudo-recombination and interspecific recombination in mixed infections and new species are emerging frequently. Association of satellite DNAs with many begomoviruses has further increased the severity of the problem. One of the primary steps in managing these diseases is to create a clear picture of diversity and geographical distribution of the causal pathogens. This chapter summarizes symptomatology, virus transmission and molecular characterization of various emerging begomoviruses in the last ten years, in different parts of India.

Chapter 7- Whitefly transmitted geminiviruses (*Geminiviridae: Begomovirus*) are economically important pathogens causing serious losses in food crops globally. Chilli pepper- infecting Begomoviruses cause serious problem throughout Indonesia. The disease was first reported in 1999 in West Java and has occurred ever since in all chilli pepper growing area, causing severe to complete loss of chilli pepper crops. Losses have been so extensive in some areas, such as southern part of Sumatra, that chilli pepper was not grown at that time period. Identification of the causal agent was initiated based on biological and molecular characterization. The virus was tentatively called *pepper yellow leaf curl Indonesia virus* and sweetpotato whitefly, *Bemisia tabaci* (Hemiptera: Aleyrodidae), was demonstrated as the only agent responsible for its transmission and dispersal. The epidemics of the disease was associated with the occurrence of a long and severe dry season which effcet the development of whitefly population. Since resistant chilli pepper cultivars are not available, the major control measures have primarily involved numerous applications of insecticides and modifications of cultural practices.

Chapter 8- Begomoviruses are one of the most important plant pathogens all over the world, especially in tropical and sub-tropical parts. The viruses from this group infect a range of crops viz. vegetables, cash crops, ornamentals, tropical fruits and even weeds which sometimes can act as latent hosts of the virus. Among these, solanaceous plants are important as food crops and are grown in a large part of the world and also cultivated throughout India. Hot and humid climate in some parts of India is suitable for spread of begomoviruses, especially affecting communities dependent on produce from these crops as sole source of very valuable protein component of diet. Many begomoviruses have been detected and characterized from important solanceous crops like tomato, chili, potato and tobacco. Due to global warming, green house cultivation, appearance and increase in population of whitefly biotype 'B' and its intrusion into new areas, begomovirus diseases are emerging as serious problem in more temperate areas of India. Human activities like introduction of new crops, exchange of germplasm and movement of infected plant materials from one place to another have also contributed significantly in the spread of these diseases. Moreover, begomoviruses are evolving very fast due to mutations, pseudo-recombination and interspecific recombination in mixed infections and new species are emerging frequently. Association of satellite DNAs with many begomoviruses has further increased the severity of the problem. One of the primary steps in managing these diseases is to create a clear picture of diversity and geographical distribution of the causal pathogens. This chapter summarizes symptomatology, virus transmission and molecular characterization of various emerging begomoviruses in the last ten years, in different parts of India.

Chapter 9- Tomato yellow leaf curl disease (TYLCD) caused by *Tomato yellow leaf curl virus* (TYLCV) is one of the most devastating viral diseases worldwide. TYLCD was first reported in Japan in 1996. Both TYLCV-Israel and TYLCV-Mild strains have become widely distributed and have caused serious reductions in tomato and *Eustoma grandiflorum* crops throughout Japan. *Bemisia tabaci* is both the insect vector of TYLCV and a serious insect pest. An invasive population of the Q biotype of *B. tabaci* was identified in 2004, and has largely displaced the B biotype. This chapter provides an overview of TYLCV and *B. tabaci*, including indigenous begomoviruses and distinct populations of *B. tabaci* in Japan.

Chapter 10- Whitefly-transmitted Gemini viruses of the genus *Begomovirus* are important pathogens of a wide range of crop eco-systems. Geminiviruses are a family of plant viruses that contain a circular single-stranded DNA (ssDNA) and replicate through a rolling-circle mechanism, en-capsidated in geminate particles. The Geminiviridae family includes a large number of viruses. The particles of gemini viruses are quasi-isometric. They are called gemini (twin) viruses, because they are usually found in pairs. Each particle has a diameter of just 15 – 20 nm. This group of viruses consists of approximately 38 species of extremely narrow host ranges. The group is peculiar and noted for its nucleic acid and is grouped as whitefly transmitted (WFT) viruses. Gemini viruses belong to the smallest virus particles able to multiply without a helper virus. They have a circular DNA with a molecular weight of 0.7–0.8x106 (2,500 base pairs). The genome of Gemini viruses consists of two molecules of DNA of almost equal size, but different sequences. The isolated circular DNA alone is not infectious. In infected host cells, the nucleus holds the chief amount of viral DNA. Symptoms vary from mosaic, yellow leaf, curling, rolling of leaves, thickening of veins, reduction of fruit set, yellow mottling, crinkling, severe stunting, reduced yields and at times death of plants. On infection typical mosaic-like leaf patterns of light and dark green occur. The infection spreads often over the whole leaf beginning at the veins. Leaves that had been infected during their development are usually deformed or volute.

Chapter 11- Viruses of the family *Geminiviridae* are responsible for many economically important crop diseases worldwide. The majority of these diseases are caused by geminiviruses that are whitefly transmitted and are placed in the genus *Begomovirus*. Within the last ten years there has been a fundamental shift in our understanding of the nature of begomoviruses. It has become evident that, in contrast to the situation in the New World where all native viruses have genomes consisting of two components, in the Old World the majority of begomoviruses are monopartite and most of these associate with single-stranded DNA satellites. Two classes of such molecules were identified at the turn of the Century. The alphasatellites are satellite-like and have not been extensively investigated. Their precise function in begomovirus infections remains unclear but they are nevertheless of evolutionary interest. The betasatellites are true satellites and are, in many cases, an essential factor for begomoviruses to infect their hosts. These, at first glance, "simple" molecules have proven to be complex with respect to their interactions with both their helper begomoviruses and host plants. This chapter will outline the advances made in studying the alpha- and betasatellites over the 10 years since their discovery. This greater understanding of how they interact with both the host and the helper virus may provide useful information for their control as well as an insight into plant developmental processes.

CONTRIBUTORS

A.A. Zaidi, Plant Virology Lab., Institute of Himalayan Bioresource Technology (CSIR), Palampur, 176061, Himachal Pradesh, India

Bikash Mandal, Plant Virology Unit, Division of Plant Pathology, Indian Agricultural Research Institute, New Delhi-110012, India.

Bong Choon Lee, National Agricultural Research Center for Kyushu Okinawa Region, 2421 Suya, Koshi, Kumamoto 861-1192, Japan, and National Institute of Crop Science, RDA, Milyang, Korea.

Henryk Czosnek The Otto Warburg Minerva Center for Agricultural Biotechnology & the Robert H. Smith Institute for Plant Science and Genetics in Agriculture. The Hebrew University of Jerusalem, Rehovot, Israel.

Hirosato Kodama, Okinawa Prefectural Plant Protection Center, Naha, Okinawa, Japan .

Imran Amin, Agricultural Biotechnology Division, National Institute for Biotechnology and Genetic Engineering, Jhang Road, PO Box 577, Faisalabad, Pakistan.

Junichi Shirakawa, National Agricultural Research Center for Kyushu Okinawa Region, 2421 Suya, Koshi, Kumamoto 861-1192, Japan.

Keisuke Kijima, Okinawa Prefectural Agricultural Research Center, Itoman, Okinawa, Japan.

Kenkichi Kanmiya, Kurume University, Kurume, Fukuoka, Japan.

Kusum Dhakar Department of Science, Mody Institute of Technology and Science, Lakshmangarh, (Sikar) Rajasthan-332311. India.

M.S. Khan, Plant Health Clinic, Biotech Park, Sector-G, Jankipuram, Lucknow-226 021, U.P., India.

Masahiro Taniguchi, Okinawa Prefectural Plant Protection Center, Naha, Okinawa, Japan.

Masato Ikegami, Department of Life Science, Graduate School of Agricultural Science, Tohoku University, 1-1 Tsutsumidori-Amamiyamachi, Aoba-ku, Sendai, Miyagi 981-8555, Japan .

Masatoshi Onuki, National Agricultural Research Center for Kyushu Okinawa Region, 2421 Suya, Koshi, Kumamoto 861-1192, Japan.

Muhammad Ilyas, Agricultural Biotechnology Division, National Institute for Biotechnology and Genetic Engineering, Jhang Road, PO Box 577, Faisalabad, Pakistan.

Muhammad Saeed, Agricultural Biotechnology Division, National Institute for Biotechnology and Genetic Engineering, Jhang Road, PO Box 577, Faisalabad, Pakistan.

Neetu Sharma, CSK, Himachal Pradesh Agricultural University Krishi Vigyan Kendra, Bara, Hamirpur, India.

Noor Aidawati, Department of Plant Pests and Diseases, University of Lambung Mangkurat, Jalan Jendral Ahmad Yani Km 36, Banjarbaru 70714, South Kalimantan, Indonesia.

P. Sharma, Department of Microbiology, Bundelkhand University Jhansi (U.P) 284128, India.

P.K. Seth, Plant Health Clinic, Biotech Park, Sector-G, Jankipuram, Lucknow-226 021, U.P., India.

Pradeep Sharma, Division of Crop Improvement, Directorate of Wheat Research, Post Box No. 158, Agarsain Road, Karnal 132001, India.

Rajarshi K. Gaur, Department of Science, Mody Institute of Technology and Science, Lakshmangarh, (Sikar) Rajasthan-332311, India.

Rob W. Briddon, Agricultural Biotechnology Division, National Institute for Biotechnology and Genetic Engineering, Jhang Road, PO Box 577, Faisalabad, Pakistan.

S.K. Raj, Plant Molecular Virology Lab, National Botanical Research Institute (NBRI), Rana Pratap Marg, Lucknow-226 001, U. P., India.

S.K. Snehi, Plant Molecular Virology Lab, National Botanical Research Institute (NBRI), Rana Pratap Marg, Lucknow-226 001, U. P., India.

Shahid Mansoor, Agricultural Biotechnology Division, National Institute for Biotechnology and Genetic Engineering, Jhang Road, PO Box 577, Faisalabad, Pakistan.

Shigenori Ueda, National Agricultural Research Center for Kyushu Okinawa Region, 2421 Suya, Koshi, Kumamoto 861-1192, Japan.

Shinichiro Okazaki, Oita Prefectural Agricultural, Forestry and Research Center, Usa, Oita, Japan.

Sri Hendrastuti Hidayat, Department of Plant Protection, Bogor Agricultural University, Jalan Kamper Darmaga, Bogor 16680, West Java, Indonesia.

Sri Sulandari, Department of Plant Protection, Gadjah Mada University, Jalan Flora no. 1 Bulaksumur, Yogyakarta 55281, Indonesia.

V. Bajpai, Department of Science, Mody Institute of Technology and Science, Lakshmangarh, (Sikar) Rajasthan-332311, India.

Vipin Hallan, Plant Virology Lab., Institute of Himalayan Bioresource Technology (CSIR), Palampur, 176061, Himachal Pradesh, India.

Yogesh Kumar, Plant Virology Lab., Institute of Himalayan Bioresource Technology (CSIR), Palampur, 176061, Himachal Pradesh, India.

In: Emerging Geminiviral Diseases and their Management ISBN 978-1-61668-620-8
Editors: P. Sharma, R. K. Gaur et al. pp. 1-35

Chapter 1

CHALLENGES FOR TRANSGENIC RESISTANCE TO CONTROL GEMINIVIRAL DISEASES

Muhammad Ilyas, Imran Amin, Shahid Mansoor, Rob W. Briddon and Muhammad Saeed [*]
National Institute for Biotechnology and Genetic Engineering, Faisalabad, Pakistan.

ABSTRACT

Phytopathogenic viruses are a major biotic constraint to agricultural productivity and, throughout the warmer parts of the world, many of the most important diseases of crops are caused by arthropod-borne viruses of the family *Geminiviridae*. With the increasing concerns over the use of insecticides as well as the adoption of agricultural practices that favour the build-up of both vector and virus populations including, for example, widespread monoculture and overlapping cropping seasons, as well as the lack of suitable sources of natural resistance (so called host-plant resistance) in many crop species the problems due to these viruses are increasing. The advent of transgensis for plants in the 1980s paved the way to a virtually unlimited source of virus resistance for application in agriculturally important plant species. This chapter will review the efforts aimed at developing transgenic resistance to geminiviruses, the drawbacks of some of these approaches and possible strategies to overcome these problems are also discussed.

INTRODUCTION

There are approximately 450 species of plant-pathogenic viruses, which cause a range of diseases in susceptible hosts (Soosaar et al., 2005). Amongst these the geminiviruses are an emerging group of plant pathogens that are responsible for increasing problems of both staple food and cash crops in developing countries (Seal et al., 2006b, Varma & Malathi, 2003). Until the 1990s, problems with geminiviruses were relatively localised, but recently they have become of worldwide concern, due the spread of both the viruses and their vectors as a result

[*] Correspondence concerning this article should be addressed to: Muhammad Saeed, E-Mail: msaeed@nibge.org; saeed_hafeez@yahoo.com.

of the global trade in agricultural products and due to increasingly intensive agricultural practices.

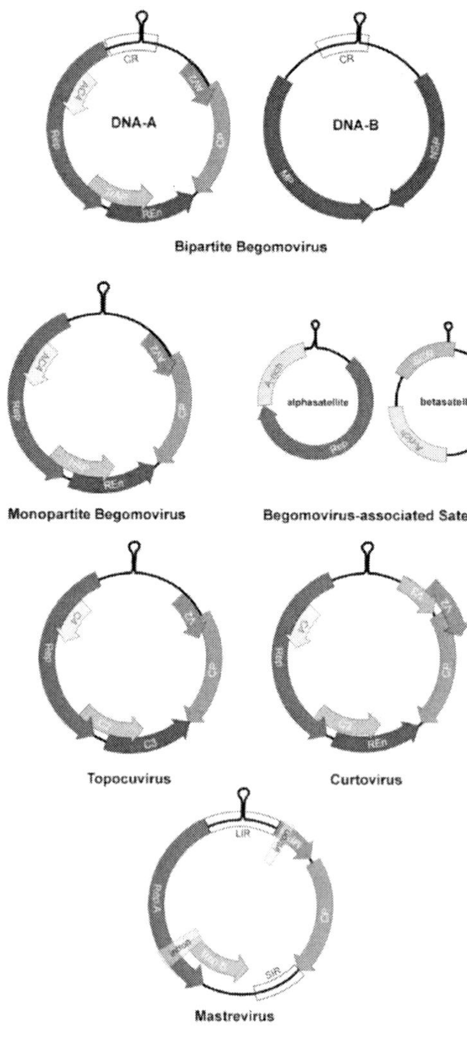

Figure 1. Typical genome organization of viruses in four genera of the *Geminiviridae* and structures of the satellites (betasatellite and alphasatellite) associated with begomovirus. The position and orientation of virion-sense (V) and complementary-sense (C) genes and the predicted hairpin structure containing the nonanucleotide sequence (TAATATTAC) are shown. Mastreviruses contains two noncoding regions referred to as the large and small intergenic regions (LIR and SIR, respectively) and two introns (boxes). The DNA-A and DNA-B components of bipartite begomoviruses contain an almost identical sequence of approximately 200 nucleotides, referred to as the common region (CR), which is usually located within the intergenic region. The begomoviruses originating from the New World, all of which are bipartite, do not encode an AV2 gene. The alpha- and betasatellites contain a region of sequence rich in adenine (A-rich), and the betasatellites contain a region of highly conserved sequence known as the satellite conserved region (SCR). The gene products (where their function is known) are indicated as coat protein (CP), replication-associated protein (Rep), transcriptional activator protein (TrAP), replication enhancer protein (REn), movement protein (MP) and nuclear shuttle protein (NSP). The only product encoded by betasatellites is known as βC1 whereas the alphasatellites encode a Rep.

The viruses of the family *Geminiviridae* are single-stranded (ss) DNA viruses that are at present classified into four genera, *Mastrevirus*, *Topocuvirus*, *Curtovirus* and *Begomovirus,* based on their genome organization (Figure 1), host ranges, and insect vectors (Stanley et al., 2005). The genomes (or genomic components) of geminiviruses consist of ssDNA circles that replicate in the nuclei of infected cells as double-stranded (ds), replicative-form intermediates by a rolling-circle mechanism. The encapsidated form of the virus DNA is referred to as the virion-strand. Transcription occurs from the replicative form with genes diverging from a non-coding region that harbours cis-acting control elements and a predicted hairpin structure containing, within the loop, the sequence TAATATTAC (known as the nonanucleotide sequence). Virion-strand rolling-circle DNA replication initiates at the 3' adenine residue in the nonanucleotide sequence and this is, by convention, designated as position 1 for numbering the nucleotide positions on the sequence of the genome (genomic component). A further convention is to name open reading frames (ORFs; genes) as either encoded on the virion-sense strand (V) or complementary-sense strand (C) and, for bipartite begomoviruses, to prefix them with A or B when they are encoded on the DNA-A or DNA-B component, respectively.

The virion-sense strand of all geminiviruses encodes the coat protein [CP; ORF (A)V1] that is the only structural protein of these viruses, forming the distinctive geminate particles, interacts with the insect vectors to facilitate transmission and is involved in movement in host plants (Hanley-Bowdoin et al., 1999) (Figure 1). For the New World begomoviruses this is the only virion-sense encoded gene. However, the Old World begomoviruses encode an additional gene, (A)V2, the product of which is involved in movement in host plants. The V2 of the mastreviruses encodes the movement protein (MP) that mediates cell-to-cell movement in host plants. In addition to the CP, the curtoviruses encode two additional genes. The V2 encodes a protein involved in regulating ssDNA/dsDNA levels, whereas the V3 is involved in virus movement in hosts. The genes encoded by the topocuvirus *Tomato pseudo curly top virus* have not been investigated. However, sequence similarities suggest that they mirror the functions of genes encoded by begomoviruses.

The dicot-infecting geminiviruses (with the exception of the dicot-infecting mastreviruses) encode four genes on the complementary-sense strand (Figure 1). The replication-associated protein [Rep, encoded by ORF (A)C1] is a rolling-circle replication initiator protein with DNA nicking and joining activity. It is a sequence specific DNA binding protein that nicks the virion-strand ahead of the 3' adenine of the nonanucleotide sequence to initiate rolling circle replication and resolves new ssDNA circles by cutting and ligating. The protein recognizes small repeated sequences, known as “iterons”, just upstream of the hairpin structure which are, for the most part, species specific. The specificity of a particular Rep for its cognate iterons is such that, in most instances, the Rep from one virus species will not recognize the iterons of a second species to initiate replication. The second complementary-sense gene [ORF (A)C2] encodes a protein that may act as a transcription factor that up-regulate host-encoded genes and that may also be suppressor of RNA interference (RNAi – a host defense mechanism that recognizes and degrades foreign and aberrant RNAs (reviewed by Baulcombe, 2004). For many begomoviruses the (A)C2 protein is also involved in up-regulating the late (virion-sense) genes and is known as the transcriptional activator protein (TrAP). The replication enhancer protein [REn, ORF (A)C3] interacts with host factors, to create a cellular environment suitable for viral replication, and with Rep, to up-regulate its activity. The (A)C4 protein (ORF C4) is for many viruses a pathogenicity determinant and in some cases a suppressor of RNAi. The complementary-sense coding strategy of the mastreviruses is unusual amongst geminiviruses. The Rep is expressed from a spliced transcript of the RepA and RepB (ORFs C1 and C2, respectively) genes. The RepA protein, translated from a non spliced transcript, interacts with host-encoded factors to create an environment suitable for virus replication and also is involved in up-regulating the expression of late, virion-sense genes. The DNA B component of bipartite begomoviruses encodes two additional genes; in the virion-sense the nuclear shuttle protein (NSP, ORF BV1) and in the complementary-sense the MP (ORF BC1). These act in a concerted manner to mediate cell-to-cell movement of the virus in plants (Stanley et al., 2005).

Geminiviruses cause problems to agriculture throughout the warmer parts of the world although a few, such as *Wheat dwarf virus* in northern Europe and China, also occur in more temperate regions. There are three prominent examples of geminiviruses causing problems to agricultural crops that stand out for entirely different reasons. Cassava is a subsistence crop across much of sub-Saharan Africa and is severely affected by bipartite begomoviruses that cause cassava mosaic disease (CMD) (Patil & Fauquet, 2009). During the 1990s a particularly severe epidemic of CMD spread throughout Uganda leading to food shortages and famine-related deaths (Thresh & Otim-Nape, 1994). Recent reports have indicated that the CMD pandemic has since affected at least nine countries in East and Central Africa, covering an area of 2.6 million square kilometres and causing an estimated annual economic loss of US$1.9–2.7 billion (Patil & Fauquet, 2009). Tomato leaf curl disease affects tomato production throughout all tropical and sub-tropical areas and is caused by a number of distinct begomovirus species. However, one particular species, the monopartite begomovirus *Tomato yellow leaf curl virus* (TYLCV) was first identified in the Middle East but has since spread throughout the Mediterranean region, the Americas, Australia, eastern China and Japan (Cohen & Lapidot, 2007). This rapid spread has been attributed to the global trade in agricultural products. In contrast to the previous two examples, cotton leaf curl disease (CLCuD) affects a fibre rather than a food crop. Nevertheless the disease severely affected the economy of Pakistan, which depends heavily on cotton exports, during the 1990s (Briddon & Markham, 2001). Losses were estimated at US$5 billion in the 5 years between 1992 and 1997. The disease is caused by a complex of monopartite begomoviruses that associate with a specific betasatellite. Although losses due to CLCuD were reduced in the late 1990s by the introduction of cotton varieties with conventional resistance to the viruses, in the early 2000s a resistance breaking strain appeared and the country is again suffering heavy losses (Mansoor et al., 2003).

Strategies to control viruses of agricultural importance, including geminiviruses, are usually preventive by eradicating the vectors using chemical insecticidess (Seal et al., 2006a). These strategies are effective in reducing the infectious potential of vectors (by lowering the number of vectors per plant) but often the virus has already been transmitted to the plant before the insect vector is killed. Secondly because of their acute toxicity and adverse effects on the environment, concerns have been raised on the use of agrochemicals to control virus vectors. Due to the severity of viral epidemics, the difficulties of implementing efficient management practices, and the increasing demand for sustainable and environment-friendly agricultural practices, there is a great need to develop varieties that are resistant to pathogenic viruses. These factors necessitate that natural resistance be augmented with genetically-engineered resistance that should provide more durable resistance to geminiviruses.

NATURAL RESISTANCE TO VIRUSES

At this time the best strategy to reduce crop losses due to viruses is either through the physical separation of the host and the pathogen or through the deployment of genetic resistance (so called host plant resistance) to viruses so as to prevent or limit the extent of the infection. In practice, the former is achieved through the use of virus-free seed or planting material or by employing physical barriers or pesticides to deter/control the vectors of viruses. However, at present, the most effective and sustainable approach to controlling diseases caused by viruses is through the deployment of host plant resistance against the viruses or their vectors. Disease resistance genes that confer resistance to plant pathogens been isolated from a number of plant species. Analysis of the resistance genes has revealed that they contain some common motifs despite having diverse origins and pathogen specificities (Lehmann, 2002).

For geminiviruses only limited sources of resistance are available in commercial cultivars of affected crops. For example, tomato varieties harbouring the partially dominant Ty-1 resistance gene, derived from *Solanum. chilense* (Zamir et al., 1994), are commonly grown throughout the Mediterranean where TYLCV and various related tomato-infecting begomovirus species are endemic. This resistance provides a strong limitation of virus accumulation and a reduction in symptom severity but does not provide immunity (Michelson et al., 1997). Recently García-Andrés et al. (2009) presented field and experimental data which indicates that the widespread use of Ty-1 resistance has contributed to the emergence of TYLCV variants that fit better in the resistant genotypes.

Farnham and Baulcombe (2006) have provided an intriguing possibility with respect to developing resistance to viruses. They illustrated a possible way of increasing the range of viruses recognised by a resistance gene. A known (cloned) resistance gene (they used the Rx gene isolated from potato which provided resistance against some strains of the potexvirus *Potato virus X* [PVX]; (Bendahmane et al., 1999) was randomly mutated *in vitro* and then the mutations were screened for altered virus specificity. They found that this system could generate Rx mutants which provided resistance to additional strains of PVX and the unrelated carlavirus *Poplar mosaic virus*. There is thus the possibility of using know resistance genes to extend their activities. With respect to geminiviruses, no natural host plant resistance genes have yet been cloned.

NON-CONVENTIONAL RESISTANCE TO PLANT-INFECTING VIRUSES

The advent of transgenesis offers many ways of obtaining virus resistant plants. It provides the ability to produce crop varieties inherently resistant to pathogen infection. The tools of genetic transformation have made it possible to introduce selected genes into plants to confer novel phenotypes. Genetic transformation is an attractive alternative for the genetic improvement of crops because it overcomes many of the shortcomings of conventional approaches by allowing the insertion of specific virus resistance traits directly into desirable, widely cultivated (so called "elite") varieties. The strategies which have been investigated for the usefulness in providing transgenic resistance against phytopathogenic viruses, including geminiviruses, can be grouped under the terms-

(i) Pathogen derived resistance (PDR) - in which a nucleic acid sequence, which may or may not encode a functional protein, derived from the pathogen is used as the source of resistance.
(ii) Non pathogen derived resistance (NPDR) - in which the source of sequence for resistance is other than the pathogen.

The concept of PDR is based upon the phenomenon of cross protection (also known as "mild strain interference") in which plants infected with less virulent strains often develop resistance against highly virulent strains of that virus (reviewed by Lecoq, 1998). Although in many cases there appears to be no cross-protection between geminiviruses, the interaction in most cases being either benign or synergistic (Alves-Júnior et al. 2009, Vanitharani et al., 2004), there is some evidence to suggest that the phenomenon might be useful for protection of cassava against begomoviruses that cause cassava mosaic disease in Africa (Owor et al., 2004). Sanford and Johnson (1985) have defined PDR as a strategy in which entire genes, or sequences of the pathogen's genome (either structural or non-structural), are used to transform host plants so that protection against the pathogen from which the sequences were derived, or a range of closely related pathogens, is achieved. The advent of plant transformation, and the concept of PDR in particular, opened new avenues in the mid 80s for the development of virus-resistant crop varieties. PDR can be protein-mediated or nucleic acid-mediated. Protein mediated resistance generally offers low but broader spectrum resistance whereas nucleic acid- mediated resistance often protects against very high levels of inoculum but is highly specific.

The majority of PDR strategies involve silencing of viral genes. This occurs either by repression of transcription (referred to as "transcriptional gene silencing" [TGS]) which is usually induced by DNA methylation or by mRNA degradation (referred to as "post transcriptional gene silencing" [PTGS]. TGS results from histone modifications that create an environment of heterochromatin around a gene that makes it inaccessible to the transcriptional machinery. TGS is achieved by expression of double-stranded RNAs derived from promoter rather than coding regions of genes. These phenomena are thought to be evolutionarily ancient and occur widely in eukaryotes. Recently they have been grouped under the term RNA interference (RNAi). RNAi was first reported to be induced by the introduction of double-stranded (ds)RNA in *Caenorhabditis elegans* cells in 1998. It is involved in the inhibition of viruses and silencing of transposable elements in plants, insects, fungi and nematodes and is mediated by small (21-24 nucleotides) interfering (si)RNAs. The dsRNA precursors for siRNA may arise from dsRNA viral replication intermediates, bidirectional transcription of a locus, transcription of extended inverted duplications or the activity of an RNA dependent RNA polymerase (RdRp) acting on a suitable RNA template. RNaseIII-type enzymes, termed Drosha and Dicer (DCR), in animals, and DCR-LIKE (DCL), in plants, catalyze processing of large dsRNA precursors into 21- to 24-nt duplexes (Bartel, 2004, Baulcombe, 2004). Distinct pathways are involved in the synthesis of siRNAs. Heterochromatin-associated siRNAs (predominantly 24-nt) form through the activities of RNA-dependent RNA polymerase (RDR) 2, RNA polymerase IV, and DCL3 and require ARGONAUTE (AGO) 4 for activity to direct methylation of DNA and histone H3. Formation of post-transcriptionally active siRNAs from exogenous (invading pathogens and transgenes) sources may involve RDR1 or RDR6 and, for some viruses, DCL2 (Baulcombe, 2004).

Double-stranded siRNAs are bound by AGO in a multiprotein complex that is referred to as the RNA-induced silencing complex (RISC). The strand in which the 5′ end is less thermodynamically stable in the duplex becomes preferentially incorporated into RISC. The incorporated siRNA provides RISC with the specificity to recognise target mRNAs which then cleaves the target near the centre of the siRNA. Several lines of evidence suggest that RNA silencing is a general, adaptive antiviral defence mechanism in plants (Baulcombe, 2004, Voinnet, 2005).

Several examples of PDR designed to be protein-mediated for distinct plant-infecting RNA viruses in a range of plant species were successful even though the underlying mechanism of resistance remained undefined. Lindbo et al (1993) dissected the resistance mechanism in transgenic tobacco lines produced to express the CP of *Tobacco etch virus*. They found that the mRNA of the transgene, rather than the expressed protein, was involved in the resistance, indicative of RNA silencing rather than a protein-mediated mechanism. Similar work also demonstrated that plant viruses can induce RNA silencing in their hosts. Long dsRNAs are generated, either as replication or transcription intermediates, when viruses infect plants, leading to the establishment of RNA silencing (Ratcliff et al., 1999). This is a major defence mechanism used by plants to defend against invading viruses.

Another important development in RNAi has been the identification of micro (mi)RNAs. These were first identified by Lee et al. (1993). miRNAs are small non-coding RNAs that are expressed in most cases by RNA polymerase II as primary miRNAs which are >1 kb long. In the nucleus, genes encoding primary miRNA (pri-miRNA) are much longer than the mature form. pri-miRNA contain a 5′ 7-methyl guanosine cap and 3′ poly adenosine tail similar to mRNAs. However miRNA embedded in repetitive elements such *Alus* can be transcribed by RNA polymerase III. The hairpin structure in pri-miRNA is normally recognized and cleaved by the nuclear RNase III enzyme Drosha and its cofactor DiGeorge syndrome critical region 8 (DGCR8) to short 60-70 nucleotides stem-looped pre-miRNA. However in some cases pri-miRNAs bypass Drosa processing. pre-miRNA are exported to the cytoplasm via the nuclear transport receptor exportin-5 and the cofactor RanGTP (Kim, 2005) In the cytoplasm they are processed by the endonuclease RNase III enzyme Dicer into ~22 nucleotides long double-stranded miRNA. In common with siRNAs the strand with less stable 5′ hydrogen bonding is then loaded into the RISC and targets this to mRNAs with partially complementary sequences (usually located in the 3′ untranslated regions of gene transcripts), triggering mRNA cleavage or inhibition of translation. It has recently been suggested that under stress conditions, the mode of miRNA regulation can change and, by association with other proteins, can turn into an activator of gene expression. The exact criteria for target recognition are currently not clear. However, pairing of the 5′ 7–8 nucleotides of the miRNA (seed region) to multiple sites in the 3′ untranslated region of a target mRNA is in many cases sufficient to trigger translational inhibition (Leung & Sharp, 2007).

Experimentally PTGS can be triggered in transgenic plants by expression of dsRNA homologous to viral sequences. These can be generated in two ways.

(a) By transcription of a transgene with sequences in sense or antisense orientation leading to silencing of the transgene and homologous viral gene.

(b) By transcription of a hairpin structure in which virus derived sequences are cloned in sense and antisense orientation, thus leading to a double-stranded, hairpin RNA.

Separation of the sense and antisense sequences by an intron has been shown to improve the efficiency of silencing (Smith et al., 2000).

The down-regulation of gene expression using anti-sense RNA (asRNA) sequences was used before the discovery of the mechanisms of gene silencing. An asRNA molecule is complementary to a particular mRNA and will base pair with a homologous mRNA forming an RNA duplex which is degraded into siRNA leading to silencing. Such asRNA strategies have been successfully used since 1991 against geminiviruses. More recently the expression of hairpin RNAs has been the norm. The antisense approach may be a better option than the expression of hairpin dsRNA, which may be more prone to TGS, since the short RNAs from a hairpin transcript are generated constitutively (whereas, at least in theory, siRNAs from a sense or asRNA are generated only in the presence of the target) increasing the chances of transgene methylation leading to TGS (Aufsatz et al., 2002, Zilberman et al., 2004). TGS may become more severe with increasing numbers of cell cycles (Fojtova et al., 2003) and lead to the inactivation of the transgene and subsequent loss of protection against virus infection.

The advantage of PTGS-based resistance strategies is that they potentially have fewer deleterious effects on plant development than those in which a protein is expressed from the transgene. Similarly, the fact that these approaches do not require expression of proteins could make them safer for use in plants which are used as food or fodder crops. Since the silencing based approach is homology dependent it is unlikely to be able to provide a particularly broad spectrum resistance. A virus would need only relatively small variations in the sequence of the targeted region to avoid the resistance, as has been demonstrated also for miRNA-mediated virus resistance (Duan et al., 2008). In addition to this the majority of viruses, including geminiviruses, express proteins which are suppressors of silencing and these can potentially undermine RNAi based resistance (Levy et al., 2008, Sharma & Ikegami, 2008).

Transgenic Resistance Against Geminiviruses

There are relatively few reports investigating the use of transgenic resistance to geminiviruses in their natural hosts. The majority of studies have been carried out in experimental plants such as *Nicotiana benthamiana* and *N. tabacum*, probably because many of the plant species affected by these viruses are recalcitrant to transformation. Nevertheless, some significant advances have been made even for these recalcitrant species over the last few years. The reports published to date are listed in Table 1, for PDR studies, and Table 2, for non-PDR studies. Additionally the strategies investigated in efforts to obtain transgenic resistance against geminiviruses are summarised diagrammatically in Figure 2.

Pathogen Derived Resistance to Geminiviruses

Protein Mediated Resistance

The first report of transgenic resistance against a plant virus involved the expression of the CP of *Tobacco mosaic virus* (Abel et al., 1986) and this strategy was subsequently also tried for geminiviruses. Tomato plants expressing the CP of TYLCV exhibited delayed symptoms development and later on showed recovery of symptoms which was dependent on the expression level of the CP (Kunik et al., 1994).

Table 1. List of published studies reporting transgenic resistance to geminiviruses that have used the pathogen derived approach

Protein mediated resistance			
Transgene	**Plant species**	**Targeted virus***	**Reference**
MP of TGMV	*N. benthamiana*	ACMV	(von Arnim & Stanley, 1992)
CP of TYLCV	*S. lycopersicum*	TYLCV	(Kunik et al., 1994)
Full-length or truncated Rep of ACMV	*N. tabacum* protoplasts and *N. benthamiana*	ACMV	(Hong & Stanley, 1996)
Truncated Rep of TYLCSV	*N. benthamiana*	TYLCSV	(Noris et al., 1996)
Truncated Rep of TYLCSV	*S. lycopersicum*	TYLCSV	(Brunetti et al., 1997)
Truncated MP of ToMoV	*N. tabacum*	ToMoV, CaLCV	(Duan et al., 1997)
Mutated Rep of ACMV	*N. benthamiana*	ACMV	(Sangaré et al., 1999)
Mutated Rep of BGMV	*N. tabacum* suspension cells	BGMV	(Hanson & Maxwell, 1999)
Wild type or mutated NSP or MP of BDMV	*S. lycopersicum*	ToMoV	(Hou et al., 2000)
Truncated Rep of TYLCSV	*N. benthamiana*	TYLCSV	(Brunetti et al., 2001)
Truncated Rep of ToLCNDV	*N. benthamiana*	ToLCNDV, PHYVV, PYMV, ACMV	(Chatterji et al., 2001)
Rep of ToMoV	*S. lycopersicum*	ToMoV	(Polston & Hiebert, 2001)
Truncated Rep of TYLCV	*N. benthamiana*	TYLCSV, TYLCV	(Lucioli et al., 2003)
Truncated Rep of TYLCV	*S. lycopersicum*	TYLCV	(Antignus et al., 2004)
Mutated Rep of MSV	*Zea mays*	MSV	(Shepherd et al., 2005)
Full-length Rep and truncated Rep of MYMV	*N. tabacum*	MYMV	(Shivaprasad et al., 2006)
Truncated and mutate Rep of MSV	*Digitaria sanguinalis* and *Zea mays*	MSV	(Shepherd et al., 2007b)
RNA interference			
Transgene	**Plant**	**Targeted virus**	**Reference**
Antisense Rep of TGMV	*N. tabacum*	TGMV	(Day et al., 1991)
Antisense Rep of TGMV	*N. tabacum*	BCTV	(Bejarano & Lichtenstein, 1994)
Antisense Rep of TYLCSV	*N. benthamiana*	TYLCSV	(Bendahmane & Gronenborn, 1997)
Antisense Rep-REn-TrAP and MP of BGMV	*Phaseolus vulgaris*	BGMV	(Aragao et al., 1998)
Truncated CP of ToMoV	*N. tabacum*	ToMoV	(Sinisterra et al., 1999)
Antisense Rep, REn and TrAP of CLCuKV	*N. tabacum*	CLCuVs	(Asad et al., 2003)
Bidirectional promoter of the DNA A of MYMV	Blackgram (*Vigna mungo*)	MYMV	(Pooggin et al., 2003)
Chemically synthesized siRNA to Rep of ACMV	*N. tabacum* protoplasts	ACMV	(Vanitharani et al., 2003)
Rep of ACMV	*Manihot esculenta*	ACMV, EACMCV, SLCMV	(Chellappan et al., 2004)
Antisense Rep of ToLCNDV	*S. lycopersicum*	ToLCNDV	(Praveen et al., 2005)
Antisense Rep, TrAP and REn of ACMV	*Manihot esculenta*	ACMV	(Zhang et al., 2005)

Table 1 (Continued)

RNA interference			
Transgene	**Plant**	**Targeted virus**	**Reference**
Hairpin RNA derived from conserved regions in Rep of TYLCV and TYLCSV	*S. lycopersicum* and *N. benthamiana*	TYLCSV, TYLCV	(Abhary et al., 2006)
Hairpin RNA derived from TrAP of ToLCV	*S. lycopersicum*	ToLCV	(Bian et al., 2006)
Hairpin RNA derived from Rep of TYLCV	*S. lycopersicum*	TYLCV	(Fuentes et al., 2006)
Hairpin RNA derived from Rep of BGMV	*Phaseolus vulgaris*	BGMV	(Bonfim et al., 2007)
Antisense AV2 of ToLCNDV	*N. tabacum*	ToLCNDV	(Mubin et al., 2007)
Hairpin RNA derived from Rep and AC4 of ToLCNDV	*S. lycopersicum*	ToLCNDV	(Ramesh et al., 2007)
Bidirectional promoter of DNA A of ACMV	*Manihot esculenta*	ACMV	(Vanderschuren et al., 2007)
Hairpin RNA derived from CP of TYLCV	*S. lycopersicum*	TYLCV	(Zrachya et al., 2007)
Hairpin RNA derived from CP and Rep of TYLCSV	*N. benthamiana*	TYLCSV	(Gharsallah et al., 2008)
Hairpin RNA derived from Rep of TYLCSV	*S. lycopersicum*	TYLCSV	(Tamarzizt et al., 2009)
Hairpin RNA derived from Rep of ACMV	*Manihot esculenta*	ACMV	(Vanderschuren et al., 2009)
DNA interference			
Transgene	**Plant**	**Targeted virus**	**Reference**
DI DNA of ACMV DNA-B	*N. benthamiana*	ACMV	(Stanley et al., 1990)
DI DNA of BCTV	*N. benthamiana*	BCTV	(Frischmuth & Stanley, 1994, Stenger, 1994)

*The virus acronyms used are *African cassava mosaic virus* (ACMV), *Bean golden mosaic virus* (BGMV), *Beet curly top virus* (BCTV), *Cabbage leaf curl virus* (CaLCV), *Cotton leaf curl virus* (CLCuV), *East African cassava mosaic Cameroon virus* (EACMCV), *Maize streak virus* (MSV), *Mungbean yellow mosaic virus* (MYMV), *Pepper huasteco yellow vein virus* (PHYVV), *Potato yellow mosaic virus* (PYMV), *Sri Lankan cassava mosaic virus* (SLCMV), *Tomato golden mosaic virus* (TGMV), *Tomato leaf curl virus* (ToLCV) *Tomato leaf curl New Delhi virus* (ToLCNDV), *Tomato yellow leaf curl Sardinia virus* (TYLCSV), *Tomato yellow leaf curl virus* (TYLCV), *Tomato mottle virus* (ToMoV).

Although the precise mechanism remains unclear, it may be that inappropriate timing of expression, as well as overexpression, of the CP may sequester viral genomic ssDNA preventing it returning to the replication cycle, by acting as a template for complementary-strand synthesis, and/or prevent other viral movement associated proteins from gaining access to the ssDNA to allow its movement into the next cell.

TYLCV is a monopartite begomovirus and for such viruses the CP is essential for infectivity. For bipartite begomoviruses the CP is not essential for infectivity, although viruses lacking the CP have longer latent periods between inoculation and the appearance of symptoms, consistent with the CP having an important, if not essential, role in virus movement.

Table 2. List of published studies reporting transgenic resistance to geminiviruses that have used the non-pathogen derived approach

Non-pathogen derived resistance			
Strategy	**Plant**	**Targeted virus***	**Reference**
Cell death induced by dianthin from *Dianthus caryphyllus*	*N. benthamiana*	ACMV	(Hong et al., 1996)
Cell death induced by barnase from *Bacillus amyloliquefaciens*	*Manihot esculenta*	ACMV	(Zhang et al., 2003)
Artificial zinc-finger protein	*Arabidopsis thaliana*	BSCTV	(Sera, 2005)
Peptide aptamers inhibiting Rep	*N. tabacum* protoplasts	TGMV	(Lopez-Ochoa et al., 2006)
Ribozyme-mediated resistance	Yeast	MYMIV	(Ushasri et al., 2007)
Trapping of viral DNA by GroEL	*S. lycopersicum*	TYLCV	(Akad et al., 2007) (Edelbaum et al., 2009)
Recombinant-antibody inhibiting Rep	*N. benthamiana*	TYLCV	(Safarnejad et al., 2009)

*The virus acronyms used are *African cassava mosaic virus* (ACMV), *Tomato golden mosaic virus* (TGMV), *Beet severe curly top virus* (BSCTV), *Tomato yellow leaf curl virus* (TYLCV) and *Mungbean yellow mosaic India virus* (MYMIV).

For this reason it has been assumed that a CP-mediated resistance strategy against bipartite begomoviruses would not be successful. Consistent with this assumption transgenic beans expressing the CP of *Bean golden mosaic virus* and *N. benthamiana* plants expressing the *African cassava mosaic virus* (ACMV) CP did not show resistance (Azzam et al., 1996, Frischmuth & Stanley, 1998). The study of Frischmuth and Stanley (1998) also highlighted one of the dangers of expressing full-length, non-mutated genes in plants; the possibility of recombination between the transgene and the episomal virus. Although some resistance to *Tomato mottle virus* was (ToMoV) achieved by transgenic expression of the CP modified by the deletion of 30 nucleotides in the 5' end in tobacco. Sinisterra et al. (1999) were unable to show expression of the CP. Resistance in this case correlated with CP gene transcript levels, suggestive of RNA mediated effect (RNA silencing) rather than a protein-mediated effect.

One possibility that has not yet been investigated is that transgenic expression of the CP could interfere with insect transmission. For geminiviruses the CP determines vector specificity and is presumed to interact with specific receptors in the digestive tract of vector (insects) to mediate acquisition of virions (Briddon et al., 1990, Rosell et al., 1999). Thus transgenic expression of CP could potentially be used to competitively block virus receptors in insects, thereby reducing the rate of transmission.

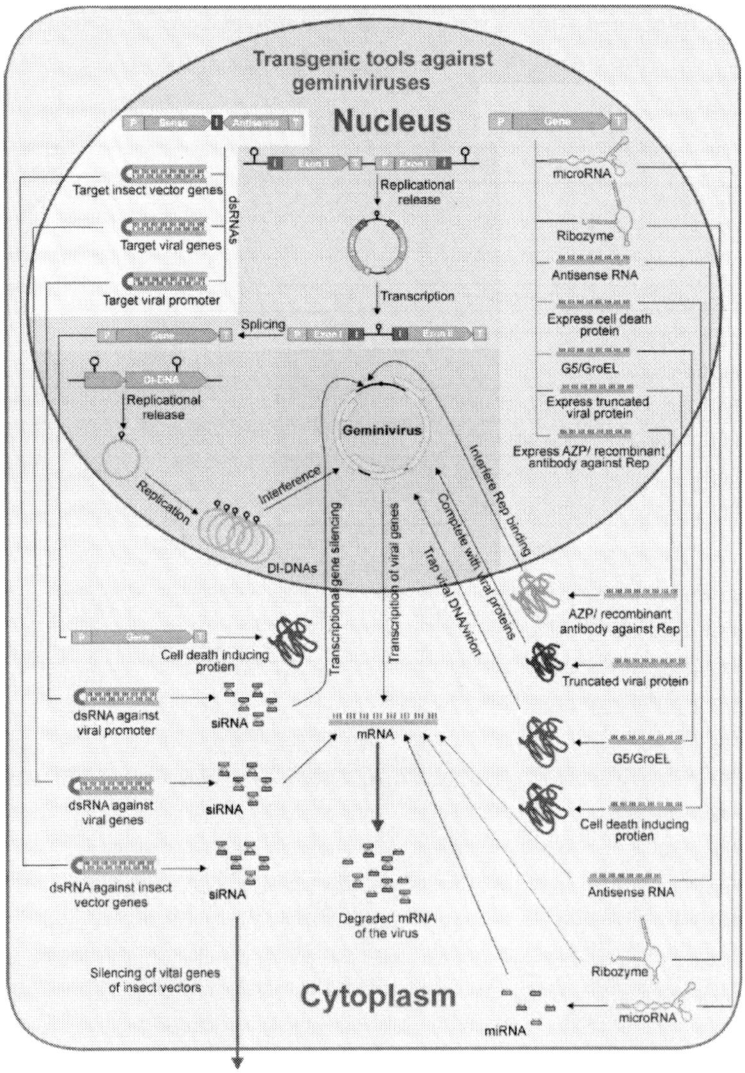

Figure 2. Ttransgenic approaches for the production of plant varieties resistance to geminiviruses. microRNA, ribozyme and antisense RNA expressed from transgenes are produced in the nucleus and are transported to the cytoplasm where they target the mRNAs from the virus. mRNAs of cell death inducing protein, G5, GroEL, truncated viral proteins, artificial zinc finger (AZPs) and recombinant antibodies expressed from their transgenes are also transported to the cytoplasm where they perform their respective functions. In "InPAct" system a circular DNA molecule with gene of interest, promoter (P), terminator (T) and introns (I) is released by replication from the designed DNA structure integrated into plant genome. Gene of interest with introns and a hairpin structure in between is expressed from the circular DNA molecule. Hairpin structure is removed along with introns during splicing and this processed mRNA is transported to the cytoplasm where it is translated into the desired protein which performs its function. Hairpin RNAs produced from expression cassettes containing virus gene sequence in sense and antisense orientation separated by an intron are also transported to the cytoplasm where they are processed to siRNAs which enter the gene silencing pathway. Circular defective interfering DNA (DI-DNA) molecules of geminiviruses are produced from partial tandem repeats which are integrated into the plant genome, either by replicational release or recombination. These DI-DNAs replicate and compete with invading geminiviruses.

Consistent with this hypothesis it has been shown that purified *Maize streak virus* (MSV) virions treated with formaldehyde (making them non-infectious to plants) significantly reduced transmission by the vector *Cicadulina mbila* when mixed with infectious virions and fed to the insects through a membrane (Briddon, 1990). Presumably the non-infectious virions competed with the infectious virions for receptors which mediated acquisition, resulting in viruliferous vectors harbouring a reduced virus inoculum for onward transmission.

Tobacco plants expressing the MP gene of ToMoV with a spontaneous mutation that suppressed symptom induction (MP is a symptom determinant for ToMoV) were identified and shown to exhibit varying resistance responses to inoculation with ToMoV and *Cabbage leaf curl virus* (Duan et al., 1997). In the most highly resistant lines no virus replication could be detected following insect transmission of the viruses in leaves distal to the site of inoculation. In a similar study Hou et al. (2000) produced tomato lines expressing the MP and NSP of *Bean dwarf mosaic virus*. They also identified a line expressing a mutated MP which did not express a virus-like phenotype, in contrast to lines with an intact MP. Upon *Agrobacterium*-mediated inoculation with infectious clones of ToMoV, lines expressing the mutated MP, as well as the lines expressing full-length MP and NSP showed a significant delay in the appearance of symptoms of virus infection in comparison to non-transgenic plants. The results suggested that the transgenically expressed MP and NSP proteins acted as trans-dominant negative mutants, interfering with the normal function of the movement proteins of the invading virus. The results also showed that this approach can potentially yield resistance to quite distantly related viruses. However, since it is not desirable to express full-length virus genes in plants, due to the possibility of recombination with the invading virus possibly yielding a fitter, better adapted progeny, it seems sensible to utilise deletion mutants which maintain the ability to suppress virus infection.

The majority of studies investigating the use of protein-mediated PDR against geminiviruses have utilised Rep due to its crucial role in virus replication. Interfering with virus replication, rather than movement or transmission, is perceived as the most likely strategy to deliver effective resistance. Overexpression of geminivirus Rep in plants usually results in a necrotic response, possibly due to interference by this protein in cell cycle. Nevertheless, some transgenic lines expressing apparently full-length Rep have been produced. Hanley-Bowdoin et al. (1990) produced tobacco lines expression the *Tomato golden mosaic virus* (TGMV) Rep and showed that some lines could complement TGMV mutants carrying a disrupted Rep gene. Although Hong and Stanley (1996) could recover some *N. benthamiana* lines apparently expressing full-length Rep of ACMV, this would not complement an ACMV Rep mutant. Since neither study determined the sequence of the integrated transgene, it would seem possible that transformation has selected for plants carrying mutated Rep gene sequences. Bendahmane & Gronenborn (1997) were unable to generate *N. benthamiana* plants that expressed the Rep of TYLCV. The study of Hong and Stanley (1996) further showed that the lines expressing apparently full-length ACMV Rep and lines expressing just the N-terminal Rep sequences were resistant to ACMV infection with a higher proportion of transgenic plants remaining asymptomatic or showing mild and/or delayed symptoms. However, no resistance was found when these lines were challenged with TGMV or *Beet curly top virus* (BCTV).

Expression of a truncated Rep (tRep; with a deletion of C-terminal NTP-binding site) of *Tomato leaf curl Sardinia virus* (TYLCSV) was very effective for resistance (Brunetti et al.,

1997, Noris et al., 1996) and was shown to correlate with expression levels of the transgene. Rep ATPase activity is required for virus replication (Desbiez et al., 1995) and tRep may thus act as a trans-dominant negative mutant, forming non-functional Rep multimers. Alternatively it may repress transcription of the Rep gene of the virus since N-terminal sequences involved in DNA-binding are maintained or may adversely affect the integrity of the viral DNA by introducing nicks at cryptic motifs.

Using a transient assay in tobacco protoplasts Chatterji et al. (2001) showed that a tRep derived from *Tomato leaf curl New Delhi virus* (ToLCNDV) could interfere with the replication of the homologous virus as well as three unrelated viruses. The effect was shown to be mediated by DNA binding, for the homologous virus, and by the formation of (presumably) non-function Rep oligomers for the homologous and heterologous viruses (acting as a trans-dominant negative mutant).

Lucioli et al. (2003) showed that resistance in *N. benthamiana* expressing tRep derived from TYLCSV provided resistance against the homologous virus and the unrelated TYLCV. The mechanism of resistance in the case of TYLCSV was shown to be due to repression of Rep gene transcription whereas resistance to TYLCV was shown to occur by the formation of dysfunctional Rep complexes.

Recently the first transgenic resistance to a monocot-infecting geminivirus, the mastrevirus MSV was reported (Shepherd et al., 2007a). Using the tRep approach they produced transgenic maize that showed a significant delay in symptom development, less severe symptoms and higher plant survival rates upon challenge with MSV. Significantly this study was conducted in a developing country and the transformed maize variety is widely grown in Africa, meaning that potentially this resistance can rapidly be introduced in farmers' fields to control this serious pathogen of a subsistence crop.

These studies demonstrate that PDR resistance to geminiviruses based upon expression of viral proteins has great promise. The use of non-function (or partially functional) gene products is desirable, to both avoid possible deleterious effects on plants and recombination with an invading virus, but has the potential to yield effective and possible wide-spectrum resistance.

RNA Interference-mediated Resistance

As the mechanism of RNAi has been elucidated, it has been developed into a widely used technique for studying gene function and, judging by the numbers of publications on the subject, the preferred strategy for trying to engineer virus resistance (Tenllado et al., 2004). In theory geminiviruses should lack the particular vulnerability of RNA viruses, whose genomes can be degraded by PTGS, because of their DNA genomes and replication taking place in nucleus. In fact this is not the case. The first evidence that virus-specific siRNAs are induced by geminivirus infections came from an analysis of RNA extracts from TYLCV infected tomato plants (Lucioli et al., 2003) that showed siRNAs corresponding to the viral Rep gene, of both sense and antisense polarity. The susceptibility of geminiviruses to RNA silencing was subsequently confirmed by the demonstration that synthetic siRNAs, designed to target the coding region of ACMV Rep, were able to interfered with Rep mRNA and thus lead to a reduction in viral DNA replication in cultured cells (Vanitharani et al., 2003).

Prior to the identification of RNAi as a general defense mechanism against viruses in plants, Day et al. (1991) produced tobacco plants expressing an asRNA construct containing sequences of the TGMV Rep. The transgenic plants showed a significant reduction in viral DNA levels upon infection by TGMV, contrasting with both non-transformed plants and plants transformed with a sense construct. This resistance was later shown to possibly be sequence specific (Bejarano & Lichtenstein, 1994). Although the Rep asRNA plants significantly reduced the replication of BCTV, this was not the case for ACMV. The authors concluded that a block of sequence in the BCTV Rep gene with high levels of similarity to the transgene likely allowed the resistance to act against this virus whereas the ACMV Rep gene contains more dispersed similarity.

Subsequently both sense and antisense RNA technologies have been investigated for their usefulness in delivering resistance against geminiviruses. Sense RNA constructs targeting the Rep of TYLCSV (Bendahmane & Gronenborn, 1997), ToMoV (Polston & Hiebert, 2001) or the Rep/TrAP/REn sequences of *Cotton leaf curl Khokhran virus* (CLCuKV; Asad et al., 2003) and asRNA constructs targeting the Rep/REn/TrAP/MP of *Bean golden yellow mosaic virus* (BGYMV; Aragao et al., 1998), AV2 gene of ToLCNDV (Mubin et al., 2007) and Rep/TrAP/REn sequences of CLCuKV (Asad et al., 2003) were shown to deliver plants that showed delayed appearance of symptoms and virus resistance, shown by a reduction in virus DNA levels, upon infection with the cognate virus. Significantly the BGYMV study achieved resistance in bean (*Phaseolus vulgaris*), the natural host of the virus; something that has not often been achieved.

A major driving force in the push to geminivirus resistance has been the need to alleviate problems due to begomoviruses in cassava across Africa. Cassava is a subsistence crop upon which a large number of people across central Africa depend that was, and continues to be, severely affected by a cassava mosaic begomovirus (CMB) epidemic during the 1990s (Patil & Fauquet, 2009). Transgenic cassava plants expressing the full-length Rep gene of ACMV in sense orientation were produced and exhibited significant broad-spectrum resistance against a number of CMBs including ACMV, *East African cassava mosaic Cameroon virus* (EACMCV) and *Sri Lankan cassava mosaic virus*. The plants also showed resistance to dual infection by ACMV and EACMCV which, in the field, results in a synergistic interaction that has been responsible for severe disease. Further analysis showed that the mechanism involved in the resistance was RNAi (Chellappan et al., 2004). Similarly Zhang et al. (2005) produced transgenic cassava plants resistant to ACMV infection by expressing the Rep, TrAP and REn genes of ACMV in anti-sense orientation. The transgenic lines showed reduced viral DNA replication. Vanderschuren et al. (2007) reported that cassava plants expressing an intron-spliced hairpin construct containing CR sequences of ACMV showed accelerated recovery from infection upon challenge with ACMV and that recovery correlated with the appearance of transgene-derived siRNA. Overall viral symptoms were attenuated and plants had less viral DNA accumulation than in those of wild type plants. The authors suggested that the resistance was due to TGS, but did not proceed to prove this.

Another significant study used an intron-spliced hairpin construct containing CP gene sequences of TYLCV to produce transgenic tomato lines (Zrachya et al., 2007). When challenged with TYLCV the transgenic lines showed a seven week delay in symptom appearance, in comparison to non-transgenic lines, as well as a significant reduction in virus DNA levels. The authors speculated that the ultimate loss of resistance and appearance of symptoms was due to the action of virus encoded suppressors of silencing.

Resistance using RNAi has also been achieved by targeting non-coding regions of geminiviruses. Transient expression of *Mungbean yellow mosaic virus* (MYMV) IR sequences as an intron spliced hairpin resulted in complete recovery in blackgram (*Vigna mungo*) plants infected with MYMV (Pooggin et al., 2003). Similarly an intron spliced hairpin construct containing sequences of the IR conserved between TYLCV, TYLCSV and *Tomato yellow leaf curl Malaga virus* yielded a broad spectrum resistance when transiently expressed in tomato and *N. benthamiana* plants challenged with these viruses by *Agrobacterium*-mediated inoculation or whitefly transmission. No virus could be detected in plants which were challenged with virus, that had earlier been inoculated with the hairpin construct, using PCR. A positive correlation between resistance and the accumulation of TYLCV-specific siRNAs was observed in silenced plants (Abhary et al., 2006).

A major drawback with RNAi technology for producing resistance to geminiviruses is that once viral DNA is produced, viral DNA binding proteins, including the CP and NSP, may serve to shield the viral DNA from the effects of the silencing machinery. Therefore, although much of the viral mRNA may be degraded, sufficient viral proteins may continue to be produced to enable continued accumulation of viral DNA. Bian et al., 2006 proposed that showed that transgene carrying homology to the *Tomato leaf curl virus* (ToLCV) was silenced upon virus inoculation while the virus managed to escape silencing possibly by de novo synthesized unmethylated DNA. In addition to this, and in common with all other viruses, geminiviruses encode suppressor proteins that can block the RNA-silencing pathway used by plants to suppress viral accumulation (Sharma & Ikegami, 2008). These factors suggest that possibly complete immunity against geminiviruses in plants will not be achieved using the RNAi approach.

DNA Interference

Plants infected with geminiviruses frequently contain virus-derived subgenomic DNAs that likely originate from errors in genome replication. Accumulation of such defective molecules, which are characteristically either half or quarter genome (or genomic component) in size (thus ~1.4 or ~0.7kbp), during virus infection was found to lead to a delay in symptom appearance and attenuation of symptoms. Due to this ability to delay and attenuate infection symptoms they behave as defective interfering (DI)-DNA (Stanley et al., 1990). Sequence analysis suggests that DI-DNAs are formed by deletion, duplication, inversion, rearrangement and sometimes by insertion of non-viral DNA sequences. For bipartite begomoviruses, DI-DNAs are preferentially produced from the DNA-B component for reasons that remain unclear. Mutational analysis has shown the minimum sequence requirements for DI-DNA replication by ToLCV is the IR containing the Rep-binding domains.The role of the host plant in the formation of DI-DNAs is also important as DI-DNAs are more readily formed in some experimental hosts than in their natural hosts (reviewed by Patil & Dasgupta, 2006). Satellites associated with geminiviruses also give rise to defective DNA (Briddon et al., 2003, Usharani et al., 2004). The mechanism of symptom attenuation by geminivirus DI-DNA is likely to be a competition for rate-limiting amounts of virus-encoded proteins, such as Rep, leading to a reduction in the accumulation of viral DNA and concomitant reduced expression of viral pathogenicity (symptom) determinants.

The ability of DI-DNA to ameliorate symptoms has been investigated as a strategy for delivering resistance to geminiviruses. Plants transformed with a tandem repeat construct of subgenomic defective molecule derived from ACMV DNA-B showed ameliorated symptoms, in comparison to untransformed plants, upon infection by ACMV (Stanley et al., 1990). The disproportionate reduction in the level of full-length virus genomic components (DNA A 20% and DNA B 70%) indicated that the episomally replicating sub-genomic DNA had been amplified at the expense of DNA B. This phenomenon was virus specific since no resistance phenotype was observed when the transgenic plants were challenged with either BCTV or TGMV. The authors speculated that effect of subgenomic DNAs could be enhanced by increasing the copy number and that this could be achieved by reducing the size of the defective DNA while ensuring that all sequences required for replication remained intact. However, BCTV DI-DNA-mediated resistance was shown to depend upon the size of DI DNA (Frischmuth et al., 1997). Although smaller sub-genomic DNA interfered with the replication of BCTV it did not efficiently suppress the virus accumulation, possibly due to an inability of the smaller DNA molecule to be moved cell-to-cell.

The association with ssDNA satellites is a common feature of most begomoviruses originating from the Old World (Briddon & Stanley, 2006) but no efforts have yet been made to investigate their possible use in providing resistance. The alphasatellites (formerly known as DNA 1) do in some cases ameliorate symptoms (slightly) and can reduce both virus and betasatellite DNA levels. The mechanism here is likely to be somewhat similar to DI-DNA-mediated inference although, since these satellite-like molecules encode their own Rep, the competition is likely to be for cellular resources rather than limiting virus encoded proteins such as Rep. The fact that alphasatellites show little, if any, helper virus specificity (Briddon et al., 2004) would overcome some of the limitations of DI-DNAs. In this respect the betasatellites show even more promise. Although there is some helper virus specificity, which is poorly understood at this time (Saunders et al., 2008), betasatellites are able to interact with more than one begomoviruses (in some cases many distinct viruses). Thus with the symptom inducing βC1 gene deleted these might be useful as broad spectrum interfering DNAs. Vectors for gene silencing based upon betasatellites, with the βC1 gene deleted, have been reported (Tao & Zhou, 2004) but so far there are no reports of investigations into whether such molecules can deliver resistance in plants.

Non Pathogen Derived Resistance to Geminiviruses

As outlined earlier, silencing may compromise PDR that is protein-mediated and virus–encoded suppressors can interfere with resistance based upon the RNA silencing pathway. This suggests that, although RNAi is a valuable tool in generating resistance, greater efforts should be made to develop and utilise non-PDR strategies to confer resistance against viruses.

Virus Induced Cell Death

As part of their defences against pathogens, plants often respond with a hypersensitive response characterised by local cell death at the site of infection that acts to limit invasion (Mur et al., 2008). Genetic engineering has made it possible to develop this specific defense response into a more general strategy for developing resistance to viruses. Ribosome-inactivating proteins (RIPs) are naturally occurring plant toxins which inhibit protein synthesis by inactivating ribosomes (Narayanan et al., 2005). Dianthin, a potent RIP isolated from *Dianthus caryophyllus*, has been exploited to engineer transgenic resistance to ACMV, in *N. benthamiana* (Hong et al., 1996). To achieve this, dianthin was expressed from the ACMV virion-sense promoter. The virion-sense promoter of bipartite begomoviruses, such as ACMV, controls late gene expression and is inducible (transactivated by) TrAP. This strategy ensured the controlled expression of the toxin, facilitating the regeneration of phenotypically normal plants, and ensured transgene expression was localized to virus-infected cells. When challenged with ACMV, transgenic plants produced atypical necrotic lesions on inoculated leaves, indicative of dianthin expression. Viral DNA accumulation was significantly reduced in these tissues, and plants exhibited attenuated systemic symptoms from which they recovered. The resistance was specific to isolates of ACMV but not active against other begomoviruses, suggesting that TrAP homologues from other viruses were unable to efficiently transactivate the ACMV promoter. Although the TGMV homologue of TrAP is able to transactivate expression from the ACMV virion-sense promoter *in planta*, it appears to do so much less efficiently than ACMV TrAP (Saunders & Stanley, 1995). It was suggested that the slight reduction in TGMV DNA accumulation that was observed in leaf discs containing the virion-sense promoter-dianthin transgene may have resulted from a low level of transactivation. Another study showed that the TrAP of ACMV can efficiently transactivate the TGMV virion-sense promoter in tobacco protoplast (Sunter et al., 1994), suggesting that the TGMV promoter might be more promiscuous in its ability to interact with heterologous TrAP. It is thus possible that, with the selection of a less specifically induced promoter, this strategy could yield RIP-mediated resistance with a broader specificity. Possibly, with our greater understanding of the activities of TrAP and the sequence elements with which it interacts, we may be able to engineer promoters susceptible to up-regulation by the TrAPs of a greater range of viruses

The *Bacillus amyloliquefaciens* gene *barnase* encodes an RNase that is specifically inhibited by the product of *barstar*. Zhang et al. (2003) outlined a novel approach for obtaining resistance by cloning *barnase* under the virion-sense promoter of a begomovirus and *basrstar* under the complementary-sense promoter. In the absence of virus infection the expression of *barstar* should exceed that of *barnase* and thus yield no RNase activity. However, upon infection with a virus expressing a TrAP which up-regulates the virion-sense promoter controlling barnase, RNase activity would result in cell death, inhibiting virus infection in much the same way as the earlier diathin mediated strategy.

One problem with strategies that express "toxins" is that leaky expression can lead to cell toxicity. To tackle this problem a novel gene expression system called "InPAct" (for In Plant Activation) has been developed and is being marketed commercially by Farmacule (Australia; http://www.farmacule.com) and has been described in Shepherd et al. (2009). The main innovation of the InPAct system is that instead of it depending on promoter transactivation, it directly exploits the extremely specific ability of geminivirus Rep to initiate replication so as to ensure that gene expression cassettes will only be functional in the presence of this protein or during virus infection. The InPAct vectors, called ''split gene cassettes'', are arranged such that the toxin gene is split into two exons and cannot be expressed in the absence of Rep (Chanson et al., 2007). The split gene cassette is flanked by two viral intergenic regions (which contain the virion strand origin of replication and the Rep binding and nicking sites), which are in turn embedded in an intron. Upon viral infection, the integrated cassette serves as a template for rolling circle replication, allowing replicative release and amplification of circular ssDNA forms. Removal of the intergenic region-containing intron during post-transcriptional processing, results in the reconstitution of a translatable in-frame transcript of the toxin gene. Apart from allowing the non-leaky expression of antiviral cytotoxic compounds, the InPAct system will also be useful for virus-induced expression of non-lethal resistance genes and in molecular farming.

One potential obstacle in resistance approaches based on initiating cell death is the recent identification of a gene encoded by begomoviruses that is a suppressor of cell death (Hussain et al., 2007, Mubin et al., 2009). The TrAP of both monopartite and bipartite begomoviruses has been shown to counter cell death initiated by the V2 protein and NSP in tobacco and tomato which could compromise cell death mediated resistance.

GroEL Mediated Protection

Circulatively transmitted viruses, such as the geminiviruses, must translocate from the site of uptake by their insect vectors (the digestive tract) to the site of excretion (the salivary glands) before they are available for reintroduction to plants (Briddon et al., 1990). This translocation occurs via the insect's circulative system (the haemocoel). Interaction of circulatively transmitted plant viruses with GroEL homologues, produced by insect endosymbiotic bacteria, appears to be part of the mechanism by which these viruses survive in the haemolymph (blood) of their vector once they cross the gut barrier *en route* to the salivary glands. GroELs are proteins belonging to the chaperonin family and are involved in post-translational protein folding. A GroEL homologue isolated from *B. tabaci*, the whitefly vector of begomoviruses, was shown to bind viruses of the genera *Begomovirus*, *Luteovirus* and *Tosposvirus* but did not bind viruses of the genera *Closterovirus*, *Cucumovirus*, *Nepovirus*, *Potyvirus*, *Trichovirus* and *Tobamovirus* (Akad et al., 2004). The authors concluded that the specificity was for viruses with capsid proteins with a basic isoelectric point that are rich in arginine residues.

It was realised that this specific virus-binding property of GroEL could potentially be used to provide resistance in plants. TYLCV is a phloem limited monopartite begomovirus that is strongly bound by GroEL (Akad et al., 2004). Transgenic expression of GroEL in tomato under the control of a phloem specific promoter (Akad et al., 2007) and in *N. benthamiana* under the control of 35S promoter (Edelbaum et al., 2009) was shown to protect plants from TYLCV. Inoculated transgenic plants were either asymptomatic or mildly symptomatic, although all plants were infected. GroEL may immobilize virions at important sites of the virus infection cycle such as exit from the phloem, movement within the cell, replication in the cytoplasm or nuclei, and cell-to-cell and long-distance translocation. Since the protein has little virus specificity, binding many different viruses, it may provide a broad spectrum resistance. However, it is clear that this is only likely to be an enhanced tolerance, rather than immunity.

A potential drawback with the GroEL mediated strategy is that it protects the plant without substantially reducing virus titre. Resulting asymptomatic plants remain a good source of virus inoculum for onward transmission by vectors. However, an advantage of the strategy may be that it does not block virus replication and transmission and hence may not select resistance breaking variants. Possibly any GroEL evading variant may be less fit, since it may be more susceptible to degradation during the insect transmission part of the virus' infection cycle, due to a lack of interaction with the protective GroEL.

Peptide Aptamer-mediated Resistance

This is a novel and potentially exciting approach for obtaining broad-spectrum resistance to viruses, not just geminiviruses. Peptide aptamers are recombinant proteins that bind to and inactivate target proteins (Hoppe-Seyler et al., 2004). Although they resemble single-chain antibodies, they differ in the way that they are isolated. Whereas single-chain antibodies are usually selected *in vitro* using phage display, peptide aptamers are identified and isolated from an *in vivo* yeast two hybrid interaction assay.

Although peptide aptamers have yet to be investigated *in planta* for their ability to deliver resistance to geminiviruses, they have shown potential for delivering resistance to a range of tospoviruses (Rudolph et al., 2003). This study showed that a 29 amino acid peptide, selected for its ability to bind the nucleocapsid proteins of a range of tospoviruses, yielded resistance to a range of distinct topsoviruses when expressed in *N. benthamiana* as a fusion with a carrier protein (in this case β-glucuronidase was used to act as both a carrier and a marker for expression). This study showed that peptide aptamers have the potential to yield a broad spectrum resistance that is limited only, at least in theory, by the number of targets against which it is screened for during selection.

For geminiviruses peptide aptamers have been identified that bind to the Rep proteins of a number of geminiviruses and their potential for delivering broad-spectrum resistance in plants has been demonstrated using a protoplast co-transfection assay (Lopez-Ochoa & Hanley-Bowdoin, 2007, Lopez-Ochoa et al., 2006). These interfering peptide aptamers possibly act through several distinct mechanisms. They might inhibit the DNA binding and cleavage/ligation activities catalyzed by Rep (Orozco & Hanley-Bowdoin, 1998), compete for Rep/REn or Rep/pRBR binding (Arguello-Astorga et al., 2004), or block Rep oligomerisation (Orozco et al., 2000).

Antibody Mediated Resistance

Full-length and single-chain variable fragment (scFv) antibodies against both structural and non-structural viral proteins have been investigated quite widely for their ability to deliver resistance to RNA viruses. However, there is only a single report investigating the use of a scFv antibody against geminiviruses (Safarnejad et al., 2009). *N. benthamiana* plants transformed with scFv antibodies, selected for their ability to bind the Rep protein of TYLCV using phage display (Safarnejad et al., 2008), remained symptomless upon challenge with the virus. However, low level virus replication could still be detected in such plants. Some transformed lines, although initially symptomless upon challenge, ultimately showed attenuated symptoms. The authors concluded that, in these lines, the levels of scFv expression were too low and virus replication ultimately flooded plants with sufficient Rep to overcome the resistance.

Since antibody-antigen interactions are highly specific, a major drawback to their use will be that they can only provide resistance to a narrow range of viruses, and will not be useful where virus diversity is high. Also, such resistance will likely be easily circumvented by recombination that changes the antigenic site recognised by the antibody, unless very highly conserved epitopes can be identified and targeted.

Ribozyme-mediated Resistance

Ribozymes (an abbreviation of ribonucleic acid enzyme) are small nucleolytic RNAs that have potential uses in the down regulation of gene expression (Doherty & Doudna, 2001, Wu et al., 2009). They act as molecular scissors by catalytically cleaving target RNAs and consist of a conserved catalytic domain, responsible for catalysis, and a flanking variable recognition domain that is responsible for the recognition of target sequence (Haseloff & Gerlach, 1988, Uhlenbeck, 1987). The main advantage of ribozyme is that suitable changes in the recognition domain can be engineered so as it can target any RNA (Ushasri et al., 2007). Chilakamarthi et al. (2007) explored the *trans*-cleaving property of a hammerhead ribozyme targeted to the Rep mRNA of *Mungbean yellow mosaic India virus*. The ribozyme showed approx. 33% cleavage activity on synthetic Rep transcript within 1 h under *in vitro* conditions. The *in vivo* efficiency of ribozyme was evaluated in *Saccharomyces cerevisiae*, as it can act as a surrogate host for replication of the MYMIV-DNA and lacks RNAi machinery. In the presence of the ribozyme, growth of the yeast cells that were dependent on geminiviral replication was inhibited by 30% and cellular generation time was increased by 2 h. Analysis showed up to 50% reduction in the Rep mRNA level in presence of the ribozyme in comparison to mutant, non-catalytic controls. Viral DNA replication was shown to decrease by approx. 65% due to the down-regulation of replication initiator protein by the ribozyme. Despite these promising results, and the successful application of ribozymes for resistance to RNA viruses (Huttner et al., 2001) and viroids (Yang et al., 1997), this strategy has yet to be investigated for providing resistance to geminiviruses in plants.

Artificial Zinc Finger Protein-mediated Resistance

Zinc-finger motifs are present in transcription factors and function as parts of DNA-binding and protein-protein interaction domains. A variety of approaches have been developed which allow the design of artificial zinc finger proteins (AZPs) with novel specificities. AZPs have a wide array of potential uses such as tools in research and as drugs to specifically target gene expression (Negi et al., 2008).

Sera (2005) designed an AZP targeting the replication origin of the curtovirus *Beet severe curly top virus* (BSCTV). Specifically the iterons, which are implicated in Rep recognition of the origin, were used as the target. *In vitro* DNA binding assays indicated that the AZP efficiently blocked binding of the viral Rep to the replication origin and transgenic *Arabidopsis* plants expressing the AZP showed phenotypes strongly resistant to virus infection, with 84% of the plants showing no symptoms. Subsequently AZP technology was applied to TYLCV by designing an AZP which could block the TYLCV Rep binding activity (Takenaka et al., 2007). In competitive binding assays the effectiveness of the AZP in inhibition of the Rep binding was demonstrated *in vitro*. This work was further extended by Koshino-Kimura et al. (2008) who produced transgenic *Arabidopsis* plants expressing AZP protein specific to TYLCV and they further extended their work to natural host of TYLCV, tomato (Koshino-Kimura et al., 2009). The advantage of AZP is that due to its high affinity and specificity it can provide effective resistance at low expression level. Furthermore for the virus to overcome AZP-mediated resistance would require mutations in both the iteron binding domain (known as the "iteron related domain" [IRD]) of Rep and of the iterons. Although theoretically less likely to occur than single mutations, since the IRD and iterons are located side by side on the viral genome, these regions are frequently exchanged among geminiviurses by recombination (Lefeuvre et al., 2007). Since this is a sequence specific interaction, the spectrum of the resistance will be narrow. A likely future advance, which is already being used for other targets (Durai et al., 2005), will be the linking of the AZP to a nuclease so as to yield a protein that not only interferes with Rep recognition, but also degrades (cuts) the viral DNA, with the potential to yield a more effective resistance.

ssDNA Binding Protein-mediated Resistance

Gene 5 protein (G5), a ssDNA binding protein from *Escherichia coli* phage M13, is a small ssDNA binding protein which binds to DNA in a highly cooperative manner without pronounced sequence specificity. During synthesis of the single-stranded (progeny) viral DNA, it prevents the conversion of ssDNA into the dsDNA replicative form (Salstrom & Pratt 1971). Padidam et al. (1999) investigated whether the G5 protein could restore the accumulation of ToLCNDV ssDNA. ToLCNDV DNA A that lacked the CP gene was modified to express G5 under the control of virion-sense promoter. The modified virus was associated with the accumulation of wild type levels of ssDNA and high levels of dsDNA. The accumulation of ssDNA was apparently due to stable binding with G5. G5-expressing ToLCNDV did not spread efficiently in *N benthamiana* plants, and inoculated plants developed only very mild symptoms. They proposed that the G5 expression interfered with the function of the NSP and thereby impaired virus spread. They suggested that the expression of G5 in transgenic plants may afford a novel way of controlling geminiviruses and that such resistance may be effective against all geminiviruses. Despite these promising results this strategy has yet to be investigated for providing resistance to geminiviruses in plants.

Resistance to Vectors

In 2002, a functional genomics project was initiated, which generated several thousand expressed sequence tags (ESTs) of *Bemisia tabci*; the vector of begomoviruses. This project provided the basic information to design experiments aimed at understanding and manipulating whitefly gene expression. Using sequences generated by this project Ghanima et al. (2007) provided evidence that the RNAi mechanism discovered in many organisms is also active in *B. tabaci*. By injecting long dsRNA molecules specifically directed against genes uniquely expressed in the midgut and salivary glands in the body cavity, the expression of the targeted mRNA were significantly inhibited. Gene expression levels in RNAi-silenced whiteflies were reduced up to 70% compared to whiteflies injected with buffer or with a green florescent protein-specific dsRNA. These findings open up the possibility of expressing dsRNA constructs in plants that target insect genes (Baum et al., 2007). As well as providing a possible means of protecting plants from insects, by down-regulating genes essential for insect survival, the system could also potentially be used to interfere with insect transmission of viruses by silencing insect genes involved in the up-take and circulation of the virus in the insect.

Conclusion

Stanford & Johnston proposed the idea of PDR in 1985 and the proof of the theory was demonstrated almost immediately for TMV, an RNA virus (Abel et al., 1986). More importantly the first transgenic crop resistant to an RNA virus was approved for commercial release as long ago as 1998 (Gonsalves et al., 1998). This *Papaya ring spot virus* resistant transgenic cultivar saved the papaya crop in Hawaii from total collapse. Indeed transgenic resistant against RNA viruses is the third agronomic trait in commercially grown transgenic crops followed by herbicide tolerance and insect resistance. Unfortunately this success for RNA viruses has so far not been achieved for geminiviruses. Although the proof of the concept of PDR for geminiviruses was demonstrated in the early 1990s (Day et al 1991) and the volume of research papers describing PDR for these viruses are not less than those dealing with RNA viruses, only recently have the first transgenic plants undergone field trials (Aragao & Faria, 2009). However, at this time, there are no geminivirus resistant crops being grown commercially in farmer's fields, despite the fact these viruses are of increasing importance in tropical and sub-tropical regions.

It is unclear why transgenic resistant against geminiviruses has met with limited success despite the wide range of strategies which have been investigated. Homology-dependant strategies such as RNAi may not be effective due to the diversity of geminiviruses and their ability to rapidly evolve by mutation and recombination. The fact that their DNA genomes replicate in the nucleus and are thus not susceptible to PTGS, as well as their genomes encoding multiple suppressors is also likely to play a part. Lucioli et al. (2008) have suggested that more efforts should be put into developing more effective non-PDR to overcome this problem.

The available evidence suggests that any control strategy for geminiviruses that relies on just a single mechanism of action will ultimately succumb to infection. This has been the case in Pakistan where natural host-plant resistance to the viruses causing CLCuD became ineffective approximately 5 years after its introduction (Mansoor et al., 2003). Although the precise mechanism of resistance breaking by the virus is unclear in this case, it would seem that recombination, between the highly diverse begomoviruses causing the disease in this region, and mutation had a part to play (Amrao et al.,, 2010), Similarly in the case of Ty-1 resistance in tomato, an incomplete resistance providing tolerance rather than immunity will select for the most aggressive (well adapted) virus in an environment with a high level of virus diversity (Garcia-Andres et al., 2009). The apparent stability of transgenic resistance to BGMV in beans over several years of field trials reported recently by Aragao & Faria (2009) is in an area with a very low diversity of viruses affecting beans, which may explain this apparent success. All these findings suggest that, particularly in areas with high levels of virus diversity, relying on a single mechanism of resistance, be that natural host-plant or transgenic resistance, is unlikely to stand the test of time. In areas with high virus diversity in a particular crop, such as cotton in southern Asia and cassava in sub-Saharan Africa, relying on a homology-based resistance strategy, such as RNAi, would seem particularly destined for failure. In situations like this, the wisest course of action must be to build several layer of resistance which will be much more difficult for viruses to circumvent. The foundation of any such strategy should be the best available host-plant resistance supplemented by at least two transgenic strategies of differing mechanisms of action; for example an RNAi based resistance supplemented with a non-PDR resistance such as GroEL. In this way the hurdles the virus has to overcome are far greater and the resistant is much more likely to be stable.

In the future additional resistance strategies will likely become available, which will provide further diversity to the arsenal of tools at our disposal. For example, Eybishtz et al. (2009) have shown that silencing a single host gene in tomato lines with resistance to TYLCV made them susceptible. Possibly overexpression of the gene, encoding a hexose transported, could be used to improve TYLCV resistance in tomato.

References

Abel, P. P., Nelson, R. S., De, B., Hoffmann, N., Rogers, S. G., Fraley, R. T. & Beachy, R. N. (1986). Delay of disease development in transgenic plants that express the tobacco mosaic virus coat protein gene. *Science 232*, 738-743.

Abhary, M. K., Anfoka, G. H., Nakhla, M. K. & Maxwell, D. P. (2006). Post-transcriptional gene silencing in controlling viruses of the tomato yellow leaf curl virus complex. *Archives of Virology 151*, 2349-2363.

Akad, F., A., Eybishtz, D., Edelbaum, R., Gorovits, O., Dar-Issa, N. & H. Czosnek (2007). Making a friend from a foe: expressing a *GroEL* gene from the whitefly *Bemisia tabaci* in the phloem of tomato plants confers resistance to *Tomato yellow leaf curl virus*. *Archives of Virology 152*, 1323-1339.

Akad, F., Dotan, N. & Czosnek, H. (2004). Trapping of *Tomato yellow leaf curl virus* (TYLCV) and other plant viruses with a GroEL homologue from the whitefly *Bemisia tabaci*. *Archives of Virology 149*, 1481 - 1497.

Alves-Júnior, M., Alfenas-Zerbini, P., Andrade, E. C., Esposito, D. A., Silva, F. N., da Cruz, A. C. F., Ventrella, M. C., Otoni, W. C. & Zerbini, F. M. (2009). Synergism and negative interference during co-infection of tomato and *Nicotiana benthamiana* with two bipartite begomoviruses. *Virology 387*, 257-266.

Amrao, L., Amin, I., Shahid, S., Briddon, R.W. & Mansoor, S. (2010). Cotton leaf curl disease in resistant cotton is associated with a single begomovirus that lacks an intact transcriptional activator protein. *Virus Research*, in press.

Antignus, Y., Vunsh, R., Lachman, O., Pearlsman, M., Maslenin, L., Hananya, U. & Rosner, A. (2004). Truncated Rep gene originated from *Tomato yellow leaf curl virus*-Israel [Mild] confers strain-specific resistance in transgenic tomato. *Annals of Applied Biology 144*, 39-44.

Araga˜o, F. J. L., Ribeiro, S. G., Barros, L. M. G., Brasileiro, A. C. M., Maxwell, D. P., Rech, E. L. & Faria, J. C. (1998). Transgenic beans (*Phaseolus vulgaris* L.) engineered to express viral antisense RNAs show delayed and attenuated symptoms to bean golden mosaic geminivirus. *Molecular Breeding 4*, 491–499.

Arguello-Astorga, G., Lopez-Ochoa, L., Kong, L.-J., Orozco, B. M., Settlage, S. B. & Hanley-Bowdoin, L. (2004). A Novel motif in geminivirus replication proteins interacts with the plant retinoblastoma-related protein. *Journal of Virology 78*, 4817-4826.

Asad, S., Haris, W. A., Bashir, A., Zafar, Y., Malik, K. A., Malik, N. N. & Lichtenstein, C. P. (2003). Transgenic tobacco expressing geminiviral RNAs are resistant to the serious viral pathogen causing cotton leaf curl disease. *Archives of Virology 148*, 2341-2352.

Aufsatz, W., Mette, M. F., van der Winden, J., Matzke, A. J. & Matzke, M. (2002). RNA-directed DNA methylation in Arabidopsis. *Proceedings of the National Academy of Sciences USA 99*, 16499-16506.

Azzam, O., Diaz, O., Beaver, J. S., Gilbertson, R. L., Russell, D. R. & Maxwell, D. P. (1996). Transgenic beans with the bean golden mosaic geminivirus coat protein gene are susceptible to virus infection. *Annual Report Bean Improvement Corporate 39*, 276-277.

Bartel, D. P. (2004). MicroRNAs: Genomics, biogenesis, mechanism, and function. *Cell 116*, 281-297.

Baulcombe, D. (2004). RNA silencing in plants. *Nature 431*, 356-363.

Bejarano, E. R. & Lichtenstein, C. P. (1994). Expression of TGMV antisense RNA in transgenic tobacco inhibits replication of BCTV but not ACMV geminiviruses. *Plant Molecular Biology 24*, 241-248.

Bendahmane, A., Kanyuka, K. & Baulcombe, D. C. (1999). The Rx gene from potato controls separate virus resistance and cell death responses. *Plant Cell 11*, 781-791.

Bendahmane, M. & Gronenborn, B. (1997). Engineering resistance against *Tomato yellow leaf curl virus* (TYLCV) using antisense RNA. *Plant Molecular Biology 33*, 351-357.

Bian, X.-Y., Rasheed, M. S., Seemanpillai, M. J. & Rezaian, M. A. (2006). Analysis of silencing escape of *Tomato leaf curl virus*: an evaluation of the role of DNA methylation. *Molecular Plant-Microbe Interactions 19*, 614-624.

Blevins, T., Rajeswaran, R., Shivaprasad, P. V., Beknazariants, D., Si-Ammour, A., Park, H. S., Vazquez, F., Robertson, D., Meins, F., Jr., Hohn, T. & Pooggin, M. M. (2006). Four plant Dicers mediate viral small RNA biogenesis and DNA virus induced silencing. *Nucleic Acids Research 34*, 6233-6246.

Bonfim, K., Faria, J. C., Nogueira, E. O., Mendes, E. A. & Aragao, F. J. (2007). RNAi-mediated resistance to *Bean golden mosaic virus* in genetically engineered common bean (*Phaseolus vulgaris*). *Molecular Plant-Microbe Interactions 20*, 717-26.

Borchert, G. M., Lanier, W. & Davidson, B. L. (2006). RNA polymerase III transcribes human microRNAs. *Nature Structural & Molecular Biology 13*, 1097-101.

Briddon, R. W. (1990). The molecular biology of geminivirus transmission by insects. Ph.D. Thesis, University of East Anglia, Norwich, UK.

Briddon, R. W., Bull, S. E., Amin, I., Mansoor, S., Bedford, I. D., Dhawan, P., Rishi, N., Siwatch, S. S., Abdel-Salam, A. M. & Markham, P. G. (2003). Diversity of DNA β; a satellite molecule associated with some monopartite begomoviruses. *Virology 312*, 106-121.

Briddon, R. W., Bull, S. E., Amin, I., Mansoor, S., Bedford, I. D., Rishi, N., S., S., Zafar, Y., Abdel-Salam, A. M. & Markham, P. G. (2004). Diversity of DNA 1: a satellite-like molecule associated with monopartite begomovirus-DNA β complexes. *Virology 324*, 462 - 474.

Briddon, R. W. & Markham, P. G. (2001). Cotton leaf curl virus disease. *Virus Research 71*, 151-159.

Briddon, R. W., Pinner, M. S., Stanley, J. & Markham, P. G. (1990). Geminivirus coat protein replacement alters insect specificity. *Virology 177*, 85-94.

Briddon, R. W. & Stanley, J. (2006). Sub-viral agents associated with plant-infecting single-stranded DNA viruses. *Virology 344*, 198-210.

Brunetti, A., Tavazza, M., Noris, E., Tavazza, R., Caciagli, P., Ancora, G., Crespi, S. & Accotto, G. P. (1997). High expression of truncated viral Rep protein confers resistance to *Tomato yellow leaf curl virus* in transgenic tomato plants. *Molecular Plant-Microbe Interactions 10*, 571-579.

Brunetti, A., Tavazza, R., Noris, E., Lucioli, A., Accotto, G. P. & Tavazza, M. (2001). Transgenically expressed T-Rep of *Tomato yellow leaf curl Sardinia virus* acts as a *trans*-dominant-negative mutant, inhibiting viral transcription and replication. *Journal of Virology 75*, 10573-10581.

Chanson, A. H., Becker, D. K., Harding, R. M. & Dale, J. L. (2007). Effect of Rep on recombinant protein production from a *Tobacco yellow dwarf virus*-based episomal expression vector, In: 5th International Geminivirus Symposium, 3rd International ssDNA Comparative Virology Workshop, Conference Proceedings, 2007, p. 82.

Chatterji, A., Beachy, R. N. & Fauquet, C. M. (2001). Expression of the oligomerization domain of the replication-associated protein (Rep) of *Tomato leaf curl New Delhi virus* interferes with DNA accumulation of heterologous geminiviruses. *Journal of Biological Chemistry 276*, 25631-25638.

Chellappan, P., Masona, M. V., Vanitharani, R., Taylor, N. J. & Fauquet, C. M. (2004). Broad spectrum resistance to ssDNA viruses associated with transgene-induced gene silencing in cassava. *Plant Molecular Biology* 56, 601-611.

Chilakamarthi, U., Mukherjee, S. K. & Deb, J. K. (2007). Intervention of geminiviral replication in yeast by ribozyme mediated downregulation of its Rep protein. *FEBS Letters 581*, 2675-2683.

Cohen, S. & Lapidot, M. (2007). Appearance and expansion of TYLCV: A historical point of view. In: H. Czosnek (Ed), Tomato yellow leaf curl virus disease, pp. 3-12. Springer.

Day, A. G., Bejarano, E. R., Buck, K. W., Burrell, M. & Lichtenstein, C. P. (1991). Expression of an antisense viral gene in transgenic tobacco confers resistance to the DNA virus *Tomato golden mosaic virus*. *Proceedings of the National Academy of Sciences USA 88*, 6721-6725.

Desbiez, C., David, C., Mettouchi, A., Laufs, J. & Groneborn, B. (1995). Rep protein of tomato yellow leaf curl geminivirus has an ATPase activity required for viral DNA replication. *Proceeding of National Academy of Sciences U.S.A 92*, 5640-5644.

Doherty, E. A. & Doudna, J. A. (2001). Ribozyme structures and mechanisms. *Annual Review of Biophysics and Biomolecular Structure* 30, 457-475.

Drake, J. W., Charlesworth, B., Charlesworth, D. & Crow, J. F. (1998). Rates of spontaneous mutation. *Genetics 148*, 1667-86.

Duan, C. G., Wang, C. H., Fang, R. X. & Guo, H. S. (2008). Artificial MicroRNAs highly accessible to targets confer efficient virus resistance in plants. *Journal of Virology 82*, 11084-11095.

Duan, Y.-P., Powell, C. A., Webb, S. E., Purcifull, D. E. & Hiebert, E. (1997). Geminivirus resistance in transgenic tobacco expressing mutated BC1 protein. *Molecular Plant-Microbe Interactions 10*, 617-623.

Duffy, S. & Holmes, E. C. (2008). Phylogenetic evidence for rapid rates of molecular evolution in the single-stranded DNA begomovirus *Tomato yellow leaf curl virus*. *Journal of Virology 82*, 957-65.

Durai, S., Mani, M., Kandavelou, K., Wu, J., Porteus, M. H. & Chandrasegaran, S. (2005). Zinc finger nucleases: custom-designed molecular scissors for genome engineering of plant and mammalian cells. *Nucleic Acids Research 33*, 5978-5990.

Edelbaum, D., Gorovits, R., Sasaki, S., Ikegami, M. & Czosnek, H. (2009). Expressing a whitefly GroEL protein in *Nicotiana benthamiana* plants confers tolerance to *Tomato yellow leaf curl virus* and *Cucumber mosaic virus*, but not to *Grapevine virus A* or *Tobacco mosaic virus*. *Archives of Virology 154*, 399-407.

Farnham, G. & Baulcombe, D. C. (2006). Artificial evolution extends the spectrum of viruses that are targeted by a disease-resistance gene from potato. *Proceedings of the National Academy of Sciences USA 103*, 18828-18833.

Fire, A., Xu, S., Montgomery, M. K., Kostas, S. A., Driver, S. E. & Mello, C. C. (1998). Potent and specific genetic interference by double-stranded RNA in Caenorhabditis elegans. *Nature 391*, 806-811.

Fojtova, M., Van Houdt, H., Depicker, A. & Kovarik, A. (2003). Epigenetic switch from posttranscriptional to transcriptional silencing is correlated with promoter hypermethylation. *Plant Physiol*ogy *133*, 1240-50.

Frischmuth, T., Engel, M. & Jeske, H. (1997). Beet curly top virus DI DNA-mediated resistance is linked to its size. *Molecular Breeding 3*, 213-217.

Frischmuth, T. & Stanley, J. (1991). *African cassava mosaic virus* DI DNA interferes with the replication of both genomic components. *Virology 183*, 539-544.

Frischmuth, T. & Stanley, J. (1994). *Beet curly top virus* symptom amelioration in *Nicotiana benthamiana* transformed with a naturally occurring viral subgenomic DNA. *Virology 200*, 826-830.

Frischmuth, T. & Stanley, J. (1998). Recombination between viral DNA and the transgenic coat protein gene of African cassava mosaic geminivirus. *Journal of General Virology 79*, 1265-1271.

Fuentes, A., Ramos, P. L., Fiallo, E., Callard, D., Sanchez, Y., Peral, R., Rodriguez, R. & Pujol, M. (2006). Intron-hairpin RNA derived from replication associated protein C1 gene confers immunity to *Tomato yellow leaf curl virus* infection in transgenic tomato plants. *Transgenic Research 15*, 291-304.

Garcia-Andres, S., Tomas, D. M., Navas-Castillo, J. & Moriones, E. (2009). Resistance-driven selection of begomoviruses associated with the tomato yellow leaf curl disease. *Virus Research 146*, 66-72.

Ghanima, M., Kontsedalova, S. & Czosnek, A. H. (2007). Tissue-specific gene silencing by RNA interference in the whitefly Bemisia tabaci (Gennadius). *Insect Biochemistry and Molecular Biology 37*, 732-738.

Gharsallah Chouchane, S., Gorsane, F., Nakhla, M. K., Salus, M., Martin, C. T., Maxwell, D. P., Marrakchi, M. & Fakhfakh, H. (2008). Evaluation of two gene-silencing constructs for resistance to tomato yellow leaf curl viruses in *Nicotiana benthamiana* plants. *Acta Virologica 52*, 143-9.

Gonsalves, D., Ferreira, S., Manshardt, R., Fitch, M., & Slightom, J. (1998). Transgenic virus resistant papaya: New hope for control of papaya ringspot virus in Hawaii. APS*net* Feature, American Pythopathological Society, www.apsnet.org/education/feature/papaya

Hamilton, W. D. O., Bisaro, D. M. & Buck, K. W. (1982). Identification of novel DNA forms in *Tomato golden mosaic virus* infected tissue. Evidence for a two component genome. *Nucleic Acids Research 19*, 4901-4912.

Hamilton, W. D. O., Bisaro, D. M., Coutts, R. H. A. & Buck, K. W. (1983). Demonstration of the bipartite nature of the genome of a single-stranded DNA plant virus by infection with the cloned DNA components. *Nucleic Acids Research 11*, 7387-7396.

Hanley-Bowdoin, L., Elmer, J. S. & Rogers, S. G. (1990). Expression of functional replication protein from *Tomato golden mosaic virus* in transgenic tobacco plants. *Proceedings of the National Academy of Sciences, USA 87*, 1446-1450.

Hanley-Bowdoin, L., Settlage, S. B., Orozco, B. M., Nagar, S. & Robertson, D. (1999). Geminviruses: models for plant DNA replication, transcription, and cell cycle regulation. *Critical Reviews in Plant Sciences 18*, 71-106.

Hanson, S. F. & Maxwell, D. P. (1999). *Trans*-dominant inhibition of geminiviral DNA replication by Bean golden mosaic geminivirus *rep* gene mutants. *Phytopathology 89*, 480-486.

Haseloff, J. & Gerlach, W. L. (1988). Simple RNA enzymes with new and highly specific endoribonuclease activities. *Nature 334*, 585-591.

Herr, A. J., Jensen, M. B., Dalmay, T. & Baulcombe, D. C. (2005). RNA polymerase IV directs silencing of endogenous DNA. *Science 308*, 118-120.

Hong, H., Saunders, K., Hartley, M. R. & Stanley, J. (1996). Resistance to geminivirus infection by virus-induced expression of dianthin in transgenic plants. *Virology 220*, 119-127.

Hong, J. & Stanley, J. (1996). Virus resistance in *Nicotiana benthamiana* conferred by *African cassava mosaic virus* replication-associated protein (AC1) transgene. *Molecular Plant-Microbe Interactions 4*, 219-225.

Hoppe-Seyler, F., Crnkovic-Mertens, I., Tomai, E. & Butz, K. (2004). Peptide aptamers: specific inhibitors of protein function. *Current Molecular Medicine 4*, 529-38.

Hou, Y. M., Sanders, R., Ursin, V. M. & Gilbertson, R. L. (2000). Transgenic plants expressing geminivirus movement proteins: abnormal phenotypes and delayed infection by *Tomato mottle virus* in transgenic tomatoes expressing the *Bean dwarf mosaic virus* BV1 or BC1 proteins. *Molecular Plant-Microbe Interactions 13*, 297-308.

Hussain, M., Mansoor, S., Iram, S., Zafar, Y. & Briddon, R. W. (2007). The Hypersensitive Response to Tomato leaf curl New Delhi virus Nuclear Shuttle Protein Is Inhibited by Transcriptional Activator Protein. *Mol Plant Microbe Interact 20*, 1581-1588.

Huttner, E., Tucker, W., Vermeulen, A., Ignart, F., Sawyer, B. & Birch, R. (2001). Ribozyme genes protecting transgenic melon plants against potyviruses. *Current Issues in Molecular Biology 3*, 27-34.

Kawasaki, H. & Taira, K. (2004). Induction of DNA methylation and gene silencing by short interfering RNAs in human cells. *Nature 431*, 211 217.

Khvorova, A., Reynolds, A. & Jayasena, S. D. (2003). Functional siRNAs and miRNAs exhibit strand bias. *Cell 115*, 209-216.

Kim, V. N. (2005). MicroRNA biogenesis: coordinated cropping and dicing. *Nature Review of Molecular Cell Biology 6*, 376-385.

Koshino-Kimura, Y., Takenaka, K., Domoto, F., Aoyama, Y. & Sera, T. (2008). Generation of plants resistant to tomato yellow leaf curl virus by using artificial zinc-finger proteins. *Nucleic Acids Symposium Series (Oxford) 52*, 189-90.

Koshino-Kimura, Y., Takenaka, K., Domoto, F., Ohashi, M., Miyazaki, T., Aoyama, Y. & Sera, T. (2009). Construction of plants resistant to TYLCV by using artificial zinc-finger proteins. *Nucleic Acids Symposium Series* (Oxford) *53*, 281-282.

Krek, A., Grun, D., Poy, M. N., Wolf, R., Rosenberg, L., Epstein, E. J., MacMenamin, P., da Piedade, I., Gunsalus, K. C., Stoffel, M. & Rajewsky, N. (2005). Combinatorial microRNA target predictions. *Nature Genetics 37*, 495-500.

Kunik, T., Salomon, R., Zamir, D., Navot, N., Zeidan, M., Michelson, I., Gafni, Y. & Czosnek, H. (1994). Transgenic tomato plants expressing the *Tomato yellow leaf curl virus* capsid protein are resistant to the virus. *Biotechnology 12*, 500-504.

Lecoq, H. (1998). Control of plant virus diseases by cross protection. In: A. Hadidi, R.K. Khetarpal and H. Koganezawa (eds) Plant Virus Disease Control, pp. 33-40. APS Press, St. Paul, MN, USA.

Lee, R. C., Feinbaum, R. L. & Ambros, V. (1993). The *C. elegans* heterochronic gene lin-4 encodes small RNAs with antisense complementarity to lin-14. *Cell 75*, 843–854.

Lee, Y., Kim, M., Han, J., Yeom, K. H., Lee, S., Baek, S. H. & Kim, V. N. (2004). MicroRNA genes are transcribed by RNA polymerase II. *EMBO Journal 23*, 4051-60.

Lefeuvre, P., Lett, J. M., Reynaud, B. & Martin, D. P. (2007a). Avoidance of protein fold disruption in natural virus recombinants. *PLoS Pathogens 3*, e181.

Lefeuvre, P., Martin, D. P., Hoareau, M., Naze, F., Delatte, H., Thierry, M., Varsani, A., Becker, N., Reynaud, B. & Lett, J. M. (2007b). Begomovirus 'melting pot' in the south-west Indian Ocean islands: molecular diversity and evolution through recombination. *Journal of General Virology 88*, 3458-68.

Lehmann, P. (2002). Structure and evolution of plant disease resistance genes. *Journal of Applied Genetics 43*, 403-414.

Leung, A. K. & Sharp, P. A. (2007). microRNAs: a safeguard against turmoil? *Cell 130*, 581-5.

Levy, A., Dafny-Yelin, M. & Tzfira, T. (2008). Attacking the defenders: plant viruses fight back. *Trends in Microbiology 16*, 194-7.

Lindbo, J. A., Silva-Rosales, L., Proebsting, W. M. & Dougherty, W. G. (1993). Induction of a highly specific antiviral state in transgenic plants: implications for regulation of gene expression and virus resistance. *Plant Cell 5*, 1749-1759.

Llave, C., Xie, Z., Kasschau, K. D. & Carrington, J. C. (2002). Cleavage of Scarecrowlike mRNA targets directed by a class of *Arabidopsis* miRNA. *Science 297*, 2053–2056.

Lopez-Ochoa, L. & Hanley-Bowdoin, L. (2007). Use of peptide aptamers for broad-based geminivirus disease resistance, in: Conference abstract, American Society of Plant Biologists (ASPB) Plant Biology and Botany Joint Congress, Chicago, USA, 2007, www.botonyconference.org.

Lopez-Ochoa, L., Ramirez-Prado, J. & Hanley-Bowdoin, L. (2006). Peptide aptamers that bind to a geminivirus replication protein interfere with viral replication in plant cells. *Journal of Virology 80*, 5841-53.

Lucioli, A., Noris, E., Brunetti, A., Tavazza, R., Ruzza, V., Castillo, A. G., Bejarano, E. R., Accotto, G. P. & Tavazza, M. (2003). *Tomato yellow leaf curl Sardinia virus* Rep-derived resistance to homologous and heterologous geminiviruses occurs by different mechanisms and is overcome if virus-mediated transgene silencing is activated. *Journal of Virology 77*, 6785-6798.

Mansoor, S., Amin, I., Hussain, M., Zafar, Y. & Briddon, R. W. (2006). Engineering novel traits in plants through RNA interference. *Trends in Plant Science 11*, 559-565.

Mansoor, S., Amin, I., Iram, S., Hussain, M., Zafar, Y., Malik, K. A. & Briddon, R. W. (2003). The breakdown of resistance in cotton to cotton leaf curl disease in Pakistan. *New Disease Reports 52*, 784.

Martin, D. P. & Rybicki, E. P. (2001). Sequence diversity and virulence in *Zea mays* of *Maize streak virus* isolates. *Virology 288*, 247-255.

Martin, D. P., van der Walt, E., Posada, D. & Rybicki, E. P. (2005). The evolutionary value of recombination is constrained by genome modularity. *PLoS Genetics 1*, e51.

Mattick, J. S. & Makunin, I. V. (2005). Small regulatory RNAs in mammals. *Human Molecular Genetics* 14 Spec No 1, R121-32.

Michelson, I., Zeidan, M., Zamir, D. & Czosnek, H. (1997). Localization of *Tomato yellow leaf curl virus* (TYLCV) in susceptible and tolerant nearly isogenic tomato lines. *Acta Horticulture 447*, 407-414.

Mubin, M., Amin, I., Amrao, L., Briddon, R. W. & Mansoor, S. (2010). The hypersensitive response induced by the V2 protein of a monopartite begomovirus is countered by the C2 protein. *Molecular Plant Pathology 11*, 245-254.

Mubin, M., Mansoor, S., Hussain, M. & Zafar, Y. (2007). Silencing of the AV2 gene by antisense RNA protects transgenic plants against a bipartite begomovirus. *Virology Journal 4*, 10.

Mur, L. A. J., Kenton, P., Lloyd, A. J., Ougham, H. & Prats, E. (2008). The hypersensitive response; the centenary is upon us but how much do we know? *Journal of Experimental Botany 59*, 501-520.

Narayanan, S., Surendranath, K., Bora, N., Surolia, A. & Karande, A. A. (2005). Ribosome inactivating proteins and apoptosis. *FEBS Letters 579*, 1324-1331.

Negi, S., Imanishi, M., Matsumoto, M. & Sugiura, Y. (2008). New redesigned zinc-finger proteins: design strategy and its application. *Chemistry - A European Journal 14*, 3236-3249.

Noris, E., Accotto, G. P., Tavazza, R., Brunetti, A., Crespi, S. & Tavazza, M. (1996). Resistance to tomato yellow leaf curl geminivirus in *Nicotiana benthamiana* plants transformed with a truncated viral C1 gene. *Virology 224*, 130-138.

Orozco, B. M. & Hanley-Bowdoin, L. (1998). Conserved sequence and structural motifs contribute to the DNA binding and cleavage activities of a geminivirus replication protein. *Journal of Biological Chemistry 273*, 24448-24456.

Orozco, B. M., Kong, L.-J., Batts, L. A., Elledge, S. & Hanley-Bowdoin, L. (2000). The multifunctional character of a geminivirus replication protein is reflected by its complex oligomerization properties. *Journal of Biological Chemistry 275*, 6114-6122.

Owor, B., Legg, J. P., Okao-Okuja, G., Obonyo, R., Kyamanywa, S. & Ogenga-Latigo, M. W. (2004). Field studies of cross protection with cassava mosaic geminiviruses in Uganda. *Journal of Plant Pathology 152*, 243-249.

Padidam, M., Sawyer, S. & Fauquet, C. M. (1999). Possible emergence of new geminiviruses by frequent recombination. *Virology 265*, 218-225.

Patil, B. L. & Dasgupta, I. (2006). Defective interfering DNAs of plant viruses. *Critical Reviews in Plant Sciences 25*, 47–64.

Patil, B. L. & Fauquet, C. M. (2009). Cassava mosaic geminiviruses: actual knowledge and perspectives. *Molecular Plant Pathology 10*, 685-701.

Polston, J. E. & Hiebert, E. (2001). Engineered ressitance to tomato geminiviruses. Pages 19-22 In: Proceedings of the Florida Tomato Institute. Naples, FL. P. Gilreath, ed. PRO-518, University of Florida, Gainesville.

Pooggin, M. & Hohn, T. (2004). Fighting geminiviruses by RNAi and vice versa. *Plant Molecular Biology 55*, 149-152.

Pooggin, M., Shivaprasad, P. V., Veluthambi, K. & Hohn, T. (2003). RNAi targeting of DNA virus in plants. *Nature Biotechnology 21*, 131-132.

Prasanna, H. C. & Rai, M. (2007). Detection and frequency of recombination in tomato-infecting begomoviruses of South and Southeast Asia. *Virology Journal 4*, 111.

Praveen, S., Kushwaha, C. M., Mishra, A. K., Singh, V., Jain, R. K. & Varma, A. (2005). Engineering tomato for resistance to tomato leaf curl disease using viral *Rep* gene sequences. *Plant Cell, Tissue and Organ Culture 83*, 311.

Ramesh, S. V., Mishra, A. K. & Praveen, S. (2007). Hairpin RNA-mediated strategies for silencing of tomato leaf curl virus AC1 and AC4 genes for effective resistance in plants. *Oligonucleotides 17*, 251-7.

Ratcliff, F. G., MacFarlane, S. A. & Baulcombe, D. C. (1999). Gene silencing without DNA: RNA-mediated cross-protection between viruses. *Plant Cell 11*, 1207-1215.

Rosell, R. C., Torres-Jerez, I. & Brown, J. K. (1999). Tracing the geminivirus-whitefly transmission pathway by polymerase chain reaction in whitefly extracts, saliva, hemolymph, and honeydew. *Phytopathology 89*, 239-246.

Ruby, J. G., Jan, C. H. & Bartel, D. P. (2007). Intronic microRNA precursors that bypass Drosha processing. *Nature 448*, 83-6.

Rudolph, C., Schreier, P. H. & Uhrig, J. F. (2003). Peptide-mediated broad-spectrum plant resistance to tospoviruses. *Proceedings of the National Academy of Sciences USA 100*, 4429-4434.

Safarnejad, M., Fischer, R. & Commandeur, U. (2009). Recombinant-antibody-mediated resistance against *Tomato yellow leaf curl virus* in *Nicotiana benthamiana. Archives of Virology 15*4, 457-467.

Safarnejad, M. R., Fischer, R. & Commandeur, U. (2008). Generation and characterization of functional recombinant antibody fragments against tomato yellow leaf curl virus replication-associated protein. *Communications in Agricultural and Applied Biological Sciences 73*, 311-321.

Salstrom, J., & Pratt, D. (1971). Role of coliphage M13 gene 5 in singlestranded DNA production. *Journal of Molecular Biology* 61, 489–501.

Sanford, J. C. & Johnson, S. A. (1985). The concept of parasite-derived resistance: deriving resistance genes from the parasites own genome. *Journal of Theoretical Biology 115*, 395-405.

Sangaré, A., Deng, D., Fauquet, C. M. & Beachy, R. N. (1999). Resistance to *African cassava mosaic virus* conferred by a mutant of the putative NTP-binding domain of the Rep Gene (AC1) in *Nicotiana benthamiana*. *Molecular Breeding 5*, 95-102.

Saunders, K., Briddon, R. W. & Stanley, J. (2008). Replication promiscuity of DNA-β satellites associated with monopartite begomoviruses; deletion mutagenesis of the *Ageratum* yellow vein virus DNA-β satellite localizes sequences involved in replication. *Journal of General Virology 89*, 3165-72.

Saunders, K. & Stanley, J. (1995). Complementation of *African cassava mosaic virus AC2* gene function in a mixed bipartite geminivirus infection. *Journal of General Virology 76*, 2287-2292.

Schwarz, D. S., Hutvagner, G., Du, T., Xu, Z., Aronin, N. & Zamore, P. D. (2003). Asymmetry in the assembly of the RNAi enzyme complex. *Cell 115*, 199-208.

Seal, S. E., Jeger, M. J., Van den Bosch, F., Maramorosch, K., Shatkin, A. J. & Thresh, J. M. (2006a). Begomovirus evolution and disease management. *Advances in Virus Research, 67*, 297-316.

Seal, S. E., van den Bosch, F. & Jeger, M. J. (2006b). Factors influencing begomovirus evolution and their increasing global significance: implications for sustainable control. *Critical Reviews in Plant Sciences 25*, 23 - 46.

Sera, T. (2005). Inhibition of virus DNA replication by artificial zinc finger proteins. *Journal of Virology 79*, 2614-2619.

Settlage, S. B., See, R. G. & Hanley-Bowdoin, L. (2005). Geminivirus C3 protein: replication enhancement and protein interactions. *Journal of Virology 79*, 9885-9895.

Sharma, P. & Ikegami, M. (2008). RNA-silencing suppressors of geminiviruses. *Journal of General Plant Pathology 74*, 189-202.

Shepherd, D. N., Mangwende, T., Martin, D. P., Bezuidenhout, M., Kloppers, F. J., Carolissen, C. H., Monjane, A. L., Rybicki, E. P. & Thomson, J. A. (2007a). *Maize streak virus*-resistant transgenic maize: a first for Africa. *Plant Biotechnology Journal 5*, 759-67.

Shepherd, D. N., Mangwende, T., Martin, D. P., Bezuidenhout, M., Thomson, J. A. & Rybicki, E. P. (2007b). Inhibition of maize streak virus (MSV) replication by transient and transgenic expression of MSV replication-associated protein mutants. *Journal of General Virology 88*, 325-36.

Shepherd, D. N., Martin, D. P., McGivern, D. R., Boulton, M. I., Thomson, J. A. & Rybicki, E. P. (2005). A three-nucleotide mutation altering the Maize streak virus Rep pRBR-interaction motif reduces symptom severity in maize and partially reverts at high frequency without restoring pRBR-Rep binding. *Journal of General Virology 86*, 803-13.

Shepherd, D. N., Martin, D. P. & Thomson, J. A. (2009). Transgenic strategies for developing crops resistant to geminiviruses. *Plant Science 176*, 1-11.

Shivaprasad, P. V., Thillaichidambaram, P., Balaji, V. & Veluthambi, K. (2006). Expression of full-length and truncated Rep genes from *Mungbean yellow mosaic virus-Vigna* inhibits viral replication in transgenic tobacco. *Virus Genes 33*, 365-74.

Sinisterra, X. H., Polston, J. E., Abouzid, A. M. & Hiebert, E. (1999). Tobacco plants transformed with a modified coat protein of tomato mottle begomovirus show resistance to virus infection. *Phytopathology 89*, 701-706.

Smith, N. A., S.P., S., Wang, M.-B., Stoutjesdijk, P. A., Green, A. G. & Waterhouse, P. M. (2000). Total silencing by intron-spliced hairpin RNAs. *Nature 407*, 319-320.

Soosaar, J. L. M., Burch-Smith, T. M. & Dinesh-Kumar, S. P. (2005). Mechanisms of plant resistance to viruses. *Nat Reviews Microbiology 3*, 789-798.

Stanley, J., Bisaro, D. M., Briddon, R. W., Brown, J. K., Fauquet, C. M., Harrison, B. D., Rybicki, E. P. & Stenger, D. C. (2005). Geminiviridae. In *Virus Taxonomy, VIIIth Report of the ICTV*, pp. 301-326. Edited by C. M. Fauquet, M. A. Mayo, J. Maniloff, U. Desselberger & L. A. Ball. London: Elsevier/Academic Press.

Stanley, J., Frischmuth, T. & Ellwood, S. (1990). Defective viral DNA ameliorates symptoms of geminivirus infection in transgenic plants. *Proceedings of the National Academy of Sciences, USA 87*, 6291-6295.

Stanley, J. & Townsend, R. (1985). Characterization of DNA forms associated with cassava latent virus infection. *Nucleic Acids Research 13*, 2189-2205.

Stenger, D. C. (1994). Strain-specific mobilization and amplification of a transgenic defective-interfering DNA of the geminivirus *Beet curly top virus*. *Virology 203*, 397-402.

Sunter, G. & Bisaro, D. M. (1992). Transactivation of geminivirus AR1 and BR1 gene expression by the viral AL2 gene product occurs at the level of transcription. *Plant Cell 4*, 1321-1331.

Sunter, G., Stenger, D. C. & Bisaro, D. M. (1994). Heterologous complementation by geminivirus AL2 and AL3 genes. *Virology 203*, 203-210.

Takenaka, K., Koshino-Kimura, Y., Aoyama, Y. & Sera, T. (2007). Inhibition of tomato yellow leaf curl virus replication by artificial zinc-finger proteins. *Nucleic Acids Symposium Series*, 429-30.

Tamarzizt, B.H., Chouchane, S. G., Lengliz, R., Maxwell, D. P., Marrakchi, M., Fakhfakh, H. & Gorsane, F. (2009). Use of *Tomato yellow leaf curl virus* (TYLCV) truncated Rep gene sequence to engineer TYLCV resistance in tomato plants. *Acta Virologica 53*, 99-104.

Tang, G., Reinhart, B. J., Bartel, D. P. & Zamore, P. D. (2003). A biochemical framework for RNA silencing in plants. *Genes & Development 17*, 49–63.

Tenllado, F., Llave, C. & Diaz-Ruiz, J. R. (2004). RNA interference as a new biotechnological tool for the control of virus diseases in plants. *Virus Res 102*, 85-96.

Thresh, J. M. & Otim-Nape, G. W. (1994). Strategies for controlling African cassava mosaic geminivirus. In: *Advances in Disease Vector Research*, pp. 215-236. Edited by K. F. Harris. New York: Springer Verlag.

Uhlenbeck, O. (1987). A small catalytic oligoribinucleotide. *Nature 328*, 596–600.

Usharani, K. S., Surendranath, B., Haq, Q. M. R. & Malathi, V. G. (2004). Yellow mosaic virus infecting soybean in northern India is distinct from the species infecting soybean in southern and western India. *Current Science 86*, 845-850.

Ushasri, C., Sunil, K. M. & Deb, J. K. (2007). Intervention of geminiviral replication in yeast by ribozyme mediated downregulation of its Rep protein. *FEBS letters 581*, 2675.

Vanderschuren, H., Akbergenov, R., Pooggin, M. M., Hohn, T., Gruissem, W. & Zhang, P. (2007a). Transgenic cassava resistance to African cassava mosaic virus is enhanced by viral DNA-A bidirectional promoter-derived siRNAs. *Plant Molecular Biology 64*, 549-57.

Vanderschuren, H., Alder, A., Zhang, P. & Gruissem, W. (2009). Dose-dependent RNAi-mediated geminivirus resistance in the tropical root crop cassava. *Plant Molecular Biology 70*, 265-72.

Vanderschuren, H., Stupak, M., Fütterer, J., Gruissem, W. & Zhang, P. (2007b). Engineering resistance to geminiviruses – review and perspectives. *Plant Biotechnology Journal 5*, 207-220.

Vanitharani, R., Chellappan, P. & Fauquet, C. M. (2003). Short interfering RNA-mediated interference of gene expression and viral DNA accumulation in cultured plant cells. *Proceedings of the National Academy of Sciences USA 100*, 9632-9636.

Vanitharani, R., Chellappan, P., Pita, J. S. & Fauquet, C. M. (2004). Differential roles of AC2 and AC4 of cassava geminiviruses in mediating synergism and suppression of posttranscriptional gene silencing. *Journal of Virology 78*, 9487-9498.

Varma, A. & Malathi, V. G. (2003). Emerging geminivirus problems: A serious threat to crop production. *Annals of Applied Biology 142*, 145-164.

Varsani, A., Shepherd, D. N., Monjane, A. L., Owor, B. E., Erdmann, J. B., Rybicki, E. P., Peterschmitt, M., Briddon, R. W., Markham, P. G., Oluwafemi, S., Windram, O. P., Lefeuvre, P., Lett, J. M. & Martin, D. P. (2008). Recombination, decreased host specificity and increased mobility may have driven the emergence of maize streak virus as an agricultural pathogen. *Journal of General Virology 89*, 2063-74.

Vazquez, F., Vaucheret, H., Rajagopalan, R., Lepers, C., Gasciolli, V., Mallory, A. C., Hilbert, J. L., Bartel, D. P. & Crete, P. (2004). Endogenous trans-acting siRNAs regulate the accumulation of Arabidopsis mRNAs. *Molecular Cell 16*, 69-79.

Voinnet, O. (2001). RNA silencing as a plant immune system against viruses. *Trends in Genetics 17*, 449-459.

Voinnet, O. (2005a). Induction and suppression of RNA silencing: insights from viral infections. *Nature Reviews Genetics 6*, 206-20.

Voinnet, O. (2005b). Induction and suppression of RNA silencing: insights from viral infections. *Nature Genetics 6*, 206-221.

von Arnim, A. & Stanley, J. (1992). Inhibition of *African cassava mosaic virus* systemic infection by a movement protein from the related geminivirus tomato golden mosaic virus. *Virology 187*, 555-564.

Wright, E. A., Heckel, T., Groenendijk, J., Davies, J. W. & Boulton, M. I. (1997). Splicing features in maize streak virus virion- and complementary-sense gene expression. *Plant Journal 12*, 1285-1297.

Wu, Q., Huang, L. & hang, Y. (2009). The structure and function of catalytic RNAs. *Science in China Series C: Life Sciences 52*, 232-244.

Xie, Z., Johansen, L. K., Gustafson, A. M., Kasschau, K. D., Lellis, A. D., Zilberman, D., Jacobsen, S. E. & Carrington, J. C. (2004). Genetic and functional diversification of small RNA pathways in plants. *PLoS Biology* 2, e104

Yang, X., Yie, Y., Zhu, F., Liu, Y., Kang, L., Wang, X. & Tien, P. (1997). Ribozyme-mediated high resistance against potato spindle tuber viroid in transgenic potatoes. *Proceedings of the National Academy of Sciences USA 94*, 4861-4865.

Zamir, D., Ekstein-Michelson, I., Zakay, Y., Navot, N., Zeidan, M., Sarfatti, M., Eshed, Y., Harel, E., Pleban, T., van-Oss, H., Kedar, N., Rabinowitch, H. D. & Czosnek, H. (1994). Mapping and introgression of a *Tomato yellow leaf curl virus* tolerance gene, *TY-1*. *Theoretical and Applied Genetics 88*, 141-146.

Zhang, P., Fütterer, J., Frey, P., Potrykus, I., Puonti-Kaerlas, J. & Gruissem, W. (2003). Engineering virus-induced ACMV resistance by mimicking a hypersensitive reaction in transgenic cassava plants. In: Plant Biotechnology 2002 and Beyond (Vasil, I.K., ed.),pp. 143–146. Norwell and Dordrecht: Kluwer Academic.

Zhang, P., Vanderschuren, H., Futterer, J. & Gruissem, W. (2005). Resistance to cassava mosaic disease in transgenic cassava expressing antisense RNAs targeting virus replication genes. *Plant Biotechnology Journal 3*, 385–397.

Zilberman, D., Cao, X., Johansen, L. K., Xie, Z., Carrington, J. C. & Jacobsen, S. E. (2004). Role of Arabidopsis ARGONAUTE4 in RNA-directed DNA methylation triggered by inverted repeats. *Current Biology 14*, 1214-20.

Zrachya, A., Kumar, P. P., Ramakrishnan, U., Levy, Y., Loyter, A., Arazi, T., Lapidot, M. & Gafni, Y. (2007). Production of siRNA targeted against TYLCV coat protein transcripts leads to silencing of its expression and resistance to the virus. *Transgenic Research 16*, 385–398.

In: Emerging Geminiviral Diseases and their Management
Editors: P. Sharma, R. K. Gaur et al. pp. 37-57
ISBN 978-1-61668-620-8

Chapter 2

MANAGEMENT OF TOMATO YELLOW LEAF CURL DISEASE; A CASE STUDY FOR EMERGING GEMINIVIRAL DISEASES

Henryk Czosnek
The Hebrew University of Jerusalem, Rehovot, Israel.

ABSTRACT

Although not the first virus geminivirus discovered, *Tomato yellow leaf curl virus* (TYLCV) has been one of the most studied. TYLCV impairs tomato cultures worldwide. Within less than 20 years TYLCV has spread from the Middle East to North America, Africa, Europe and Far East Asia. Molecular analyses have shown that the Tomato leaf curl disease is caused by several virus species and many strains related but different from the Middle Eastern TYLCV. This virus is an ideal object to study such topics in molecular virology as diverse as replication, expression, evolution, recombination and interaction with host factors. TYLCV is transmitted by the whitefly vector to plants in a circulative manner. The interactions between the virus, the vector and the target plant are the object of intense scrutiny. TYLCV causes a serious disease of tomato that needs to be avoided by all means. Hence ways to escape inoculation by whiteflies have been implemented, which include nets, chemical treatment to curb down whitefly populations, traps and natural enemies. Breeding for resistance has started already 30 years ago by introgressing resistance traits found in wild tomato species into the susceptible domesticated tomato. Breeding has been backed by genetic studies of resistance, discovery of QTLs and polymorphic DNA markers linked to resistance, and screening for resistance genes. Commercial lines with good tolerance to TYLCV are available; some are effective against several different TYLCV species, others are not. Altogether the study of TYLCV and the results achieved in molecular virology, genetics, plant-virus-insect interactions, IPM and breeding for resistance, can serve as model for similar studies involving other deleterious geminiviruses infecting important crops, both for basic and applied interests.

Introduction

Changes in agriculture practices, global transportation of propagative materials, vegetables, fruits and flowers, have resulted in numerous emerging infectious diseases affecting cultivated plants. The geminiviruses emerged as one of the most important groups of plant viruses. With negligible economic impact until the 1970s, today geminivirus-induced diseases are among some of the most economically important in vegetable and field crops, including beans, cassava, cotton, cucurbits, maize, pepper, and tomato. In southern Brazil, *Bean golden mosaic virus* (BGMV), a minor crop pathogen before intensification of soybean cultivation, emerged to become the limiting factor in dry season bean production. In Argentina, *Bean dwarf mosaic virus* (BDMV) threatened the collapse of bean production. African cassava mosaic disease is presently at pandemic levels on the African continent and is causing losses estimated as high as $2 billion annually (Legg and Fauquet, 2004). *Maize streak disease* is similarly a major constraint on corn production in Africa (Palmer and Rybicki, 1998). *Beet curly top disease*, which almost eliminated sugar beet production from the western United States in the early 1900s, continues to plague tomato and sugar beet production in this region (Stenger and McMahon, 1997). Finally, tomato yellow leaf curl disease limits tomato production in many regions of the world (Czosnek, 2007), and bean golden mosaic disease is distributed throughout the Americas, where it can completely decimate common bean production (Morales and Anderson, 2001).

The genome of the most important begomoviruses has been cloned and sequenced. The acquisition and transmission parameters by their inset vector have been studied. The host range of these viruses has been determined and diagnostic tools have been developed. Despite all these advances, management of the begomovirus-induced diseases is still problematic and crops are still threatened by the diseases they induce. Despite concerted efforts to contain geminiviruses and their vectors, menacing disease epidemics caused by newly emerging or re-emerging geminiviruses are becoming frequent and appearing even in new regions, previously free from such diseases.

In this chapter we will review the strategies, advances and problems of management of begomovirus induced diseases. We have taken *Tomato yellow leaf curl virus* as a case study since TYLCV is one of the best known begomoviruses and one where management has been mostly successful.

Geminiviruses

Common Features

Geminiviruses form the second largest family, of plant viruses, the *Geminiviridae* (for recent review, see Krupovic et al., 2009). They have unique twinned icosahedral particle morphology, of about 20 x 38 nm in size (Zhang et al., 2001). Begomoviruses have a genome either split between two circular single-stranded DNA molecules of approximately 2,700-2,900 nucleotides each named DNA A and DNA B (bipartite) or with a single genomic DNA A-like molecule (monopartite). Virions contain a single copy of circular ssDNA, hence, for viruses with bipartite genomes, two virions containing different genomic components are required for infection. The genome of begomoviruses share several features (see drawings of genetic maps in Rojas et al., 2005, Varma and Malathi, 2003). The coat protein (CP) is encoded by the DNA-A genomic component. This is the only protein that builds the capsid (incomplete icosahedra of 22 capsomeres each containing five ~30 kDa CP copies) and that dictates the specificity of the insect that vectors the virus (Bridon et al., 1990). The replication associated protein (Rep) is encoded by the DNA-A genome complementary strand. It has a primordial role in the initiation and termination of the rolling circle replication process of the viral genomes (Hanley-Bowdoin et al., 1999). The genes involved in virus movement are encoded by the DNA-B of bipartite genomes, while the movement function of monopartite viruses are ensured by the DNA-A. Some monopartite begomoviruses are associated with small circular satellite ssDNA (named β-DNA or DNA-1), that facilitates movement and enhance symptoms. All geminiviruses also carry an intergenic regions (IRs), which contains the origin of replication and a stem-loop structure containing the invariant nonanucleotide motif TAATATTAC involved in rolling circle replication. Geminiviruses utilize bidirectional transcription and overlapping genes for efficient coding of proteins and expression.

Geminivirus Genera

Geminiviruses have been grouped into four genera: *Mastrevirus*, *Curtovirus*, *Topocuvirus* and *Begomovirus*, depending on their vector, host range and genomic characteristics (Fauquet and Stanley, 2003). During the last two decades a large number of new begomoviruses causing mosaic and leaf curl symptoms have emerged in various parts of the world. New mastreviruses and curtoviruses have also emerged but to a lesser extent.

Mastreviruses are transmitted by leafhoppers, in most cases by a single species, in a circulative, non-propagative manner. The type species is the *Maize streak virus* (MSV). They have a monopartite genome. They infect monocotyledonous hosts, mostly grasses. Mastreviruses induce bright yellow streaks in the monocotyledonous hosts, and general yellowing and dwarfing.

Curtoviruses are transmitted by the beet leafhopper *Circulifer tenellus*. The type species is the *Beat curly top virus* (BCTV). They have a monopartite genome. They infect a wide range of dicotyledonous plants. Curtoviruses cause typical curly top symptoms in their hosts.

Topocuviruses are transmitted by the treehopper *Micrutalis malleifera*. The topocuvirus has only one member, *Tomato pseudo curly top virus* (TPCTV), which has a monopartite genome. It infects dicotyledonous plants. Analysis of the TPCTV genome reveals features typical of both mastreviruses and begomoviruses, consistent with it being a natural recombinant.

Begomoviruses are transmitted by the whitefly *Bemisia tabaci* in a circulative manner. More than 80% of the known geminiviruses belong to the genus *Begomovirus*, they mostly have bipartite genomes and infect dicotyledonous plants. Begomoviruses such as *Tomato yellow leaf curl virus* and related tomato-infecting viruses have a single genomic component. Some monopartite begomoviruses depend on a ssDNA satellite component, designated DNA β. Diverse symptoms develop in plants infected with begomoviruses, including leaf curl, vein yellowing and yellow mosaic.

TOMATO YELLOW LEAF CURL DISEASE, A RAPIDLY EXPANDING PANDEMIC

In 1959 tomatoes grown in the Jordan valley, Israel, became affected by an unknown disease and yields were greatly reduced. The plants were severely stunted and the yellowing leaflets were cupped down and inward. The disease was accompanied by large populations of whiteflies. Controlled transmission experiments indicated that the whiteflies were able to transmit the disease found in the field in a manner consistent with a viral agent. The virus was named *Tomato yellow leaf curl virus* (TYLCV) (Cohen and Harpaz, 1964). It was isolated in 1988, proven to be monopartite and sequenced in 1991 (Czosnek et al., 1988; Navot et al., 1991).

From the early 1960s, the Tomato yellow leaf curl disease has quickly spread to the entire Middle East and is presently found in many regions of Africa, America and Asia. It has been reported in the mid and late 1970s in Cyprus, Jordan, and Lebanon. It has been identified in Egypt and Turkey in the early 1980s, and in the mid-late 1990s in Iran, the Asian republics of the former USSR, Japan, Saudi Arabia and Yemen. In the early 1990s, the disease has been identified in Italy, Spain and Portugal, and later in France and Greece. In Morocco and Tunisia it was identified in the early 2000s. In East Africa, the Tomato yellow leaf curl disease was present in Sudan as early as the late 1970s. In the Réunion Island the disease was detected in the late 1990s. It has appeared in the Western Hemisphere in the mid 1990s in the Caribbean Islands, first in the Dominican Republic, then Cuba, Jamaica, Puerto Rico and the Bahamas. From there, the disease has reached the USA, identified first in Virginia in the late 1990s, then in Florida, Georgia, Louisiana, North Carolina and Mississipi. The Tomato yellow leaf curl disease has been identified in several regions of Australia (2003), Mexico (2007), in Arizona and in California (2007).

Sequencing of begomoviruses inducing the Tomato yellow leaf curl disease has shown that it was not induced by a single virus species. The TYLCV type from Israel has been found in regions as diverse as Spain (Moriones and Navas-Castillo, 2000), Japan (Kato *et al.* 1998), China (Wu, J. *et al.* 2006) and Australia (Tesoriero and Azzopardi 2006), It is interesting to note that TYLCV has not been reported in South America so far. A second TYLCV species has been described in Israel and coined *Tomato yellow leaf curl virus Mild* (TYLCV-Mld) (Antignus and Cohen, 1994). A new TYLCV-like species has been discovered in Sardinia Italy in the early 1990s and was named *Tomato yellow leaf curl Sardinia virus* (TYLCSV) (Kheyr-Pour et al., 1991). A relative of TYLCSV was identified in Sicily (Crespi et al., 1995). TYLCSV has spread since to Spain, Morocco and Tunisia (Sánchez-Campos *et al.*, 1999; El Mehrach et al., 2007; Gharsallah Chouchane *et al.*, 2006). TYLCV and TYLCSV recombinants have also been found (Monci *et al.*, 2001). The spread of TYLCV westward causing the displacement of the endogenous TYLCSV, especially in Spain (Sánchez-Campos *et al.*, 1999), and the invasion of the Americas have lead to the assumption that TYLCV-associated diseases radiate from the Middle East. However in a recent study, besides the endogenous TYLCV and TYLCV-Mld species, the Sicilian strain of TYLCSV and a Spanish strain of TYLCSV (TYLCSV-ES) have been found in Jordan and Israel demonstrating that TYLCSV has spread eastward (Anfoka et al., 2008).

Recent multiple introduction of members of the TYLCV complex in Spain provides an excellent model to analyze aspects of adaptation and evolution of an invading virus population (Sánchez-Campos *et al.*, 1999, 2002). Initial colonization with isolates of the Spanish strain of *Tomato yellow leaf curl Sardinia virus* (TYLCSV-ES) during the early 1990's, resulted in a relatively stable population in which reduced genetic diversity was observed (Sánchez-Campos *et al.*, 2002). Subsequent introductions of TYLCV and TYLCV-Mld (Navas-Castillo *et al.*, 1999; Morilla *et al.*, 2003) resulted in novel sources of variation, and conditions for recombination to occur. This was the case of the novel recombinant variant named *Tomato yellow leaf curl Málaga virus* (TYLCMalV) that emerged as a result of a genetic exchange between TYLCSV-ES and TYLCV-Mld. This natural recombinant variant showed to be better adapted ecologically than either parental virus and spread rapidly in the population (Monci *et al.*, 2002). Therefore, recombination showed to be an important force driving the evolution of the population of these viruses for adaptation to the novel ecological conditions present in the invaded area.

It seems that human activities such as commercial transfer or smuggling of infected plantlets are responsible for the rapid spread of TYLCV worldwide. The newcomer may have a better interaction with the whitefly vector than the endogenous geminiviruses. For example the new recombinant TYLCMalV became prevalent in Spain because of its better acquisition by the whiteflies and its broader host range than both of its parents (Monci *et al.* 2002). The accidental translocation of viruliferous whiteflies by winds, or by plane transportation (e.g. on ornamentals) can also be instrumental in the expansion of TYLCV. It is interesting to note that some regions have not yet been invaded by the virus such as South America or India. It is also possible that in these regions TYLCV isolates entered in unsuccessful competition with local geminiviruses.

Management of Tomato Yellow Leaf Curl Disease

Difficulties in Curbing the Spread of TYLCV

Whitefly control is a difficult task. This animal is well adapted to extreme environmental conditions. In addition, the agricultural practices favor the rapid colonization of important crops. Adults can survive extreme conditions of low and high temperature and humidity. Eggs can develop after long periods of cold and desiccation (Bosco and Caciagli, 1998; Lacy et al., 1999). Although usually sedentary, whiteflies are able to move up to several km, using wind-assisted flight (Cohen et al., 1988). Tomatoes are often grown in proximity to cotton fields, which constitute an excellent reservoir of whiteflies. Cotton harvesting often results in rapid whitefly migration to the tomato fields.

Acquisition and transmission studies in the laboratory have demonstrated that TYLCV is transmitted very efficiently by the whitefly *B. tabaci* (Czosnek, 2007). In controlled conditions, a single insect is able to infect one tomato plant (and more). Fifteen minutes of feeding on an infected tomato plant is sufficient for a whitefly to become infectious. TYLCV inoculation can occur following 15 minutes feeding period on a tomato test plant (Cohen and Nitzany, 1966). Adult whiteflies can retain the virus and spread the TYLCV disease for almost their entire adult life (Rubinstein and Czosnek, 1999).

The rapidity and efficacy with which TYLCV is transmitted to tomato plants makes whitefly control very difficult. Tomato yellow leaf curl disease outbreaks are usually managed by preventing contact between viruliferous whiteflies and the crop in the field or greenhouse. This can be achieved by using physical barriers such as nets or by curbing down the whitefly populations with insecticides. Nets need to be hermetically close since a few insects reaching the protected plants through openings may endanger the entire crop. The ability of the insecticide to kill viruliferous whiteflies before they induce damage (sucking or virus transmission) is critical for the efficacy of the insecticide.

Chemical Control of the Whitefly Vector

A number of different classes of chemicals have been used to reduce whitefly populations in the field to the minimum. The insecticide has to be extremely efficient in the field because of the very low number of insects needed to inoculate tomato. The most effective and widely used class of insecticides is the neonicotinoids, of which imidacloprid is the most popular. Laboratory experiments have shown that to be efficient in preventing inoculation of TYLCV, the insecticide has to induce the death of all whiteflies before the minimum period of 15 min during which they are able to transmit the virus. However the mean 50% mortality point was 35-40 min by 3 days after imidacloprid application. Increasing the concentration of the compound given to the soil did not lower these values (Rubinstein et al., 1999). Management of TYLCV or any other whitefly-transmitted geminivirus by the use of imidacloprid is effective only if the insecticide is delivered on a large area. Invasion of a treated field by large amount of viruliferous whiteflies originating from an untreated field will lead to a high level of infection.

Due to heavy applications, resistance to neonicotinoids has developed in several locations around the world (Nauen and Denholm, 2005). On the long-range, insecticide-resistant whiteflies populations, or resistant biotypes, may displace the susceptible resident whiteflies. A survey conducted in Israel in 1999-2000 has provided evidence for the presence of a new invasive *B. tabaci* Q biotype, alongside the endogenous B biotype (Horowitz et al., 2005). Most of the Q biotype populations were highly resistant to pyriproxyfen, while all the B biotype populations were susceptible to this insecticide.

Natural Enemies

Biological control of field populations of *B. tabaci* has been attempted by the introduction, conservation and augmentation of natural enemies such as *Eretmocerus* spp. (Goolsby et al. 2005). However the use of commercial use of natural enemies is still embryonic because its efficacy is dependent on many external factors such as temperature and humidity, on the host plant and on the size of the target whitefly population.

Cultural Practices

It is important to eliminate the potential virus reservoirs and reduce whitefly breeding host plants before tomato planting. For instance, tomato field can be selected to be isolated from cotton and bean. When possible, a mandatory whitefly host-free period of several weeks should be implemented. Such a strategy was proven successful to curb the incidence fo TYLCV in the Dominican Republic (Salati et al., 2002). Seedlings used to be transplanted in the field and greenhouses need to be virus-free. An application of amidocloprid in the nursery a week before plants are set in the field will protect plants for the first two weeks. Yellow sticky traps are helpful for suppressing adult populations, especially in greenhouses. However, their effect is negligible in open field where vast populations colonize tomato plants at the peak of the culture season.

Physical Barriers

Greenhouses covered with 50-mesh screens have been used in the Mediterranean Basin since 1990 to protect tomato crops from whitefly-transmitted TYLCV (Berlinger et al., 2002). Combined with a few insecticide sprays, the incidence of TYLCV in the covered structures was only 1%, compared with 100% disease incidence in unscreened greenhouses. Ultraviolet-absorbing plastic films are used in Israel as greenhouse covers or insect-proof nets (Antignus et al., 1996). These UV-absorbing films disturb the whitefly vision, and reduce its penetration and movement in greenhouses. TYLCV incidence in tomato grown under the UV-absorbing sheets was only 1%. It is important to note that besides the higher production cost, the use of nets and screens may also result in increased temperature and humidity inside greenhouses, requiring a good ventilation system.

Diagnosis

Since the isolation and sequencing of TYLCV (Czosnek et al., 1988; Navot et al. 1991), several diagnostic tools have been developed, both for mass screening and for more specific characterization (Accoto and Noris, 2007). Because of the conserved sequence of the coat protein and of the capsid shape, polyclonal and monoclonal antibodies are not able to specifically recognize TYLCV and distinguish between the various TYLCV species. Molecular hybridization was efficiently used after the virus was cloned. Radioactive probes were used to Southern-blot detect TYLCV in plants and viruliferous whiteflies (Czosnek et al., 1989) as well as in plant and insect tissues squashed on membranes (Navot et al., 1989). Non-radioactive labels were introduced to facilitate routine diagnosis (Zilberstein et al., 1989; Crespi et al., 1991) and virus quantification in plants and insects (Caciagli and Bosco, 1997). Both in Spain and Italy epidemiological studies on TYLCSV and TYLCV were conducted using tissue printing on membranes that were consecutively hybridized with two species-specific probes (Sanchez-Campos *et al.*, 1999; Davino *et al.*, 2006).

The advent of PCR has revolutionized begomovirus diagnostics. Specific primers allowed amplifying the full-length of the TYLCV genome (Navot et al., 1992). Degenerate primers have been designed to amplify TYLCV species (Accotto et al., 2000) as well as the DNA-A and DNA-B of unknown begomoviruses (Rojas et al., 1993). Print-PCR protocols have been successfully employed. For example, squashes of plant or whitefly tissue on small pieces of nylon membranes or Whatman 3MM paper were directly used in amplification reactions to study TYLCV acquisition and transmission, omitting the DNA extraction step (Atzmon *et al.*, 1998; Navas-Castillo et al., 1998).

Multiplex PCR, sometimes combined with restriction fragment length polymorphism (RFLP) has allowed to amplifying DNA fragments belonging to several species of TYLCVs in a single reaction, using several pathogen-specific pairs of primers. The virus is identified by the size of the amplified DNA fragment or after the PCR product has been digested by an indicative restriction enzyme. This method has been useful in detecting TYLCV and TYLCV-Mld in tomato plants and viruliferous insects (Lefeuvre et al., 2007). Multiplex PCR-RFLP has allowed to discriminated between TYLCV and TYLCSV and their recombinant, the *Tomato yellow leaf curl Malaga virus* (TYLCMalV) (Davino et al., 2008). Multiplex PCR has allowed to show that, besides the endogenous TYLCV and TYLCV-Mld species, the Sicilian strain of TYLCSV and a Spanish strain of TYLCSV (TYLCSV-ES) have been found in Jordan and Israel (Anfoka et al., 2008).

The use of real-time quantitative PCR (qPCR) to quantify the pathogen associated with plants (Korimbocus et al., 2002) and insects (Lett et al., 2001) has increased during the last few years. The distribution and amounts of TYLCSV were measured in infected tomato plants and in viruliferous whiteflies (Mason et al., 2008). The method allowed detection of as few as 10 viral copies in a sample. The real-time PCR was 2200-fold more sensitive than membrane hybridization. qPCR was instrumental in measuring the amounts of TYLCV in resistant (R) and susceptible (S) tomato genotypes, indicating that young S leaves contained three orders of magnitude more virus than R leaves. Upon silencing of a Permease I-like gene preferentially expressed in R plants, the amounts of TYLCV in infected R plants was similar to that in infected S plants (Eybishtz et al., 2009).

Rolling circle amplification (RCA), a method which utilizes the DNA polymerase from bacteriophage Φ29, has been proposed for diagnosis of geminiviruses (Haible *et al.*, 2006). It is able to amplify the geminiviral full-length genome as well as defective and satellite ssDNA molecules. Combined with restriction fragment length polymorphism (RFLP), RCA allowed discriminating between different mastreviruses (Schubert et al., 2007).

Breeding for Resistance to TYLCV

Wild Tomato Species as Source for Resistance Genes in Breeding Programs

Up to about 20 years ago, all the domesticated tomato cultivars (*Solanum lycopersicum*) were susceptible to TYLCV. Breeding programs for resistance to TYLCV have started in the middle 1970s and several commercial varieties with adequate resistance have been released since (Lapidot et al., 1997). Early efforts to identify sources of resistance to TYLCV within the domesticated tomato were unsuccessful. It was then observed that some accessions of several wild relatives of tomato, such as *S. pimpinellifolium*, *S. peruvianum*, *S. chilense* and S. *habrochaites* remained symptomless upon whitefly-mediated inoculation of TYLCV; in some cases, virus was barely detectable by Southern blot hybridization (Cohen and Nitzany 1966; Zakay et al., 1991). Crosses between *S. lycopersicum* and various accessions of wild tomato species followed by the genetic analysis of segregation of resistance in F2, backcrosses and

F3 populations allowed to determine the number of loci associated with resistance to TYLCV. The results indicated that any one of several loci (or combinations of several loci) could provide good levels of resistance to the virus. In *S. pimpinellifolium* accession LA121 an incomplete dominant locus controlled resistance (Pilowsky and Cohen 1974) while accession LA1478 provided a dominant locus (Kasrawi 1989). In contrast, the inheritance of resistance to TYLCV in accession PI 126935 of *S. peruvianum* is controlled by three to five recessive factors (Pilowsky and Cohen 1990). Breeding for resistance consisted in introgressing resistance from the wild tomato species into the domesticated tomato. Selection in the field was initially based on symptom appearance and yield. Controlled whitefly-mediated inoculation in the laboratory prior to planting and TYLCV diagnosis with molecular tools (Navot et al., 1989) increased the efficacy of prediction and selection and avoided most escapes (Vidavsky et al., 1998). The first attempts at breeding for TYLCV-resistant tomato plants were made in the early 1970s using *S. pimpinellifolium* accession LA 121 as the resistant host (Pilowsky and Cohen, 1974). Because the resistance level provided by LA 121 was found to be insufficient, efforts were shifted to accession PI 126935 of *S. peruvianum*, which was found to express a higher level of TYLCV resistance. In 1986, the first commercial TYLCV-resistant tomato hybrid named TY20 was released (Pilowsky and Cohen, 1990), followed by several highly TYLCV-resistant lines (Lapidot *et al.*, 1997). Additional accessions of wild tomato species such as *S. chilense* and S. *habrochaites*, were found to be resistant (low virus amounts and absence of symptoms) in controlled whitefly-mediated inoculation experiments (Zakay et al. 1991). In *S. chilense* resistance is determined by a semi-dominant locus and several minor loci, different for each accession (Zamir et al. 1994; Agrama and Scott 2006). Breeding projects in the Mediterranean region have also used *S. cheesmanii*, *S. peruvianum* and S. *pimpinellifolium* as wild genitors (Laterrot and Moretti 1996). TYLCV resistance introgressed from *S. habrochaites* accessions LA1777 and LA386 was found to be under the control of a major dominant locus (Vidavsky and Czosnek 1998). S. *habrochaites* has been back-crossed with *S. lycopersicum*, yielding tolerant and resistant lines. One of the resistant lines was back-crossed with S. *lycopersicum*, to produce the tolerant hybrid Favi-9 (Vidavsky and Czosnek 1998).

Mapping of Loci Associated with Resistance to TYLCV

Breeding for resistance to TYLCV has consisted in introgressing the resistant traits found in wild tomato species in the cultivated tomato. As a result, the resistant tomato contains chromosomal fragments from the wild species on a background of the domesticated tomato, identifiable with polymorphic DNA markers (Ji et al. 2007).

Resistance introgressed from *S. chilense* accession LA1969 was found to be controlled by a major partially dominant locus on the short arm of tomato chromosome 6, named *Ty-1*, and at least two minor loci mapped to chromosomes 3 and 7 (Zamir et al., 1994). TYLCV resistance derived from *S. pimpinellifolium* Hirsute-INRA was found to be mediated by a single dominant gene (Kasrawi, 1989). A quantitative trait locus (QTL), accounting for about 28% of the variation in symptom severity, was mapped on chromosome 6, at a location different from that of Ty-1 (Chagué et al. 1997). A dominant locus in line H24 developed in Taiwan and India, originating from S. *habrochaites*, was mapped to the short arm of chromosome 11. This locus, coined *Ty*-2, conferred resistance to the Indian Tomato leaf curl virus (TLCV) but not to the Mediterranean TYLCV (Hanson et al., 2000, 2006). The accession from *S. chilense*, LA2779 was found to be resistant to the Mediterranean TYLCV as well as to the bipartite *Tomato mottle virus* (ToMoV) (Agrama and Scott, 2006). A major partially dominant gene, termed *Ty-3*, was mapped to chromosome 6, near the *Ty-1* locus (Ji et al., 2007). Another TYLCV resistance locus, termed *Ty-4*, was mapped on the long arm of Chromosome 3 (Ji et al., 2008). While approximately 60% of the variance in the TYLCV resistance in a segregating population was explained by the *Ty-3* locus, *Ty-4* accounted for only 16%. A QTL was mapped using the breeding highly resistant line TY172, which originating from *S. peruvianum* (Lapidot et al., 1997; Friedmann et al., 1998). The major QTL, termed *Ty-5*, maps to chromosome 4 and accounts for approximately 45% of the variation in symptom severity among segregating plants. Three minor QTLs were mapped to chromosomes 1, 7, 9 and 11, and contributed 12% to the variation in symptom severity (Anbinder et al., 2009). The loci conferring resistance to TYLCV in *S. habrochaites* have not been localized yet.

Pile up of Resistance Sources

Breeding usually uses a single wild tomato as resistance source. Pyramiding resistances from different wild tomato species might improve the degree of resistance of the domesticated tomato to TYLCV, shorten the time for breeding commercially valuable tomato resistant to TYLCV, with higher levels of resistance and higher yields than each of the resistant parents.

TYLCV-resistant lines that originated from different wild tomato genitors (*S. chilense*, *S. peruvianum*, *S. pimpinellifolium*, and *S. habrochaites*) were crossed among themselves and with a susceptible commercial cultivar (Vidavski et al., 2008). The parental lines and the F1 hybrids were inoculated using viruliferous whiteflies. Control, non-inoculated plants of the same lines and hybrids were exposed to non-viruliferous whiteflies. The inoculated and control plants were transplanted to the field. Resistance was assayed by comparing symptom severity and yield of inoculated plants to those of the control non-inoculated plants of the same variety. Results showed that the F1 hybrids between the resistant lines and the susceptible line suffered major yield reduction because of infection, but all hybrids were more resistant than the susceptible parent, indicating a dominant effect. All F1 hybrids resulting from a cross between two resistant parents, showed a relatively high level of resistance, which in most cases was similar to that displayed by the more resistant parent. In some cases, the hybrids displayed better levels of resistance than both parents. The F1 hybrid between a line with resistance from *S. habrochaites* and a line with resistance from *S. peruvianum* exhibited the lowest yield loss and the mildest level of symptoms. Therefore, pyramiding the chromosomal regions associated with resistance from different origins improves resistance to TYLCV (and other begomoviruses). The combination of breeding with the use of molecular markers linked to the different sources of resistance will allow pyramiding of the genes, distinguishing between the different sources of resistance and combining all TYLCV resistances from the three main wild tomato species.

Resistance to Multiple Begomoviruses

The wild tomato species used as TYLCV resistance source also provide resistance to other begomoviruses. *S. chilense* (accessions LA1932, LA 1938 and LA 2779), which was used as source for resistance to the monopartite Mediterranean TYLCV was also used as a source of resistance to the New World bipartite begomovirus *Tomato mottle virus* (ToMoV) (Scott and Schuster 1991). The ToMoV-resistance introgressed from LA 1932 was governed by two genes. The genes were additive: a hybrid between a resistant and a susceptible parent (heterozygous resistance) had intermediate resistance (Griffiths and Scott 2001). Accessions of *S. habrochaites* (PI 390658 and PI 390659) and *S. peruvianum* (PI 127830 and PI 127831) were resistant to another tomato geminivirus present in India, the *Tomato leaf curl virus* (Muniyappa et al., 1991). The resistance mechanism in these wild species was subsequently associated with the presence of exudates from trichome glands on the leaf surface (Channarayappa et al., 1992). Other wild tomato species used as source of TYLCV resistance (*S. pennelli* accession LA2560 and *S. habrochaites* PI127826) produce whitefly repellent volatiles that may be used in breeding programs (Bleeker et al., 2009).

The tomato lines with improved resistance to TYLCV (Zamir et al., 1994; Vidavski and Czosnek, 1998; Friedmann et al., 1998; Ji et al., 2007a) also showed resistance to several bipartite begomoviruses in Central America (Mejía et al. 2005) and India (Maruthi et al., 2003), and to monopartite viruses in South Africa (Pietersen and Smith, 2002). Hence, the TYLCV resistant genotypes and hybrids present a broad-range of resistance and could be used as genitors in programs aimed at breeding tomato adapted to local market and consumer preferences.

Genetic Engineering for Resistance

The genetics of resistance to TYLCV and other begomoviruses is complicated (see above) and likely involves multiple gene systems, some recessive others dominant, with partial additivity. Genetic engineering technologies have been applied to cope with the TYLCV disease problem. The transgene is usually dominant and behaves as a single gene.

A variety of strategies have been devised based on the pathogen-derived resistance concept (Sanford and Johnston, 1985), which involves the expression of functional as well as dysfunctional pathogen genes such as coat protein, replicase, and movement protein (Wilson, 1993). The begomovirus *CP* gene was one of the first TYLCV genes evaluated for the ability to generate pathogen-derived resistance. Tomato plants expressing the TYLCV CP showed a delay in symptoms, a recovery from infection, and resistance upon repeated inoculations (Kunik et al., 1994). Expression antisense RNA of the TYLCSV *Rep* gene provide a good level of resistance in *Nicotiana benthamiana* upon agroinfection (Bendahmane *et al.* 1997). The expression of sense and antisense constructs of truncated TYLCSV Rep in *N. benthamiana* showed resistance upon agroinfection (Noris *et al.*, 1996). When a truncated TYLCV-Mld Rep gene was expressed in tomato, the transgenic plants were resistant to whitefly-mediated inoculation of TYLCV-Mld, but were susceptible to inoculation with TYLCV (Antignus et al., 2004). A 2/5 TYLCV *Rep* gene construct conferred high levels of resistance and often immunity in both transformed *N. tabacum* and *S. lycopersicum* (Yang et al., 2004).

RNA-mediated virus resistance proved to be more efficient than protein-mediated resistance but was shown to be highly-sequence dependent and therefore with a narrower-range potential. It was shown to trigger existing plant antiviral mechanisms leading to the degradation of viral RNA, and was collectively named RNA-mediated gene silencing. It was found that some viruses establish themselves by expressing viral proteins which interfere with plant host silencing mechanisms (Voinnet et al., 1999). Transgenic plants have been developed which exploited the mechanism of silencing via double-stranded RNA sequences (Smith et al., 2000). However, transgenic RNA-mediated resistance can be overcome by strong silencing suppressors of non-related viruses in mixed infections (Mitter et al., 2001). Using a transformation cassette consisting of the 3'-end of the *Rep* (sense and antisense orientation) as the arms of the hairpin, and a castor bean catalase as the intron, the treated tomato plants showed immunity to TYLCV upon whitefly-mediated inoculation (Fuentes et al., 2006). A similar approach was used with the 5'-end of the TYLCV CP gene and a maize ubiquitin intron, inducing resistance to TYLCV in tomato (Zrachya *et al.* 2006). The intron-airpin approach was used to express multiple TYLCV sequences in tobacco and tomato plants (Abhary *et al.*, 2006). The treated plants were symptomless and TYLCV DNA was not detected. In general, silencing was specific of the TYLCV strain; silencing using TYLCV sequences conferred resistance to TYLCV but not to other TYLCV species or strains, such as TYLCSV (Noris et al., 2004).

We have devised a new strategy based on the biology of whitefly-begomovirus interactions. Some, and perhaps all, plant viruses transmitted by their insect vectors in a circulative manner interact in the insect haemolymph with GroEL homologues produced by the vector endosymbiotic bacteria. Binding between GroEL and virions has been shown to occur *in vivo* and *in vitro* (van den Heuvel et al., 1994; Morin et al., 1999, 2000; Akad et al., 2004). The ability of GroEL to bind the TYLCV CP was exploited to generate transgenic tomato plants expressing the whitefly GroEL in their phloem. We expected that once inoculated by their vector, phloem-limited circulative viruses will be trapped by GroEL in the plant phloem. As a result invasion of phloem-associated cells and long distance movement may be significantly inhibited, rendering the plants resistant to the virus. Indeed GroEL was immunolocalized in the vascular system of the transgenic tomato plants. For at least three generations, the plants expressing the insect GroEL remained almost symptomless upon whitefly-mediated inoculation, but contained significant amounts of virus. TYLCV-GroEL complexes were detected in the tomato phloem (Akad et al., 2007).

Discovering Genes Underlying Resistance to TYLCV

At least six different loci contributing significantly to resistance to TYLCV have been localized on tomato chromosomes (see above). However TYLCV resistance genes have not been identified so far. The number of loci discovered so far and their specific contribution to resistance led us to postulate that, as in the case of many viruses, resistance to TYLCV is the result of gene networks that respond to biochemical triggers induced by virus inoculation (Goldbach et al. 2003; Murphy et al. 2001). Hence describing the networks sustaining resistance to TYLCV and discovering the participating genes are of primary importance to understanding and combating the TYLCV disease. To achieve this goal, we took advantage of two inbred tomato lines issued from the same breeding program that used *S. habrochaites* as a source of resistance: the TYLCV-susceptible line 906-4 (S) and the TYLCV-resistant line 902 (R).

Postulating that genes involved in the resistance gene network will be expressed at higher levels in R vs. S plants, we compared the trancriptomes of R and S tomato lines, before and after inoculation. We assumed that if these genes arc located at important nodes in the network, resistance will collapse upon silencing using the Tobacco rattle virus-induced gene silencing (TRV VIGS) (Liu et al., 2002). This approach proved fertile. From the ca. 300,000 clones assayed, 69 transcripts were found to represent genes preferentially expressed in R plants. We focused our attention on membrane proteins that may be involved in signal transduction. Silencing of a *Permease I-like protein* gene, which is preferentially expressed in R plants and up-regulated upon TYLCV inoculation, led to the appearance of discase symptoms similar to those in infected S plants, and in the early accumulation of large amounts of virus (Eybishtz et al., 2009). Another membrane protein gene, the hexose transporter *LeHT1* was found to be expressed preferentially in R tomato plants. Following TYLCV inoculation, *LeHT1*-silenced R plants showed inhibition of growth, and enhanced virus accumulation and spread. In addition, a necrotic response characteristic of programmed cell death (PCD) was observed along the stem and petioles of infected *LeHT1*-silenced R plants only, but not on infected not-silenced R plants (Eybishtz et al., 2010). These results demonstrate that the hexose transporter *LeHT1* is essential for the expression of natural resistance against TYLCV and its expression correlates with inhibition of virus replication and movement. They constitute the first evidence for a necrotic response backing natural resistance to TYLCV in tomato, confirming that plant defense is organized in multiple layers. The role of other R-specific genes is presently under scrutiny.

Conclusion

Although TYLCV was not the first geminivirus identified (Goodman, 1977; Harrison et al., 1977), the increasing worldwide impact of TYLCV on the tomato industry has made this virus a frontrunner in research in the many aspects of begomovirus research and management. Today TYLCV research covers aspects as diverse as breeding tomato for resistance (classical, molecular-assisted, genetic engineered), integrated pest management, plant-virus-vector interactions, and molecular plant virology. The methodologies developed for TYLCV and the results obtained in any aspect of TYLCV research and management could be applied to research and management of other geminiviruses.

References

Abhary, M. K., Anfoka, G. H., Nakhla, M. K. & Maxwell, D. P. (2006). Post-transcriptional gene silencing in controlling viruses of the *Tomato yellow leaf curl virus* complex. *Archives of Virology,15,* 291–304.

Accotto, G. P. & Noris, E. (2007) Detection methods for TYLCV and TYLCSV. In: *The Tomato Yellow Leaf Curl Virus Disease: Management, Molecular Biology and Breeding for Resistance*. H. Czosnek Ed. Springer, The Netherlands, Pp 241-249.

Accotto, G. P., Navas-Castillo, J., Noris, E., Moriones, E. & Louro, D. (2000). Typing of tomato yellow leaf curl viruses in Europe. *European Journal of Plant Pathology, 106,* 179-186.

Agrama, H. A. & Scott, J. W. (2006) Quantitative trait loci for *Tomato yellow leaf curl virus* and *Tomato mottle virus* resistance in tomato. *Journal of the American Society for Horticultural Science,131,* 267-272.

Akad, F., Dotan, N. & Czosnek, H. (2004) Trapping of *Tomato yellow leaf curl virus* (TYLCV) and other plant viruses with a GroEL homologue from the whitefly *Bemisia tabaci*. *Archives of Virology,149,* 1481-1497.

Akad, F., Eybishtz, A., Edelbaum, D., Gorovits, R., Dar-Issa, O., Iraki, N. & Czosnek, H. (2007) Making a friend from a foe: Expressing a GroEL gene from the whitefly *Bemisia tabaci* in the phloem of tomato plants confers resistance to *Tomato yellow leaf curl virus*. *Archives of Virology,152,* 1323-1339.

Anbinder, I., Reuveni, M., Azari, R., Paran, I.,·Nahon, S., Shlomo, H.,· Chen, L., Lapidot, M. & Levin, I. (2009) Molecular dissection of *Tomato leaf curl virus* resistance in tomato line TY172 derived from *Solanum peruvianum. Theoretical and Applied Genetics,119,* 519–530.

Anfoka, G., Abhary, M., Haj Ahmad, F., Hussein, A. F., Rezk, A., Akad, F., Abou-Jawdah, Y., Lapidot, M., Vidavski, F., Nakhla, M. K., Sobh, H., Atamian, H., Cohen, L., Sobol, I., Mazyad, H., Maxwell, D. P. & Czosnek, H. (2008) Survey of tomato yellow leaf curl disease –associated viruses in the eastern mediterranean basin. *Journal of Plant Pathology,90,* 311-320.

Antignus, E. Y. & Cohen, S. (1994). Cloning of tomato yellow leaf curl virus (TYLCV) and the complete nucleotide sequence of a mild infectious clone *Phytopathology,84,* 707-712.

Antignus, Y., Mor, N., Joseph, R. B., Lapidot, M. & Cohen, S. (1996). Ultraviolet-absorbing plastic sheets protect crops from insect pests and from virus diseases vectored by insects. *Environmental Entomology,25,* 919-924.

Antignus, Y., Vunsh, R., Lachman, O., Pearlsman, M., Maslenis, L., Hananya, U. & Rosner, A. (2004). Truncated *Rep* gene originated from *Tomato yellow leaf curl virus* – Israel [Mild] confers strain-specific resistance in transgenic tomato. *Annalsof Applied Biology,144,* 39–44.

Atzmon, G., van Oss, H. & Czosnek, H. (1998). PCR-amplification of tomato yellow leaf curl virus (TYLCV) DNA from squashes of plants and whitefly vectors: application to the study of TYLCV acquisition and transmission. *European Journal of Plant Pathology,104,* 189-194.

Bendahmane, M. & Gronenborn, B. (1997). Engineering resistance against tomato yellow leaf curl virus (TYLCV) using antisense RNA. *Plant Molecular Biology,33,* 351-357.

Berlinger, M. J., Taylor, R. A. J., Lebiush Mordechi, S., Shalhevet, S. & Spharim, I. (2002). Efficiency of insect exclusion screens for preventing whitefly transmission of tomato yellow leaf curl virus of tomatoes in Israel. *Bulletin Entomological Research,92,* 367-373.

Bleeker, P. M, Diergaarde, P. J., Ament, K., Guerra, J., Weidner, M., Schutz, S., de Both, M. T. J., Haring, M. A. & Schuurink, R. C. (2009) The role of specific tomato volatiles in tomato-whitefly interaction. *Plant Physiology* DOI:10.1104/pp.109.142661.

Bosco, D. & Caciagli, P. (1998). Bionomics and ecology of *Bemisia tabaci* (Sternorrhyncha: Aleyrodidae) in Italy. *European Journal of Entomology,95,* 519-527.

Briddon, R. W., Pinner, M. S., Stanley, J. & Markham, P. G. (1990). Geminivirus coat protein gene replacement alters insect specificity. *Virology,177,* 85–94.

Caciagli, P., & Bosco, D. (1997). Quantitation over time of tomato yellow leaf curl geminivirus DNA in its whitefly vector. *Phytopathology,87*, 610-613.

Chagué, V., Mercier, J. C., Guenard, M., de Courcel, A. & Vedel, F (1997) Identification of RAPD markers linked to a locus involved in quantitative resistance to TYLCV in tomato by bulked segregant analysis. *Theoretical and Applied Genetics,95*, 671-677.

Channarayappa, C., Shivashankar, G., Muniyappa, V. & Frist, R. H. (1992). Resistance of Lycopersicon species to *Bemisia tabaci*, a tomato leaf curl virus vector. *Canadian Journal of Botany,70,* 2184-2192.

Cohen, S., & Harpaz, I. (1964). Periodic rather than continual acquisition of a new tomato virus by its vector, the tobacco whitefly (*Bemisia tabaci* Gennadius). *Entomologia Experimentalis and Appicata*, *7*, 155-166.

Cohen, S., & Nitzany, F. E. (1966). Transmission and host range of tomato yellow leaf curl virus. *Phytopathology,56,* 1127-1131.

Cohen, S., Kern, J., Harpaz, I. & Ben-Joseph, R. (1988). Epidemiological studies of the tomato yellow leaf curl virus (TYLCV) in the Jordan Valley, Israel. *Phytoparasitica* 16, 259-270.

Crespi, S., Accotto, G.P., Caciagli, P. & Gronenborn, B. (1991). Use of digoxigenin-labeled probes for detection and host range studies of tomato yellow leaf curl geminivirus. *Research in Virology* 142, 283-288.

Crespi, S., Noris, E., Vaira, A.M., & Accotto, G.P. (1995). Molecular characterization of cloned DNA from a tomato yellow leaf curl virus isolate from Sicily. *Phytopathologia Mediterranea,34,* 93-99.

Czosnek, H. (2007) Interactions of *Tomato yellow leaf curl virus* with its insect vector. In: *The Tomato Yellow Leaf Curl Virus Disease: Management, Molecular Biology and Breeding for Resistance*. H. Czosnek Ed. Springer, The Netherlands. Pp 157-170.

Czosnek, H., Ber, R., Antignus, Y., Cohen, S., Navot, N., & Zamir, D. (1988). Isolation of tomato yellow leaf curl virus, a geminivirus. *Phytopathology,78,* 508-512.

Czosnek, H., Ber, R., Navot, R., Antignus, Y., Cohen, S. & Zamir, D. (1989) Characterization of tomato yellow leaf curl virus DNA forms in the viral capsid, in infected plants and in the insect vector. *Journal of Phytopathology,125,* 47-54.

Davino, S.; Davino, M. & Accotto, G. P. (2008) A single-tube PCR assay for detecting viruses and their recombinants that cause tomato yellow leaf curl disease in the Mediterranean basin. *Journal of Virological Methods,147*, 93-98.

Davino, S., Napoli C., Davino, M. & Accotto, G.P. (2006).Spread of Tomato yellow leaf curl virus in Sicily: partial displacement of another Geminivirus originally present. *European Journal of Plant Pathology,114,* 293-299.

Elbert, A. & Nauen, R. (2000). Resistance in *Bemisia tabaci* (Homoptera: Aleyrodidae) to insecticides in southern Spain with special reference to neonicotinoids. *Pest Management Science,56,* 60-64.

El Mehrach, K., Sedegui, M., Hatimi, H., Tahrouch, S., Arifi, A., Czosnek, H., Nakhla, M. K. & Maxwell, D. P. (2007) Molecular characterization of a Moroccan isolate of *Tomato yellow leaf curl Sardinia virus* and differentiation of the *Tomato yellow leaf curl virus* complex by the polymerase chain reaction. *Phytopathologia Mediterranea,46,* 185-194.

Eybishtz, A., Peretz, Y., Sade, D., Akad, F. & Czosnek, H. (2009) Silencing of a single gene in tomato plants resistant to *Tomato yellow leaf curl virus* renders them susceptible to the virus. *Plant Molecular Biology,71,* 157-171.

Eybishtz, A., Peretz, Y., Sade, D., Gorovits, R. & Czosnek, H. (2010) *Tomato yellow leaf curl virus* (TYLCV) infection of a resistant tomato line with a silenced sucrose transporter gene *LeHT1* results in inhibition of growth, enhanced virus spread and necrosis. *Planta, 231,* 537-548.

Fauquet, C. M., Mayo, M. A., Bishop, D. H. L., Maniloff, J., Desselberger, U. & Ball, L. A. Eds. (2005) Virus Taxonomy: The Classification and Nomenclature of Viruses. *The Eight Report of the International Committee on Taxonomy of Viruses.* Academic Press, 1162 pp. Elsevier.

Fauquet, C. M. & Stanley, J. (2003). Geminivirus classification and nomenclature; progress and problems. *Annals of Applied Biology,142,* 165-189.

Friedmann, M., Lapidot, M., Cohen, S. & Pilowsky, M. (1998) A novel source of resistance to tomato yellow leaf curl virus exhibiting a symptomless reaction to viral infection. *Journal of the American Society for Horticultural Science,123,* 1004-1007.

Fuentes, A., Ramos, P. L., Fiallo, E., Callard, D., Sánchez, Y., Peral, R., Rodrıguez, R. & Pujol, M. (2006). Intron–hairpin RNA derived from replication associated protein C1 gene confers immunity to *Tomato yellow leaf curl virus* infection in transgenic tomato plants. *Transgenic Research,15,* 291–304.

Gharsallah Chouchane, S., Gorsane, F., Nakhla, M.N., Maxwell, D.P., Marrakchi, M., & Fakhfakh, H. (2006). Complete Nucleotide Sequence and Construction of an Infectious Clone of a Tunisian Isolate of *Tomato yellow leaf curl Sardinia virus. Journal of Phytopathology,154,* 626-631.

Goldbach, R, Bucher, E. & Prins, M. (2003) Resistance mechanisms to plant viruses: an overview. *Virus Research,92,* 207-212

Goodman, R. M. (1977). Single-stranded DNA genome in a whitefly-transmitted plant virus. *Virology,83,*171-179.

Goolsby, J. A., DeBarro, P. J., Kirk, A. A., Sutherst, R. W., Canas, L., Ciomperlik, M. A., Ellsworth, P. C., Gould, J. R., Hartley, D. M., Hoelmer, K. A., Naranjo, S. E., Rose, M., Roltsch, W. J., Ruiz, R. A., Pickett, C. H. & Vacek, D. C. (2005). Post-release evaluation of biological control of Bemisia tabaci biotype ''B'' in the USA and the development of predictive tools to guide introductions for other countries. *Biological Control,32,* 70–77.

Griffiths, P. D., & Scott, J. W. (2001). Inheritance and linkage of tomato mottle virus resistance genes derived from *Lycopersicon chilense* accession LA 1932. *Journal of the American Society for Horticultural Science,126,* 462-467.

Haible, D., Kober, S. & Jeske, H. (2006). Rolling circle amplification revolutionizes diagnosis and genomics of geminiviruses. *Journal of Virological Methods,135,* 9-16.

Hanley-Bowdoin, L., Settlage, S. B., Orozco, B. M., Nagar, S. & Robertson, D. (1999). Geminiviruses: models for plant DNAreplication, transcription, and cell cycle regulation. *Critical Reviews in Plant Sciences,18,* 71–106

Hanson, P. M., Bernacchi, D., Green., S, Tanksley, S. D., Muniyappa, V., Padmaja, S., Chen, H. M., Kuo G., Fang, D. & Chen, J. T. (2000) Mapping a wild tomato introgression associated with tomato yellow leaf curl virus resistance in a cultivated tomato line. *Journal of the American Society for Horticultural Science,125,*15-20.

Hanson, P., Green, S.K. & Kuo, G (2006) *Ty-2*, a gene on chromosome 11 conditioning geminivirus resistance in tomato. *Tomato Genetic Cooperative Reports,56*, 17-18

Harrison, B. D., Barker, H., Bock, K. R., Guthrie, E. J., Meredith, G. & Atkinson, M. (1977). Plant viruses with circular single-stranded DNA. *Nature,270,* 760-762.

Horowitz, A.R., Kontsedalov, S., Khasdan,V. & Ishaaya, I. (2005_. Biotypes B and Q of *Bemisia tabaci* and their relevance to neonicotinoid and pyriproxyfen resistance. *Archives of Insect Biochemistry and Physioliogy,58*, 216-225.

Ji, Y., Schuster, D. J. & Scott, J. W. (2007a) *Ty-3*, a begomovirus resistance locus near the *Tomato yellow leaf curl virus* resistance locus *Ty-1* on chromosome 6 of tomato. *Molecular Breeding,20,* 271–284.

Ji, Y., Scott, J. W., Hanson, P., Graham, E. & Maxwell, D. P. (2007b) Sources of resistance, inheritance, and location of genetic loci conferring resistance to members of the tomato-infecting begomoviruses. In: Czosnek H (ed) *The Tomato Yellow Leaf Curl Virus Disease: Management, Molecular Biology and Breeding for Resistance.* Springer, Dordrecht, Netherlands, pp. 343-362

Ji, Y., Scott, J. W., Maxwell, D. P. & Schuster, D. J. (2008) *Ty-4*, a tomato *yellow leaf curl virus* resistance gene on chromosome 3 of tomato. *Tomato Genetic Cooperative Reports,58*, 29–31.

Kasrawi, M. A. (1989). Inheritance of resistance to tomato yellow leaf curl virus (TYLCV) in Lycopersicon pimpinellifolium. *Plant Dis*. 73, 435-437.

Kato, K., Onuki, M., Fuji, S. & Hanada, K. (1998). The first occurrence of tomato yellow leaf curl virus in tomato (*Lycopersicon esculentum* Mill.) in Japan. *Annals of the Phytopathological Society of Japan,64,* 552-559.

Kheyr-Pour, A., Bendahmane, M., Matzeit, M., Accotto, G.P., Crespi, S. & Gronenborn, B. (1991). Tomato yellow leaf curl virus from Sardinia is a whitefly-transmitted monopartite geminivirus. *Nucleic Acids Research,19*, 6763-6769.

Korimbocus, J., Coates, D., Barker, I. & Boonham, N. (2002). Improved detection of *Sugarcane yellow leaf virus* using a real-time fluorescent (TaqMan) RTPCR assay. *Journal of Virological Methods,103*, 109–120.

Krupovic, M., Ravantti, J. J. & Bamford, D. H. (2009) Geminiviruses: a tale of a plasmid becoming a virus. *BMC Evolutionary Biology,9,* 112.

Kunik, T., Salomon, R., Navot, N., Zeidan, M., Michelson, I., Zamir, D., Gafni, Y. & Czosnek, H. (1994) Transgenic tomato plants expressing the tomato yellow leaf curl virus capsid protein are resistant to the virus. *BioTechnology,12,* 500-504.

Lacey, L. A., Millar, L., Kirk, A. A. & Perring, T. M. (1999). Effect of storage temperature and duration on survival of eggs and nymphs of *Bemisia argentifolii* (Homoptera: Aleyrodidae) and pupae of the whitefly parasitoid *Encarsia formosa* (Hymenoptera: Aphelinidae). *Annals of the Entomological Society of America* 92, 430-434.

Lapidot, M., Friedmann, M., Lachman, O., Yehezkel, A., Nahon, S., Cohen, S. & Pilowsky, M. (1997). Comparison of resistance to tomato yellow leaf curl virus among commercial cultivars and breeding lines. *Plant Disease,81,* 1425-1428.

Laterrot, H. & Moretti, A. (1996). Chepertylc lines. *Tomato Yellow Leaf Curl Newsletter,8,* 4.

Lefeuvre, P., Hoareau, M., Delatte, H., Reynaud, B. & Lett, J. M. (2007) A multiplex PCR method discriminating between the TYLCV and TYLCV-Mld clades of *Tomato yellow leaf curl virus*. *Journal of Virological Methods, 144,* 165-168.

Legg, J. P. & Fauquet, C. M. (2004). Cassava mosaic geminiviruses in Africa. *Plant Molecular Biology, 56,* 585–99.

Lett, J. M., Granier, M., Hippolyte, I., Grondin, M., Royer, M., Blanc, S., Reynaud, B. & Peterschmitt, M. (2001). Spatial and temporal distribution of Geminiviruses in leafhoppers of the Genus *Cicadulina* monitored by conventional and quantitative polymerase chain reaction. *Phytopathology,92,* 65–74.

Liu, Y., Schiff, M. & Dinesh-Kumar, S. P. (2002) Virus-induced gene silencing in tomato. *Plant Journal,31,* 777-786.

Mason, G., Caciagli, P., Accotto, G. P. & Noris, E. (2008) Real-time PCR for the quantitation of *Tomato yellow leaf curl Sardinia virus* in tomato plants and in *Bemisia tabaci. Journal of Virological Methods,147,* 282–289.

Morales, F. J. & Anderson, P. K. (2001). The emergence and dissemination of whitefly-transmitted geminiviruses in Latin America – Brief review. *Archives of Virology,146,* 415–441.

Muniyappa, V., Jalikop, S. H., Saikia, A. Chennarayappa, K., Shivashankar G., Bhat A. I., & Ramappa, H. K. (1991). Reaction of *Lycopersicon* cultivars and wild accessions to tomato leaf curl virus. *Euphytica,56,* 37-41.

Murphy, A. M., Gilliland, A., Wong, C. E., West, J., Singh, D. P. & Carr, J. P. (2001) Signal transduction in resistance to plant viruses. *European Journal Plant Pathology,107,*121-128

Palmer, K. E. & Rybicki, E. P. (1998). The molecular biology of mastreviruses. *Advance in Virus Research,50,* 183–234

Pietersen, G. & Smith, M. F. (2002) *Tomato yellow leaf curl virus* resistant tomato shows resistance to *Tomato curly stunt virus. Plant Disease,86,* 528–34.

Rojas, M. R., Hagen, C., Lucas, W. J. & Gilbertson, R. L. (2005) Exploiting chinks in the plant's armor: Evolution and Emergence of Geminiviruses. *Annual Reviews of Phytopathology,43,* 361–94

Maruthi, M. N, Czosnek, H., Vidavski, F., Tarba, S.-Y., Milo, J., Leviatov, S., Venkatesh, H. M., Padmaja, A. S., Kulkarni, R. S. & Muniyappa, V. (2003). Comparison of resistance to *Tomato leaf curl virus* (India) and *Tomato yellow leaf curl virus* (Israel) among *Lycopersicon* wild species, breeding lines and hybrids. *European Journal of Plant Pathology, 109,* 1-11.

Mason, G., Caciagli, P., Accotto, G. P. & Noris, E. (2008) Real-time PCR for the quantitation of *Tomato yellow leaf curl Sardinia virus* in tomato plants and in *Bemisia tabaci. Journal of Virological Methods,147,* 282–289.

Mejía, L., Teni, R. E., Vidavski, F., Czosnek, H., Lapidot, M., Nakhla M. K. & Maxwell, D. P. (2005). Evaluation of tomato germplasm and selection of breeding lines for resistance to begomoviruses in Guatemala. *Acta Horticulturae,695,* 251-256.

Mitter, N., Sulistyowati, E., Graham, M. W. & Dietzgen, R.G. (2001) Suppression of gene silencing, a threat to virus-resistant transgenic plants. *Trends in Plant Science,6,* 246-247.

Monci, F., Sanchez-Campos, S., Navas-Castillo, J., & Moriones, E. (2002). A Natural Recombinant between the Geminiviruses Tomato yellow leaf curl Sardinia virus and

Tomato yellow leaf curl virus Exhibits a Novel Pathogenic Phenotype and is Becoming Prevalent in Spanish Populations. *Virology,303,* 317-326.

Morin, S., Ghanim, M., Zeidan, M., Czosnek, H., Verbeek, M. & van den Heuvel, J. F. J. M. (1999) A GroEL homologue from endosymbiotic bacteria of the whitefly *Bemisia tabaci* is implicated in the circulative transmission of *Tomato yellow leaf curl virus. Virology,256,* 75-84.

Morin, S., Ghanim, M., Sobol, I. & Czosnek, H (2000) The GroEL protein of the whitefly *Bemisia tabaci* interacts with the coat protein of transmissible and non-transmissible begomoviruses in the yeast two-hybrid system. *Virology,276*, 404-416.

Moriones, E. & Navas-Castillo, J. (2000). Tomato yellow leaf curl virus, an emerging virus complex causing epidemics worldwide. *Virus Research,71*, 123-134.

Nauen, R. & Denholm, I. (2005). Resistance of insect pests to neonicotinoid insecticides: current status and future prospects. *Archives of Insect Biochemistry and Physiology, 58,* 200–215.

Navas-Castillo, J., Díaz, J. A., Sánchez-Campos, S. & Moriones, E. (1998). Improvement of the print-capture polymerase chain reaction procedure for efficient amplification of DNA virus genomes from plants and insect vectors. *Journal of Virological Methods,75*, 195-198.

Navot, N., Ber, R. & Czosnek, H. (1989) Rapid detection of tomato yellow leaf curl virus in squashes of plants and insect vectors. *Phytopathology,79,* 562-568.

Navot, N., Pichersky, E., Zeidan, M., Zamir, D. & Czosnek, H. (1991). Tomato yellow leaf curl virus: a whitefly-transmitted geminivirus with a single genomic component. *Virology,185*, 131-161.

Navot, N., Zeidan, M., Pichersky, E., Zamir, D. & Czosnek, H. (1992) Use of the polymerase chain reaction to amplify tomato yellow leaf curl virus DNA from infected plants and viruliferous whiteflies. *Phytopathology,82,* 1199-1202.

Noris, E., Accotto, G. P., Tavazza, R., Brunetti, A., Crespi, S. & Tavazza, M. (1996). Resistance to Tomato yellow leaf curl geminivirus in *Nicotiana benthamiana* plants transformed with a truncated viral C1 gene. *Virology,224,* 130-138.

Noris, E., Lucioli, A., Tavazza, R., Caciagli, P., Accotto, G. P. & Tavazza, M. (2004). *Tomato yellow leaf curl Sardinia virus* can overcome transgene-mediated RNA silencing of two essential viral genes. *Journal of General Virology*,85, 1745-1749.

Pilowsky, M. & Cohen, S. (1974). Inheritance of resistance to tomato yellow leaf curl virus in tomatoes. *Phytopathology,64*, 632-635.

Pilowsky, M. & Cohen, S. (1990). Tolerance to tomato yellow leaf curl virus derived from *Lycopersicon peruvianum. Plant Disease,74,* 248-250.

Rojas, M. R., Gilbertson, R. L., Russell, D. R. & Maxwell, D.P. (1993) Use of degenerate primers in the polymerase chain reaction to detect whitefly-transmitted geminiviruses. *Plant Disease, 77,* 340-347.

Rubinstein, G. & Czosnek, H. (1997) Long-term association of tomato yellow leaf curl virus (TYLCV) with its whitefly vector *Bemisia tabaci*: effect on the insect transmission capacity, longevity and fecundity. *Journal of General Virology,78,* 2683-2689.

Rubinstein, G., Morin, S. & Czosnek, H. (1999) Long-term effect of imidacloprid on mortality of the whitefly *Bemisia tabaci* caged with treated eggplant and tomato, and on transmission of tomato yellow leaf curl geminivirus (TYLCV) to tomato. *Journal of Economic Entomology,92*, 658-662.

Salati, R., Nahkla, M. K., Rojas, M. R., Guzman, P., Jaquez, J., Maxwell, D. P. & Gilbertson, R. L. (2002). *Tomato yellow leaf curl virus* in the Dominican Republic: Characterization of an infectious clone, virus monitoring in whiteflies, and identification of reservoir hosts. *Phytopathology,92,* 487-496.

Sánchez-Campos, S., Navas-Castillo, J., Camero, R., Soria, C., Diaz, J. A. & Moriones, E. (1999). Displacement of Tomato yellow leaf curl virus (TYLCV)-Sr by TYLCV-Is in tomato epidemics in Spain. *Phytopathology,89,* 1038-1043.

Sanford, J. C. & Johnston, S. A. (1985). The concept of parasite-derived resistance – deriving resistance genes from the parasite's own genome. *Journal of Theoretical Biology,113,* 395-405.

Schubert, J., Habekuß, A., Kazmaier, K. &Jeske, H (2007) Surveying cereal-infecting geminiviruses in Germany-Diagnostics and direct sequencing using rolling circle amplification. *Virus Research,127,* 61-70.

Scott, J. W. & Schuster, D. J. (1991). Screening of accessions for resistance to the Florida tomato geminivirus. *Tomato Genetic Cooperation Report,41,* 48-50.

Smith, N. A., Singh, S. P., Wang, M. B., Stoutjesdijk, P. A., Green, A. G. & Waterhouse, P. M. (2000) Total silencing by intron-spliced hairpin RNAs. *Nature,407,* 319-320.

Stenger, D. C. & McMahon, C. L. (1997). Genotypic diversity of beet curly top virus populations in the western United States. *Phytopathology*,87, 737–44

Tesoriero, L. & Azzopardi, S. (2006). Tomato yellow leaf curl virus in Australia. *Primefact,220,* 1-2.

van den Heuvel, J. F. J. M., Verbeek, M. & van der Wilk, F. (1994) Endosymbiotic bacteria associated with circulative transmission of *Potato leafroll virus* by *Myzus persicae. Journal of General Virology,75,* 2559-2565.

Varma, A. & Malathi, V. G. (2003) Emerging geminivirus problems: A serious threat to crop production. *Annals of Applied Biology*,142, 145-164.

Vidavski, F. & Czosnek, H. (1998). Tomato breeding lines immune and tolerant to tomato yellow leaf curl virus (TYLCV) issued from Lycopersicon hirsutum. *Phytopathology,88,* 910-914.

Vidavski, F., Czosnek, H., Gazit, S., Levy, D. & Lapidot, M. (2008) Pyramiding of genes conferring resistance to *Tomato yellow leaf curl virus* from different wild tomato species. *Plant Breeding,127,* 625-631.

Vidavsky, F., Leviatov, S., Milo, J., Rabinowitch, H. D., Kedar, N. & Czosnek, H. (1998) Behavior of tolerant tomato breeding lines (*Lycopersicon esculentum*) originated from three different sources (*L. peruvianum*, *L. pimpinellifolium* and *L. chilense*) upon early controlled inoculation by tomato yellow leaf curl virus. *Plant Breeding,117,* 165-169.

Voinnet, O., Pinto, Y. M. & Baulcombe, D. C. (1999) Suppression of gene silencing: a general strategy used by diverse DNA and RNA viruses of plants. *Proceedings of the National Academy of Sciences USA,96,* 14147-14152.

Wilson, T. M. A. (1993) Strategies to protect crop plants against viruses: Pathogen-derived resistance blossoms. *Proceedings of the National Academy of Sciences USA,90,* 3134-3141.

Wu, J. B., Dai, F. M. & Zhou, X. P. (2006). First Report of *Tomato yellow leaf curl virus* in China. *Plant Disease,90,* 1359.

Yang, Y., Sherwood, T. A., Patte, C. P., Hiebert, E. & Polston, J. E. (2004). Use of tomato yellow leaf curl virus Rep gene sequences to engineer TYLCV resistance in tomato. *Phytopathology*,94, 490–496.

Zakay, Y., Navot, N., Zeidan, M., Kedar, N., Rabinowitch, H., Czosnek, H. & Zamir, D. (1991) Screening of *Lycopersicon* accessions for resistance to tomato yellow leaf curl virus: presence of viral DNA and symptom development. *Plant Disease,75,* 279-281.

Zamir, D., Ekstein-Michelson, I., Zakay, Y., Navot, N., Zeidan, M., Sarffati, M., Eshed, Y., Harel, E., Pleban, T., van Oss, H., Kedar N., Rabinowitch H. D. & Czosnek, H. (1994) Mapping and introgression of a tomato Yellow leaf curl virus tolerance gene, TY-1. *Theoretical and Applied Genetics,88,* 141-146

Zhang, W., Olson, N. H., Baker, T. S., Faulkner, L., Agbandje-McKenna, M., Boulton, M. I., Davies, J. W. & McKenna, R. (2001). Structure of the *Maize streak virus* geminate particle. *Virology,279,* 471-477.

Zilberstein, A., Navot, N., Ovadia, S., Reinhartz, A., Herzberg, M. & Czosnek, H. (1989). Field-usable assay for the diagnosis of the tomato yellow leaf curl virus in squashes of plants and insects by hybridization with a chromogenic DNA probe. *Technique,1,* 118-124.

In: Emerging Geminiviral Diseases and their Management
Editors: P. Sharma, R. K. Gaur et al. pp. 59-69
ISBN 978-1-61668-620-8

Chapter 3

EMERGENCE OF TOMATO LEAF CURL JAVA DISEASE COMPLEX

***Pradeep Sharma*[1,*] *Neetu Sharma*[2] *and Masato Ikegami*[3]**

[1]Division of Crop Improvement, Directorate of Wheat Research, Karnal 132001, India;
[2]CSK HPKV Krishi Vigayan Kendra, Bara, Hamirpur, India;
[3]Graduate School of Agricultural Science, Tohoku University, 981555, Japan.

ABSTRACT

Geminiviruses are single-stranded (ss) DNA viruses that cause devastating disease of important food and fiber crops worldwide. *Tomato leaf curl Java virus* (ToLCJV) is a distinct begomovirus species associated with betasatellite DNA (DNAβ02), and a recombinant DNA (recDNA) prevalent in Indonesia. Recently, DNA satellites associated with begomoviruses have come to our attention; they are widespread throughout the world and are associated with many diseases, causing huge losses to economically important crops particularly developing countries. We have established the functional role of betasatellite DNA in the etiology of begomovirus disease. DNAβ02, ToLCJV induced severe leaf curling type of symptoms in tomato plants. The infectious clones of ToLCJV alone did not induce leaf curl symptoms in tomato plants. However, in the presence of DNAβ02, ToLCJV induced leaf curl symptoms in tomato plants. The leaf curl symptoms that DNAβ02 induced in agroinoculated tomato plants were similar to those observed in the field in Java, Indonesia. The recombinant DNA had a chimeric sequence, which arose from recombination among ToLCJV, *Ageratum yellow vein virus*-Java (AYVV-[Java]). We also found that ToLCJV and its betasatellite encode as many as three RNA silencing suppressors.

* Correspondence concerning this article should be addressed to: Pradeep Sharma, email: neprads@yahoo.com.

1. INTRODUCTION

Geminiviruses have monopartite or bipartite circular single-stranded DNA genomes, encapsidated in geminate particles. The *Geminiviridae* family consists of four genera, distinguished according to their host range, insect vector, and genome organization (Fauquet et al., 2008). The majority of the *Geminiviridae* belong to the genus *Begomovirus*, which is transmitted by the whitefly (*Bemisia tabaci*). Tomato leaf curl diseases caused losses in tomato production throughout the world (Varma and Malathi, 2003). Many viruses causing leaf curl disease in tomato has been characterized (Dry et al., 1993, Hong et al., 1995; Padidam et al., 1995; Kon et al., 2006;, Matsuda et al., 2008). For example, *Tomato leaf curl New Delhi virus*-Severe (ToLCNDV-Svr) (Padidam et al., 1995) has a bipartite genome consisting of two circular ssDNA molecules (DNA-A and DNA-B), whereas *Tomato leaf curl virus* (ToLCV) (Dry et al., 1993), *Tomato leaf curl Bangalore virus* (ToLCBV) (Krithi et al., 2002), *Tobacco leaf curl Japan virus* (TbLCJV) (Shimizu and Ikegami, 1999), and *Tomato leaf curl Philippines virus* (ToLCPV) (Matsuda et al., 2008) are monopartite. Several distinct begomoviruses have been associated with tomato leaf curl in Indonesia (Sukomato et al., 2005).

Recently, satellite DNAs associated with several monopartite begomoviruses have come to our attention, they are widespread throughout the world and are associated with many diseases, causing huge losses to economically important crops particularly in developing countries. Furthermore, infectious clones of the monopartite begomoviruses AYVV from Singapore, *Cotton leaf curl Multan virus* (CLCuMV) from Pakistan, *Bhendi yellow vein mosaic virus* (BYVMV) from India, *Tomato yellow leaf curl China virus*-[Tobacco Y10] (TYLCCNV-[Tb:Y10]), and *Eupatorium yellow vein virus* (EpYVV) were unable to induce typical symptoms. A novel group of satellite DNA molecules called DNAβ, which are associated with these viruses, is essential for inducing the characteristic disease symptoms (Briddon et al., 2001; Jose and Usha, 2003; Saunders et al., 2004; Saunders et al., 200; Zhou et al., 2003; Kon et al., 2006; Sharma and Rishi, 2007; Matsuda et al., 2008). DNAβ is approximately half the size of the genomic DNA, and except for a conserved hairpin structure and a TAATATTAC loop sequence, has little sequence similarity with the genomic DNA of the associated begomovirus. DNAβ requires a helper begomovirus DNA for replication, encapsidation, insect transmission, and movement in plants (Saunders et al., 2000). Previously, many begmoviruses such as *Ageratum yellow vein virus* – Java (AYVV-Java), *Pepper yellow vein Indonesia virus* (PeYLCIDV) and its new strains were reported from Indonesia (Sakata et al., 2008). Among all, ToLCJV is very acute problem and causing enormous damage to the tomato production (Sukomato et al., 2005).

In this Chapter, we described ToLCJV, a distinct begomovirus species which is associated with betasatellite (DNAβ02), and a recombinant DNA isolated from tomato plants exhibiting leaf curl symptoms from Indonesia. ToLCJV and its associated betasatellite encodes as many as three post transcriptional gene silencing suppressor (PTGS) (Kon et al. 2007; Sharma and Ikegami, 2008).The recombinant DNA had a chimeric genome and consisted of sequences of begomovirus genomic DNA and satellite DNAβ.

2. Etiology of the Disease

Tomato leaf curl disease is among the most important limiting factors that affect tomato cultivation in Indonesia. Infected plants exhibit severe downward leaf curling (Figure 1 left panel) and stunting of plant growth. Efforts to detect DNA-B components associated with Java virus based on RFLP analysis or PCR using universal degenerate primers PCRc1 and PBL1v2040 (Table 1) were unsuccessful. From these results, we concluded that Java virus is monopartite begomovirus (Sukomato et al., 2005; Kon et al., 2006). In addition to genomic component, betasatellite was also detected from the same tomato infected plants exhibiting leaf curl type symptoms. Table 1 shows the primers used for detection and cloning of the tomato leaf curl disease. Figure 2 shows the detection of Java virus and its betasatellites, resolved on 1% agarose gel by using polymerase chain reaction (PCR) amplification.

Figure 1. Typical symptoms showing severe downward leaf curling by Tomato leaf curl Java virus. Natural infection uder fields (right panel) and agro-inculated with DNA plus its associated betasatellite (left panel) under glasshouse conditons.

3. The Begomovirus, Tomato Leaf Curl Java Virus

The genome organizations of these viruses were similar to those of other monopartite begomoviruses originating from the Old World. ToLCJV encodes six partially overlapping open reading frames (ORFs) that are organized bidirectionally (Figure 3). The complete genome consisted of 2.7 kb (Kon et al., 2006). An identical 33-base potential stem loop-forming region was identical in the intergenic regions (IR) of both viruses. ToLCJV contains *Rep* high-affinity binding site GGTCTCAA (ntd. 2617-2624 and 2651-2659). The complete nucleotide sequences of ToLCJV were compared with those of other begomoviruses showed identities with AYVV (87%) and AYVV-Java (85%). Similarly, ORF V1 (CP) of ToLCJV showed the highest amino acid sequence identities with its counterparts from AYVV (87%), *Tomato leaf curl Laos virus* (ToLCLV) (87%), and AYVV-Java (87%). Figure 4 shows the phylogenetic relationships of ToLCJV with other selected geminiviruses. This analysis revealed that ToLCJV was distantly related to the ageratum-infecting begomovirus group (Figure 4). International Committee on Taxonomy of Viruses (ICTV) proposed that a virus with less than 90% identity for the complete genome and CP sequences should be considered a distinct species (Fauquet et al., 2005). Therefore, keeping in view-nucleotide (ntd.) identity and phylogenetic analysis, Kon et al. (2006) proposed ToLCJV to be a distinct species of begomovirus.

Table 1. Primers used to detect the *Tomato leaf curl Java virus* and its associated betasatellite

Primer name	Primer sequence
ToXba1 **ToXba2** (for Full DNA)	**For.**5'-*TCTAGA*CACGATGTTAGTCACGTGGG-3' **Rev.**5'-*TCTAGA*CTCAAAAGACTGGACCTTAC-3'
CRv1 CRc2	5'-TAATATTACCGGATGGCCGC-3' 5'-AAAAAATTATGCCA-3'
PCRc1 PBL1v2040 (for DNA B)	5'-CTAGCTGCAGCATATTTACRARWATGCCA-3' 5'-GCCTCTGCAGCARTGRTCKATCTTCATACA-3'
DNAβ01	StCla1.5'-*ATCGAT*GAATCTTTATACATGATCC-3'
DNAβ02	StCla2.5'-*ATCGAT*CGAGGAGATCAAAGCAGAAG-3'
RecDNA	StBamH15'-*GGATCC*TCTTTTGAACGAGTTTCCTG-3' StBamH25'-*GGATCC*CACATGTTAAAATAATACTTGG-3'

(R=A, G; W= A, T; K= G, T; R=A, G).

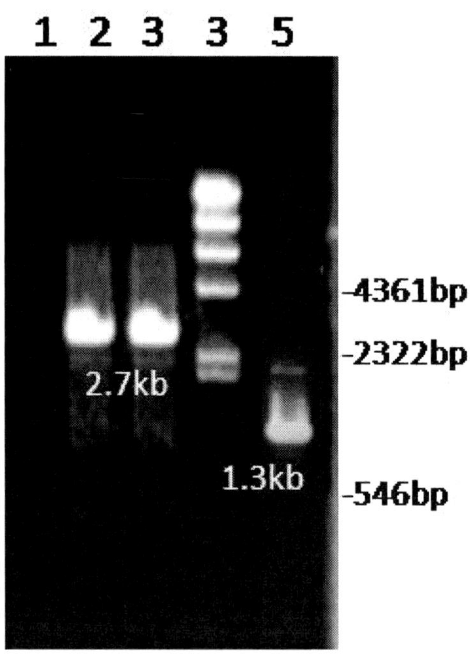

Figure 2. PCR based detection of ToLCJV genome and its associated betasatellite DNA component. PCR product were analyzed on 1% agarose gel and stained with ethiduim bromide.

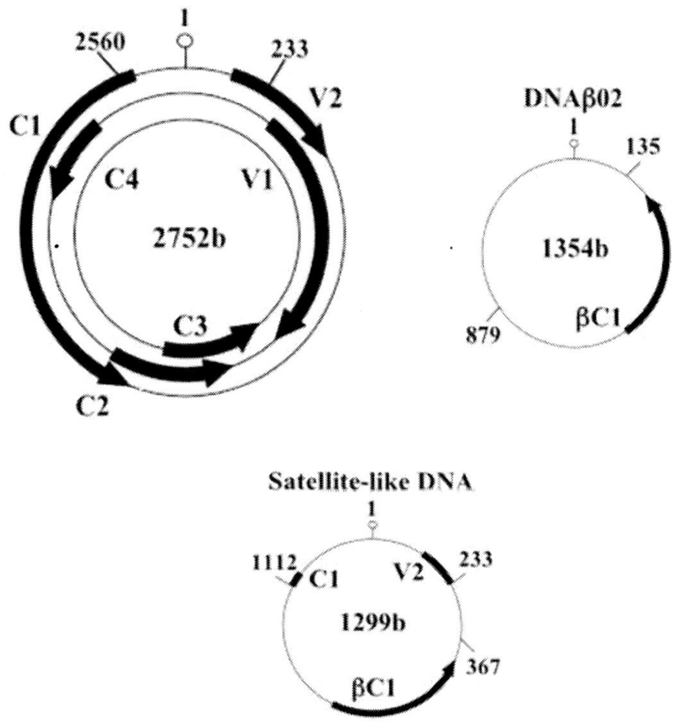

Figure 3. Genome organizations of Tomato leaf curl Java virus and its associated betasatellite DNA (DNAβ02), and recombinant betasatellite like DNA molecule. Open circles indicate the stem loop-forming region containing the conserved nonanucleotide TAATATT↓AC. Nucleotide numbering starts from the nick site (↓) within the conserved nonanucleotide. The arrows represent predicted ORFs in both orientations ('C' for complementary and 'V' for viral sense) and potentially encoded proteins are shown as black arrows. The region of recombinant satellite DNA from nt. 1112-233 originates from its cognate DNA (ntd. 2560-233).

4. THE BETASATELLITE DNA (DNAβ02)

Satellites are viruses or nucleic acids that depend on a helper virus for replication but lack nucleotide sequence homology to the helper virus. DNAβ appears to be capable of being replicated by the Rep proteins from a diverse range of begomoviruses (Ogawa et al., 2008; Briddon et al., 200;, Saed et al., 2005).Satellite codes for their own coat protein, whereas RNA satellites are packaged in coat protein encoded by the helper virus (Mayo et al., 2005). Many plant RNA viruses have RNA satellites associated with them. The majority of satellites interfere with the replication of helper virus resulting in attenuated symptoms. Some of satellites exacerbate disease symptoms induced by helper virus or produce novel symptom which is usually not associated with helper infection (Roossinck et al., 1992).The betasatellite DNA (DNAβ02) consisted of 1354 nucleotides,(accession number AB113651). DNAβ satellites have a conserved genome organization consisting of a single complementary-sense ORF (βC1) (Figure 3), an adenine rich region (~65%), and a satellite conserved region (SCR). The SCR contains a stem loop sequence TAATATT/AC which is similar to sequence in the origin of replication of geminiviruses (Briddon et al., 2003). Besides full-length molecules, naturally occurring DNAβ deletion mutants of 660 to 896 nt have been reported (Briddon et al., 2003). The putative replication associated protein-binding motifs present in ToLCJV (GGTCTCAA) was not found in DNAβ01 or DNAβ02, but similar sequences were located upstream from the stem loop structure: GGTGTGTA (nt. 1268-1275: DNAβ02), GGTAGAAA (nt. 1286-1293: DNAβ02), and GGAGAAG (nt. 1218-1224: DNAβ02).

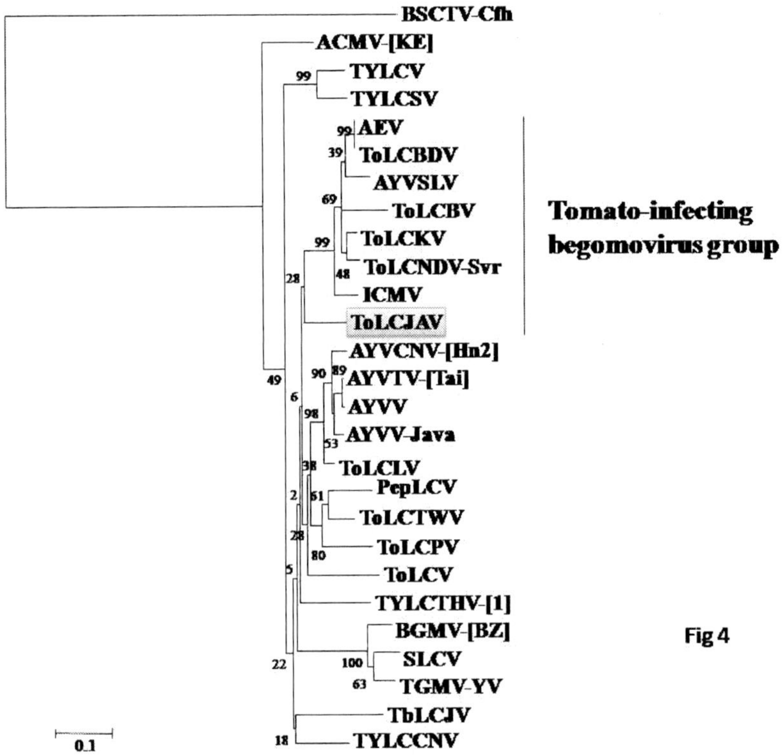

Fig 4

Figure 4. Neighbor-joining phylogenetic tree obtained from the complete nucleotide sequences of *Tomato leaf curl Java virus* and other begomovirus DNA-A components. The vertical distances are arbitrary and the horizontal distances are proportional to the calculated mutation distances. The tree was rooted on the sequence of *Beet severe curly top virus*-Cfh (BSCTV-Cfh: accession no. U02311), belonging to the distinct genus *Curtovirus*. The following sequences were obtained from GenBank and used for comparisons and phylogenetic analysis: *African cassava mosaic virus*-Kenya (ACMV-[KE]: J02057, DNA-A), *Ageratum enation virus* (AEV: AJ437618), *Ageratum yellow vein China virus*-[Hainan 2] (AYVCNV-[Hn2]: AJ495813), *Ageratum yellow vein SriLanka virus* (AYVSLV: AF314144), *Ageratum yellow vein Taiwan virus*-[Taiwan] (AYVTV-[TW]: AF307861), *Ageratum yellow vein virus* (AYVV: X74516), *Bean golden mosaic virus*-[Brazil] (BGMV-[BZ]: M88686, DNA-A), *Indian cassava mosaic virus* (ICMV: Z24758, DNA-A), *Pepper leaf curl virus* (PepLCV: AF134484), *Squash leaf curl virus* (SLCV: M38182, DNA-A), *Tobacco leaf curl Japan virus* (TbLCJV: AB028604), *Tomato golden mosaic virus*-Yellow vein (TGMV-YV: K02029, DNA-A), *Tomato leaf curl Bangalore virus* (ToLCBV: Z48182), *Tomato leaf curl Bangladesh virus* (ToLCBDV: AF188481), *Tomato leaf curl Karnataka virus* (ToLCKV: U38239), *Tomato leaf curl Laos virus* (ToLCLV: AF195482), *Tomato leaf curl New Delhi virus*-Severe (ToLCNDV-Svr: U15015, DNA-A), *Tomato leaf curl Philippines virus* (ToLCPV: AB050597), *Tomato leaf curl Taiwan virus* (ToLCTWV: U88692), *Tomato leaf curl virus* (ToLCV: S53251), *Tomato yellow leaf curl China virus* (TYLCCNV: AF311734), *Tomato yellow leaf curl Sardinia virus* (TYLCSV: X61153), *Tomato yellow leaf curl Thailand virus*-[1] (TYLCTHV-[1]: X63015, DNA-A), *Tomato yellow leaf curl virus* (TYLCV: X15656).

The nucleotide sequence of recombinant betasatellite DNA comprised of 1299 nucleotides and contained one complementary-sense ORF (βC1) (Figure 3). The region from nt.1112-233 of satellite-like DNA had the highest sequence identity with the corresponding sequences of ToLCJV (nt. 2560-233: 82%) and AYVV-Java (nt. 2560-236: 84%). The satellite-like DNA sequence from nt. 367-1110 had the highest sequence identity with of nt.135-879 of DNAβ02 (94%) (Kon et al., 2006). The iteron (GGTCTCAA) present in ToLCJV was not same in satellite-like DNA, whereas the iteron (GGAGACA) in AYVV-Java was found in recombinant betasatellite DNA (virus sense nt. 1167-1173, 1194-1200, and 1201-1207). A similar recombinant DNA component was found in *Ageratum yellow vein virus*-infected *Ageratum conyzoides* plants and *Cotton leaf curl virus*-infected cotton plants (Saunders et al., 2000; Briddon et al., 2003). These results demonstrate that potential for the emergence of new satellite-like DNAs by recombination between previously existing begomoviruses and satellite DNAβs.

The total nucleotide sequence of DNAβ02 had the highest identity (93%) with that of DNAβ01, followed by AYVV DNAβ (81%). The amino acid sequence of the βC1 product of DNAβ02 had the highest identity (89%) with that of DNAβ01. The βC1 products of DNAβ01 and DNAβ02 also had relatively high amino acid sequence identities (79% and 85%, respectively) with that of AYVV DNAβ. Kon et al. (2007) have recently demonstrated that betasatellite DNA encoded βC1 protein determines the pathogenesis.

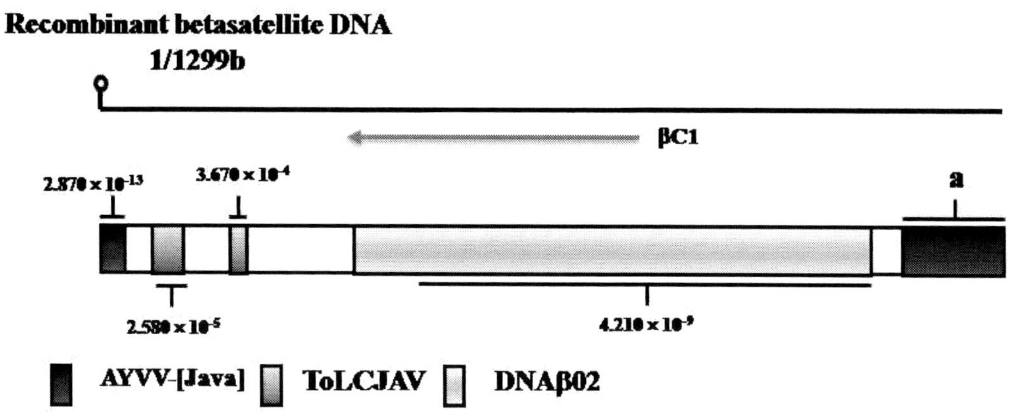

Figure 5. Recombinant regions detected within virus sequences. Potential recombination break points and the probability (*P*) that the indicated regions do not have a recombination origin are presented. The coloured patterns indicate the viral genomic DNA or satellite DNAβ origins of the recombinant region. The open circle indicates the stem loop-forming region. The βC1 ORF is indicated with an arrow.

5. Recombination Analysis

A mixed infection is one of the prerequisites for developing new recombinant viruses or strains. The emergence of new begomovirus species in nature via recombination between existing species has been demonstrated (Ogawa et al., 2008, Matsuda et al. 2008, Galvo et al., 2003). Kon et al. (2006) tested whether satellite-like DNA had a chimeric genome that arose from viral genomic DNA and satellite DNA using RDP. RDP provided evidence for the occurrence of recombination among ToLCJV, AYVV-[Java], DNAβ01, and DNAβ02 (Figure 5). In these recombination events, a region (nt. 1155-39) of recDNA was corresponds to nt. 2633-43 of AYVV-[Java] genomic DNA ($P = 2.870 \times 10^{-13}$), which included the IR. Two regions (nt. 77-123 and 190-216) of recDNA were also corresponds to nt. 77-123 and nt. 190-216 of ToLCJV genomic DNA ($P = 2.580 \times 10^{-5}$ and 3.670×10^{-4}, respectively). An additional two regions (nt. 367-883 and 450-1110) of recDNA were corresponds to nt. 135-652 of DNAβ01 ($P = 9.981 \times 10^{-5}$) and nt. 219-879 of DNAβ02 ($P = 4.210 \times 10^{-9}$), respectively. From these results, we concluded that satellite-like DNA had a chimeric genome that arose from recombination involving ToLCJV, AYVV-[Java], DNAβ01, and DNAβ02, and given the name recombinant DNA (recDNA)

6. Agro-Infectivity Analysis

Kon et al. (2006) analysed infectivity of ToLCJV, DNAβ02, and recDNA was carried out to examine infectious nature of the obtained clones in tomato and *N.benthamiana.* Plants mock-agroinoculated with *Agrobacterium* cells with the empty binary vector did not develop symptoms. Tomato plants agroinoculated with ToLCJV (clone pBToX1.4) were infected systemically, but displayed no symptoms, while co-agroinoculated with ToLCJV and DNAβ02 (clone pBToX1.4β02) showed severe downward leaf curling and crinkling symptoms developed (Figure 1 right panel). Results were further confirmed through PCR detection of ToLCJV-, and DNAβ02-specific DNAs from agroinoculated tomato plants (Kon et al., 2006), and the PCR-amplified fragment was cloned and sequenced. Similar results were obtained while tested on *N. benthamiana* plants. Consistent with these results, a change from upward to downward leaf curling in the presence of DNAβ has been demonstrated experimentally for AYVV in *N. benthamiana* plants (Saunders et al., 2004). These results further support the hypothesis that DNAβ plays an important role in symptom induction.

We tested whether recDNA is replicated by ToLCJV. Tomato plants co-agroinoculated with ToLCJV and recDNA (clone pBToX1.4Rec) displayed no symptoms. ToLCJV genomic DNA was detected in infected tomato plants, but not recDNA by PCR analysis. *N. benthamiana* plants co-agroinoculated with ToLCJV and recDNA developed mild downward leaf curling symptoms. However, both ToLCJV genomic DNA and recDNA were detected in agroinoculated tomato plants by PCR analysis. No symptoms were observed in tomato plants agroinoculated with recDNA alone. When tomato plants were agroinoculated with ToLCJV, DNAβ02, and recDNA, leaf curl symptoms developed, which also resembled symptomatologicaally in tomato plants infected with ToLCJV and DNAβ02. The presence of viral genomic DNA, DNAβ, and recDNA in agroinoculated plants was confirmed by PCR analysis (Kon et al., 2006).

7. RNA Silencing Suppressors

RNA silencing is a major means of defense against viruses (Kasschu and Carrington, 1998, Voinnet, 2001, Dunoyer et al., 2004). When viruses or transgenes are introduced into plants, they trigger a host response that is generally referred to as posttranscriptional gene silencing (PTGS). The PTGS pathway uses viral RNA as a template to generate small RNA species of 21 to 26 nt, termed small interfering RNAs (siRNAs). The siRNAs subsequently assemble into a nuclease complex known as the RNA-induced silencing complex (RISC), guiding the complex to bind and destroy homologous transcripts (Hamilton et al., 2002). The majority of well studied plant viruses have evolved the ability to evade this defense by encoding suppressors of PTGS. Different viral suppressors act at distinct steps in PTGS and can help to elucidate the PTGS pathway (Sharma and Ikegami, 2008). Viral suppressors are often determinants of pathogenicity, and their suppression activity is associated with the pathogenic determination (Brigneti et al., 1998).

Recently we reported that *Tomato leaf curl Java virus* encodes multiple suppressors (Sharma and Ikegami, 2008). We reported that the βC1 protein encoded by ToLCJV DNAβ02 produces virus-like symptoms in *Nicotiana benthamiana* and functions as a suppressor of PTGS. Furthermore, we demonstrate that C2 as a weak suppressor and C4 protein encoded by ToLCJV is not suppressor of PTGS (Kon et al., 2007). The results of the transient expression of GFP fused with *β*C1 protein suggest that this protein is localized to the cell periphery and or cytoplasmic strands.

ACKNOWLEDGMENTS

We thank Project Director, DWR, Karnal and Prof. HL Sharma, Palampur and Prof. N. Rishi for their valuable comments and encouragement during preparation of this Chapter. We acknowledge partly support by Grants-in-Aid of the Academic Frontier Promotion Program from the Ministry of Education, Culture, Sports, Science and Technology of Japan to MI, and Young Scientist's scheme from Indian council of Agricultural Research and Japan Society for the Promotion for Science grants to P.S.

REFERENCES

Briddon, R.W., Bull, S.E., Amin, I., Idris, A.M., Mansoor, S., Bedford, I.D., Dhawan, P., Rishi, N., Siwatch, S.S., Abedel-Salam, A.M., Brown, J.K., Zafar, Y., and Markham, P. G. 2003. Diversity of DNAβ, a satellite molecule associated with some monopartite begomoviruses. *Virology 312*:106-121.

Briddon, R.W., Mansoor, S., Bedford, I.D., Pinner, M.S., Saunders, K., Stanley, J., Zafar, Y., Malik, K.A., and Markham, P.G. 2001. Identification of DNA components required for induction of cotton leaf curl disease. *Virology 285*:234-243.

Brigneti, G., Voinnet,O., Li, W.X., Ji, L.H., Ding, S.W., Baulcombe, D.C. 1998. Viral pathogenicity determinants are suppressors of transgene silencing in *Nicotiana benthamiana*. *EMBO J. 17*: 6739-6746.

Dry, I. B., Rigden, J. E., Krake, L. R., Mullineaux, P. M., and Rezaian, M. A. 1993. Nucleotide sequence and genome organization of tomato leaf curl geminivirus. *J Gen. Virol. 74*:147-151.

Dunoyer, P., Lecellier, C.H., Parizotto, E.A., Himber, C., Voinnet, O. 2004. Probing the microRNA and small Interfering RNA pathways with virus-encoded suppressors of RNA silencing. *Plant Cell 16*: 1235-1250.

Engel, M., Fernandez, O., Jeske, H., and Frischmuth, T. 1998. Molecular characterization of a new whitefly-transmissible bipartite geminivirus infecting tomato in Panama. *J Gen. Virol. 79*:2313-2317.

Fauquet, C.M., Briddon, R.W., Brown, J.K., Moriones, E., Stanley, J., Zerbini, M., Zhou, X. 2008. Geminivirus strain demarcation and nomenclature. *Ach. Virol. 153*:783–821.

Fauquet, C.M., and Stanley, J. 2005. Revising the way we conceive and name viruses below the species level: A review of geminivirus taxonomy calls for new standardized isolate descriptors. *Ach. Virol 150*:2151-2179.

Galvao, R.M., Mariano, A.C., Luz, D.F., Alfenas, P.F., Andrade, E.C., Zerbini, F.M., Almeida, M.R., and Fontes, E. P.B. 2003. A naturally occurring recombinant DNA-A of a typical bipartite begomovirus does not require the cognate DNA-B to infect *Nicotiana benthamiana* systemically. *J. Gen. Virol. 84*:715-726.

Hamilton, A., Voinnet, O., Chappell, L., Baulcombe, D. 2002. Two classes of short interfering RNA in RNA silencing. *EMBO J. 21*:4671-4679.

Hong, Y.G., and Harrison, B.D. 1995. Nucleotide sequences from tomato leaf curl viruses from different countries: evidence for three geographically separate branches in evolution of the coat protein of whitefly-transmitted geminiviruses. *J. Gen. Virol. 76*:2043-2049.

Jose, J., and Usha, R. 2003. Bhendi yellow vein mosaic disease in India is caused by association of a DNAβ satellite with a Begomovirus. *Virology 305*:310-317.

Kasschau, K.D., Carrington, J.C. 1998. A counterdefensive strategy of plant viruses: suppression of posttranscriptional gene silencing. *Cell 95*:461-470.

Kirthi, N., Miya, S.P., Murthy, M.N. and Savithri, H.S.2002. Evidence for recombination among the tomato leaf curl virus strains/species from Bangalore, India. *Arch. Virol. 147*:255-272.

Kon, T., Hidayat, S.H., Hase, S., Takahashi, H., Ikegami, M. 2006. The natural occurrence of two distinct begomoviruses associated with DNAbeta and a recombinant DNA in a tomato plant from Indonesia. *Phytopathology 96*:517-525.

Kon, T., Sharma, P. and Ikegami, M. 2007. Suppressor of RNA silencing encoded by the monopartite Tomato leaf curl Java begomovirus. *Arch. Virol. 152*: 1273-1282

Matsuda, N., Sharma, P., Bajet N.J., and Ikegami, M. 2008. Molecular characterization of a new strain of *Tomato leaf curl Philippines virus* and its associated satellite DNAβ molecule: further evidence for natural recombination amongst begomoviruses. *Arch. Virol. 153*: 961-967.

Ogawa, T., Sharma, P. and Ikegami, M. 2008. The begomoviruses *Honeysuckle yellow vein mosaic* and *Tobacco leaf curl Japan viruses* associated with DNAβ satellite's cause tomato yellow dwarf disease. *Virus Res. 137*: 235-244.

Padidam, M., Beachy, R.N., and Fauquet, C.M. 1995. Tomato leaf curl geminivirus from India has a bipartite genome and coat protein is not essential for infectivity. *J.Gen. Virol. 76*:25-35.

Saeed, M., Behjatnia, S. A., Mansoor, S., Zafar, Y., Hussain, S., and Rezaian, M. A. 2005. A single complementary-sense transcript of a geminiviral DNAβ satellite is determinant of pathogenicity. *Mol. Plant Microbe Intr. 18*:7-14.

Sakata, J., Shibuya, Y., Sharma, P. and Ikegami, M. 2008. *Pepper yellow leaf curl Indonesia virus*, a new bipartite begomovirus and its new strains in leaf curl diseased tomato and yellow vein diseased ageratum in Indonesia. *Arch. Virol. 153*: 2307-2313.

Saunders, K., Bedford, I. D., Briddon, R. W., Markham, P. G., Wong, S. M., and Stanley, J. 2000. A unique virus complex causes *Ageratum* yellow vein disease. *P.N.A.S* U.S.A. *97*:6890-6895.

Saunders, K., Bedford, I. D., Yahara, T., and Stanley, J. 2003. The earliest recorded plant virus disease. *Nature 422*:831.

Saunders, K., Norman, A., Gucciardo, S., and Stanley, J. 2004. The DNAβ satellite component associated with ageratum yellow vein disease encodes an essential pathogenicity protein (βC1). *Virology 324*:37-47.

Sharma, P. and Ikegami, M. 2008. RNA silencing suppressors of geminiviruses. *J Gen. Plant Pathol. 74*: 189-202.

Sharma, P. and Rishi, N. 2007. Cotton leaf curl, an emerging whitefly-borne Begomovirus complex of cotton. *Plant Viruses 1*: 127-134

Shimizu, S., and Ikegami, M. 1999. Complete nucleotide sequence and the genome organization of tobacco leaf curl geminivirus from Japan. *Microbiol.Immunol. 43*:989-992.

Sukamto, Kon, T., Hidayat, S. H., Ito, K., Hase, S., Takahashi, H., and Ikegami, M. 2005. Begomoviruses associated with leaf curl disease of tomato in Java, Indonesia. *J. Phytopathol. 153*:562-566.

Varma, A., and Malathi, V. G. 2003. Emerging geminivirus problems: a serious threat to crop production. *Ann. Appl. Biol. 142*:145-164.

Voinnet, O. 2001. RNA silencing as a plant immune system against viruses. *Trends Gen. 17*:449–459

Zhou, X., Xie, Y., Tao, X., Zhang, Z., Li, Z., and Fauquet, C. M. 2003. Characterization of DNAβ associated with begomoviruses in China and evidence for co-evolution with their cognate viral DNA-A. *J.Gen. Virol. 84*:237-247.

In: Emerging Geminiviral Diseases and their Management ISBN 978-1-61668-620-8
Editors: P. Sharma, R. K. Gaur et al. pp. 71-103

Chapter 4

MOLECULAR STUDIES OF GEMINIVIRUS DISEASES AFFECTING CULTIVATION OF CHILLI CROPS AND THEIR POSSIBLE MANAGEMENT

S.K. Raj[1,*], M.S. Khan[2], S.K. Snehi[1] and P.K. Seth[2]
[1]National Botanical Research Institute (NBRI), Lucknow-226 001, U. P., India;
[2] Plant Health Clinic, Biotech Park, Lucknow-226 021, U.P., India.

ABSTRACT

Chilli crop is susceptible to a number of plant viruses belonging to various groups viz. *Cucumo, Poty, Tospo, Tobamo, Tobra, Tombus* and Geminiviruses etc. Among them the whitefly-transmitted geminiviruses (WTGs) pose a constraint to chilli cultivation and are highly prevalent in tropical climatic conditions due to the *Bemisia tabaci* complex. Apart from the tropical countries, the whiteflies have also gained importance in temperate regions in Europe because of their adaptability in greenhouse. Numerous WTGs have been reported on chilli from India, Pakistan, Sri Lanka, Bangladesh, China, Malaysia, Mexico, U.S.A., Thailand, Jamaica and Barbados. The most prevalent viral disease of chilli is the leaf curl and yellow mosaic.

Chilli plants often show symptoms similar to tomato leaf curl, such as yellowing, leaf curling, reduction in leaf size and stunting. Since chilli and tomato crops overlap in the field, hence it is likely that chilli may become infected with tomato begomoviruses and tomato might be infected by chilli infecting begomoviruses as the severe epidemics caused by *Pepper golden mosaic virus* in Mexico and *Pepper hausteco yellow vein virus* in Western Hemisphere has been reported. Recently infection of *Pepper leaf curl Bangladesh virus* in tomato (PepLCBV) in West Bengal, India has been reported. The association of DNA-β molecules has also been reported with bipartite begomoviruses for example: *Tomato leaf curl New Delhi virus* (ToLCNDV) causing leaf curl disease in chilli. The round year cultivation of the crops must have resulted to be susceptible for numerous economically important plant viruses and have become the challenge for the

* Correspondence concerning this article should be addressed to: S.K. Raj, E-Mail: skraj2@rediffmail.com.

plant pathologist. The leaf curl disease is of moderate to high incidence and widely distributed wherever chilli crop is grown. Recent reports indicate that there has been continuous increase in the incidence of chilli leaf curl disease resulting in the severe economic losses. This chapter deals with the detailed work done at Molecular Plant Virology Laboratory, NBRI, Lucknow, India and by various workers from India and abroad on biological, molecular and management studies of geminivirus infecting chilli crops.

INTRODUCTION

Chilli is an important and widely cultivated spice and vegetable crop. From the economic point of view, chilli is no longer considered a minor vegetable crop. Chilli is one of the most susceptible crops and has been found to be affected by various pathogens, which cause heavy losses to chilli production; viruses are considered as one of the important factors (Kang *et al.*, 1973; Villalon, 1975). Natural occurrence of several viruses have been reported on chilli by various workers and among them *Chilli leaf curl virus* (ChLCV), *Cucumber mosaic virus* (CMV) and *Chilli vein mottle virus* (CVMV) have been reported as most destructive viruses affecting chilli cultivation in terms of incidence and yield loss. The literature survey revealed that in last few years, the leaf curl disease of chilli has become the major constraint for chilli production in most of the chilli growing areas (Reddy *et al.*, 2005; Senanayake *et al.*, 2006, Khan, 2006; Quiñones *et al.*, 2002, Sri *et al.*, 2008). However, a little work has been done to diagnose the associated begomovirus (es) of leaf curl disease in India.

Chilli (*Capsicum annuum* L.), a member of family Solanaceae, is an important vegetable and spice crop of immense commercial importance. It is also called as hot pepper, cayenne pepper and sweet pepper. Several sources put the origin of chilli in Bolivia and Peru (Purseglove *et al.*, 1981; Bosland *et al.*, 1996) which was distributed after the discovery of America to other parts of the world. Brazil and Mexico is also considered as place of origin of chilli. It was introduced in India by Portuguese in Goa from Mexico during middle of 17th century and since then it had rapidly spread throughout the country. It is now grown in all parts of the world (Somos, 1984). Chilli is essentially a crop of the tropics and grown better in hotter regions. But it has a wide range of adoptability and is also grown extensively under various environments and climatic conditions in more than 60 countries (Martelli and Quacquarelli, 1983). It is cultivated over large areas in Asia, Africa South and Central America, parts of USA and Southern Europe both under tropical and subtropical conditions. The major chilli growing countries are India, Nigeria, Mexico, China, Indonesia and Korean Republic (Anonymous 2002).

Chilli is nature's wonder; its fruit appears in different sizes, shapes and colours. The fruit size varies from variety to variety and may be some times more than hundred times bigger to that of others. The shape may be elongated or round and distal pointed, blunt or sucked in. Outer skin may be corrugated, leathery or smooth. Colour varies from blue to green, orange, red, yellow, violet, cream white and near black. The great variation in shape, size, flavour and colour of the fruit, as well as in plant habit, makes it difficult to devise a practical system of classification in *C. annuum*. Smith *et al.* (1987) reviewed the genus and proposed a horticultural classification of the chilli.

There are several synonyms of hot pepper worldwide; chile, chili, and chilli. Etymologically, the word "chile" is derived from the Aztec language, and refers to *Capsicum* peppers in Central America including Mexico, and in the many parts of southwestern United States (Smith *et al.*, 1987). The word "chili" is thought to be the anglicized form of "chile" and now denominates pungent types of *Capsicum* peppers in the United States. Similarly, the words "chilli" and "chillies" are used in Middle Eastern and Asian countries in connection with pungent forms (Sanogo, 2003).

Economic Importance of Chilli

It is constituent of many foods, flavour, food color, minerals and pungency. Green chilli is eaten raw or stuffed and baked or cooked or fried and is also used as spice. The sweet pepper, bell pepper etc. have the mildest flavour and least pungency; these are eaten raw in salads and also cooked in various ways. They are used as stuffed with various vegetables, meat, and cooked. Several varieties of sauce and pickles are prepared with the green as well as the ripe and red chillies. Chilli is extensively used in vegetable soups, hot sauce and as a spicy flavouring in meat. The juice of pomegranate is boiled with capsicum to make a tasty sauce. Chilli can also be pickled with vinegar or employed to preserve onions and cucumbers. It is used to increase the pungency of ginger ale, ginger soda, ginger beer, rum, galeniealas etc. and in tobacco smoking and chewing. The dried, ripped and powered chillies are employed in various preparations of daily food (Biacs *et al.*, 1989).

Since ancient times green, fresh and ripe as well as dry chillies have been used by healers to cure a variety of ailments. They have been used extremely to relieve pain and internally to cure anything from yellow fever to the common cold. The medicinal properties of chilli includes stimulant, tonic, sialagogne, alternative, rubefacient, carminative, anticoagulant etc.

The active ingredient in chilli is a compound called capsaicin (8-methyl-*N*-vanillyl-6-nonenamide) which is beneficial to the respiratory system, blood pressure and heart. It prevents the heart disease by dilating blood vessels. Capsaicin ointment have been found to relieve the pain of arthritis and shingles when applied externally, and when taken internally, capsaicin triggers release of endorphins in the brain, which has a pain relieving effect similar to that of morphine. A single pepper has been found to contain a full day's supply of beta-carotene and nearly twice the recommended daily allowance for vitamin C, which makes the chilli an invaluable food to fight against cancer and heart disease. Recently Russian scientists have identified vitamin P in green chillies which is considered to be important as it protects from secondary irradiation injury. Chilli may also help in weight loss by speeding up the metabolism. Because of high content of Vitamin A and C, Chilli is an excellent source for balancing man's nutrition (Villalon, 1981).

Cultivation and Production of Chilli

Numerous varieties of chilli are cultivated around the world. The world production of chilli crop sums up to around 7 million tons that occupies approximately 1.5 million hectares of land. India is the world largest producer of chilli followed by China and Pakistan. India accounting for 11 lakh tons of production annually followed by China with a production of around 4 lakh tons and Pakistan also producing 3 lakh tons of chilli every year. This shows that the bulk share of chilli production is held by the Asian countries though it is produced throughout the world. India is also the largest exporter and consumer of chilli in the world and it contributes to $1/4^{th}$ share in the total quantity of chilli exported in the world. It consumes around 6.2 million tons of chillies i.e. about 90% of the total produce of the country. The crop is cultivated in almost all the states of the country, Andhra Pradesh being the leader of all. It contributes to approximately 25% of the total production. The major chilli exporting players along with their percentage share in the world total exports are: India (25%), China (24%), Spain (17%), Mexico (8%), Pakistan (7.2%), Morocco (7%), and Turkey (4.5%) (http://www.crnindia.com/commodity/chilli.html).

Diseases of Chilli

Chilli is one of the most susceptible crops and has been found to be affected by various pathogens viz. fungi, bacteria, viruses, phytoplasma, and nematodes (Persley *et al.*, 1989; Bosland *et al.*, 1994; Goldberg, 1995; Singh and Singh, 2000), which cause heavy losses to chilli production. The important diseases of chilli are mosaic, leaf curl, damping off, die back and powdery mildews. Viruses are considered as one of the important constraints to chilli production all over the world (Kang *et al.*, 1973; Villalon, 1975; Villalon, 1981).

Viral Diseases of Chilli

The virus diseases occurring on chilli show different types of symptoms like vein clearing, vein banding, mosaic, ring spot, crinkling, blistering, puckering, necrosis, wilting, stunting, curling, etc. The diseased plants mostly show mosaic, curling, crinkling and blistering symptoms on the leaves. Symptoms of *Chilli leaf curl virus* consists of abaxial and adaxial curling of the leaves accompanied by puckering and blistering of interveinal areas and thickening and swelling of the veins. Infected plants by *Chilli mosaic virus/Cucumber mosaic virus* (CMV) generally exhibit mosaic, yellow discoloration, vein clearing, shoestring, or top necrosis, and necrotic rings and spots on leaves and fruits (Tobias, 1983; Fujisawa *et al.*, 1986; Francki, 1979; Nagaraju and Reddy, 1982). Several strains of CMV are known to infect peppers (Simons, 1957). In chili peppers, yield losses up to 41% (Joshi and Dube, 1973). Twenty six species of aphids are known to transmit CMV in nature. *Tomato aspermy virus* (TAV) causes mosaic and leaf distortion in *Capsicum* sp. and has been reported from Japan, Hungary and Netherlands (Proctor, 1975; Tochihara, 1970). Chilli mosaic disease caused by *Potato virus Y* is characterized by severe mosaic mottling, vein banding and interveinal chlorosis of the infected leaves. Fruits from infected plant show chlorotic streaks and rough skin.

Natural occurrence of several viruses has been reported on chilli by various workers (Osaki and Inouye, 1981; Martelli and Quacquarelli, 1983; Green and Kim 1991). Incidence and severity of viral disease of chilli vary as reported by the several workers from different geographical region may be due to the type of insect-vectors in that particular area and their infestation and climatic conditions. Five types of viruses have been reported to infect chilli plants in Malaysia. These are *Cucumber mosaic virus* (CMV), *Chilli veinal mottle virus*

(CVMV), *Tobacco mosaic virus* (TMV), *Chilli leaf curl virus* (CLCV), *Tomato spotted wilt virus* (TSWV) and *Tobacco leaf curl virus* (TLCV). Among these, CMV and CVMV appear to be the most important viruses in terms of incidence and yield loss (Green, 1992). In Korea, ten viruses are reported and, out of them, two species of *Pepper vein chlorosis* virus and *Pepper vein mosaic virus* are reported only in Korea (Kim *et al.*, 1990 & 1991).

In India, the most important disease caused by viruses in chilli is yellow mosaic disease (Biswas *et al.*, 2005). Incidence of the disease is reported to be 12-95% resulting yield losses as high 65-75% (Bidari and Reddy, 1990; Narasimhan and Alagianagalingam, 1986). The yellow mosaic disease (YMD) has been found to be of complex nature; eight viruses have been identified in Karnataka (Bidari and Reddy, 1994) and four viruses in Punjab (Kaur and Singh, 1985). The most common ones are *Potato virus Y* (PVY), *Pepper vein banding virus* (PVBV), *Pepper veinal mottle virus* (PVMV), *Chilli veinal mottle virus* (CVMV), *Tobacco etch virus* (TEV), *Tobacco mosaic virus* (TMV), *Cucumber mosaic virus* (CMV) and *Tobacco leaf curl virus* (ToLCV) (Kaur and Singh, 1985; Rishi and Dhawan, 1989; Bidari and Reddy, 1990 & 1994; Kiranmai *et al.*, 1997; Satya Prakash *et al.*, 2002). *Pepper veinal mottle virus* (Nagaraju and Reddy, 1980), *Paper vein banding virus* (Nagaraju and Reddy, 1981), *Marigold mottle virus* (Naqvi *et al.*, 1978), *Launaea mosaic virus* (Naqvi *et al.*, 1976), *Brinjal mosaic virus* (Naqvi and Mahmood 1975), *Calliopsis mottle virus* (Joshi and Suteri 1976), *Potato virus S* and *Potato virus M* (Misra *et al.*, 1979), *Datura distortion mosaic virus* (Mali *et al.*, 1985) have also been reported time to time from India on *Capsicum annuum*. Among them, PVY and CMV are of common occurrence and cause enormous losses in Chilli (Tomlinson, 1987; Singh-Amarjit et al., 1990; Kiranmai *et al.*, 1997, Satya Prakash *et al.*, 2002). This chapter mainly deals the occurrence of geminivirus in chilli crops across the world.

Geminivirus

Geminiviruses of family *Geminiviridae* are characterized by the twinned, small (ca. 18-30 nm), quasi-icosahedral morphology of the virion particles and by their genomes consisting of one or two molecules of ssDNA 2.5-3.0 kb in size (Goodman 1977; Goodman *et al.*, 1977; Harrison *et al.*, 1977; Hatta and Franki, 1979; Reisman *et al.*, 1979; Francki *et al.*, 1980; Harrison, 1985; Lazarowitz, 1992). It has four genera: *Mastrevirus*, *Curtovirus*, *Topocuvirus* and *Begomovirus* that differ mainly with respect to genomic characteristics, type of insect vector and host range (van Regenmortel *et al.*, 2000; Harrison and Robinson, 2002; Fauquet and Stanley 2003; Fauquet *et al.*, 2003; Stanley *et al.*, 2005).

Geminiviruses Infecting Chilli

Among four genera of *Geminiviridae,* only *Curtovirus* and *Begomovirus* are known to infect chilli, which are being described as:

1) Curtovirus

Out of six species in *Curtovirus*, only *Beet curly top virus* (BCTV) is reported to infect chilli. The leafhopper transmitted BCTV causes yellow discoloration and curling of the leaves followed by decline and death of chilli plants. Symptoms, however, vary with different strains. High incidences of BCTV can cause considerable yield losses (UmaUngs *et al.*, 1977).

2) Begomovirus

The type species of *Begomovirus* is *Bean golden mosaic virus*. *Begomovirus* species are transmitted by whitefly in dicotyledonous and monocotyledonous plants. About twenty nine begomovirus species have been reported infecting chilli globally viz.: *Bean distortion dwarf virus* (Xi *et al.*, 1982), *Chino del tomato virus* (syn. Tomato golden mosaic virus; Brown and Nelson, 1988), *Pepper hausteco yellow vein virus* (syn. Pepper huasteco virus; Garzon-Tiznado *et al.*, 1993), *Pepper mild tigre virus* (Brown and Nelson, 1989), *Texas pepper virus* (Stenger *et al.*, 1990), *Serrano golden mosaic virus* (syn. Pepper golden mosaic virus; Brown and Poulos, 1990; Holguin-Pena and Vazquez, 2004; Ala-Poikela *et al.*, 2005), *Pepper leaf curl Bangladesh virus* (Fauquet *et al.*, 2003), *Pepper leaf curl Lahore virus* (Fauquet *et al.*, 2008), *Pepper leaf curl virus* (Fauquet *et al.*, 2008), *Pepper yellow vein Mali virus* (Fauquet *et al.*, 2008), *Tobacco leaf curl virus* (Osaki and Inouye, 1981; Sugiura *et al.*, 1975), *Pepper yellow leaf curl virus* (Samretwanich *et al.*, 2000), *Potato yellow mosaic Trinidad virus* (Umaharan *et al.*, 1998; Jones, 2003), *Tomato dwarf leaf curl virus* (Roye *et al.*, 1999), *Tomato leaf curl Barbados virus* (Fauquet *et al.*, 2003), *Tomato leaf curl Sinaloa virus* (Ala-Poikela *et al.*, 2005), *Tomato severe leaf curl virus* (Ala-Poikela *et al.*, 2005), *Tomato leaf curl Jodebpur virus* (Shih *et al.*, 2007), *Tomato leaf curl Gujarat virus* (Chakraborty *et al.*, 2003), *Tomato leaf curl Sulawesi virus* (Tsai *et al.*, 2009), *Bell pepper leaf curl virus* (Tahir and Haider, 2006), *Pepper yellow leaf curl Indonesia virus* (PepYLCIDV; Tsai *et al.*, 2006), *Tomato yellow leaf curl virus* (Quiñones *et al.*, 2002), *Tomato leaf curl Bangalore virus* (Reddy *et al.*, 2005), *Tomato leaf curl New Delhi virus* (Hussain *et al.*, 2004; Khan *et al.*, 2006a; Reddy *et al.*, 2005), *Tomato severe rugose virus* (Bezerra-Agasie *et al.*, 2006) a new begomovirus similar (93-94% nucleotide sequence identity) to *Cabbage leaf curl virus* (Zubiaur *et al.*, 2006), *Chilli leaf curl Multan virus* (Akhter *et al.*, 2009) and *Chilli leaf curl virus* from China (Tsai *et al.*, 2000), Pakistan (Shih *et al.*, 2003, Fauquet *et al.*, 2008), Sri Lanka (Shivanthan, 1982), Taiwan (Shih *et al.*, 2003) and Bhutan (Thinlay and Penjore, 1996) and India (Senanayake *et al.*, 2006, Chattopadhyay *et al.*, 2008, Fauquet *et al.*, 2008). The distribution of these begomoviruses is summarized in table 1.

Table 1. Begomoviruses affecting chilli and their distribution

Virus	Acronym	Distribution	Reference
Bean distortion dwarf virus	BDDV	China	Xi *et al.*, 1982
Chino del tomato virus	CdTV	Mexico, USA	Brown & Nelson, 1988
Pepper huasteco yellow vein virus	PHYVV	Mexico	Garzon-Tiznado *et al.*, 1993
Pepper mild tigre virus	PepMTV	Mexico	Brown & Nelson, 1989
Texas pepper virus	TPV	Texas, U.S.A	Stenger *et al.*, 1990
Serrano golden mosaic virus / Pepper golden mosaic virus	SGMV/ PepGMV	Nicaragua, Mexico, Baja, California	Brown & Poulos, 1990, Holguin-Pena & Vazquez, 2004, Ala-Poikela *et al.*, 2005

Pepper leaf curl Bangladesh virus	PepLCBV	Bangladesh	Fauquet *et al.*, 2003
Pepper leaf curl Lahore virus	PepLCLV	Pakistan	Fauquet *et al.*, 2008
Pepper leaf curl virus	PepLCV	Malaysia	Fauquet *et al.*, 2003
Pepper yellow vein Mali virus	PepYVMV	Mali	Fauquet *et al.*, 2008
Tobacco leaf curl virus	ToLCV	Tanzania	Osaki & Inouye, 1981
Pepper yellow leaf curl virus	PepYLCV	Thailand	Samretwanich *et al.*, 2000
Potato yellow mosaic Trinidad virus	PYMTV	Trinidad	Umaharam *et al.*, 1998, Jones, 2003
Tomato dwarf leaf curl virus	TDLCV	Jamaica	Roye *et al.*, 1999
Tomato leaf curl Barbados virus	ToLCBBV	Barbados	Fauquet *et al.*, 2003
Tomato leaf curl Sinaloa virus	ToLCSV	Nicaragua	Ala-Poikela *et al.*, 2005
Tomato severe leaf curl virus	ToSLCV	Nicaragua	Ala-Poikela *et al.*, 2005
Tomato leaf curl Jodebpur virus	ToLCJV	India	Shih *et al.*, 2006
Tomato leaf curl Gujarat virus	ToLCGV	India	Chakarborty *et al.*, 2003
Tomato leaf curl Sulavesi virus	ToLCSuV	Indonesia	Tsai *et al.*, 2009
Bell pepper leaf curl virus	BPepLCV	Pakistan	Tahir &Haider, 2006
Pepper yellow leaf curl Indonesia virus	PepYLCIV	Indonesia	Tsai *et al.*, 2006
Tomato yellow leaf curl virus	TYLCV	Cuba	Quiñones *et al.*, 2002
Tomato leaf curl Bangalore virus	ToLCBV	India	Reddy *et al.*, 2005
Tomato leaf curl New Delhi virus	ToLCNDV	Pakistan, India	Khan *et al.*, 2006a; Reddy *et al.*, 2005
Tomato severe rougase virus	ToSRV	Brazil	Bezerra-Agasie *et al.*, 2006
Cabbage leaf curl virus	CabLCV	Cuba	Zubiaur *et al.*, 2006
Chilli leaf curl Multan virus	ChLCMV	Pakistan	Akhter *et al.*, 2009
Chilli leaf curl virus	ChLCV	China, Taiwan, Sri Lanka, Pakistan, India Bhutan.	Tsai *et al.*, 2000; Shivanthan, 1982; Shih *et al.*, 2001; Shih *et al.*, 2003; Thinlay & Penjore, 1996; Senanayake *et al.*, 2006, Fauquet *et al.*, 2008

Biological Studies of Geminivirus Infecting Chilli Crops

Incidence and Natural Symptoms of Leaf Curl Disease on Chilli Cultivars Grown in India:

The natural occurrence of severe leaf curl disease was observed in various chilli cultivars growing in different locations of Uttar Pradesh: Bahraich (Matera and District Horticulture Research Station), Lucknow (Banthara Research farm of NBRI) and Kanpur (Vegetable farm of CSA University, Kalyanpur) districts of Uttar Pradesh in the years 2003-2005. The disease incidence was significant ranging from 35 to 50% in different locations which were surveyed. The incidence of leaf curl disease in two districts viz. Kanpur and Lucknow was found increasing in the three subsequent years. In the year 2003, the incidence was about 35% which reaches to 50% in the year 2005 in research farm of Vegetable Science, C. S. A. University of Agricultural & Technology at Kanpur while it was about 40% in 2003 and increased to 46% in the year 2005 at Banthara Research farm of NBRI, Lucknow. However, the disease incidence was almost constant (40-42%) in the years 2003-2005 at Matera and District Horticulture Research Station, Bahraich. The disease incidence in various cultivars showed variable degree of susceptibility/tolerance/resistance under field conditions (Khan, 2006).

Various varieties/cultivars of chilli showed leaf curl disease symptoms typical of geminivirus infection with different degrees of severity. The cultivars Surajmukhi, Agni,

Andhra Jyoti, X-235 from Bahraich; Chandramukhi, Pusa Jwala, Pusa Sadabahar, KDCS-810, Longi Jpani from Lucknow; Chanchal, Chaman, Achar-8, KDCS-810, Selection-13, Selection-54, Achar-36, Assam-5 B and Selection-56 from Kanpur shows typical symptoms of leaf curl disease (Figure 1) with high disease incidence. The symptoms were upward and downward curling of the leaves accompanied with puckering, blistering of intervenial areas and thickening/swelling of the veins (Figures 2-3). Infected plants showed stunted growth, less flowering and deformed fruits as compared to the healthy plants (Figure 3). The severely infected plants did not bear any flowers or fruits resulting in 25-30% of total yield loss in various locations (Khan, 2006).

Figure 1. A view of chilli field showing natural occurrence of severe leaf curl disease with a high incidence in chilli growing at Vegetable farm of CSA University, Kanpur, U.P., India.

Figure 2. A close view of severely infected chilli plant showing upward and downward curling of the leaves accompanied with puckering, blistering of intervenial areas and thickening/swelling of the veins (a) as compared to the healthy plants (b).

Figure 3. Infected plants of four chilli cultivars showing mild to severe leaf curl symptoms, growth stunting, less flowering (upper panel) and fruit deformation in 3 chilli cultivars (lower panel).

Virus Transmission by Whitefly (*Bemisia Tabaci*) and Host Range Study:

The presence of whiteflies (*Bemisia tabaci*) was also noticed on the infected chilli plants showing leaf curl disease and in the vicinity of growing area, therefore, causal agent of the disease was suspected as a whitefly transmitted virus (begomovirus) and whitefly transmission tests were carried out (Khan, 2006).

Virus transmission tests using whiteflies (Figure 4a &b) were performed with samples of chilli collected from Lucknow, Kanpur and was maintained in glass-house (Figure 4c). Host range studies were also conducted to a few test plants. In each whitefly transmission tests, approximately 6 viruliferous whiteflies were transferred to healthy seedlings of *Capsicum annuum*, *Lycopersicon esculentum*, *Nicotiana tabacum* cv. White Burley, *Nicotiana glutinosa* and *Solanum melongena* for inoculation. The inoculated seedlings were maintained in insect proof cages and observed periodically for symptom development for a period of 6 weeks. The disease was successfully transmitted from naturally infected *C. annuum* to healthy *C. annuum* (Figure 4d), *L. esculentum* (Figure 4e), *N. tabaccum* cv. White burley (Figure 4f) and *N. glutinosa* which developed severe and mild leaf curl symptoms by 25-30 days post inoculation. However, *Solanum melongena* did not develop any symptoms even after 6 weeks of inoculation (Khan, 2006).

Figure 4. A diagrammatic representation of transmission of chilli leaf curl disease by whitefly (*B. tabaci*) from naturally infected chilli to healthy chilli, tomato and tobacco seedling. Colonization of whiteflies on a tobacco host plant (a), their close view (b), naturally infected chilli (c) and *C. annuum* (d), *L. esculentum* (e), *N. tabaccum* cv. White burley (f) plants showing systemic leaf curl symptoms 25-30 days of post inoculation.

Studies on Chilli Leaf Curl Disease Carried in other Countries

A geminivirus causing leaf curl and distortion symptoms was isolated from pepper (*C. annuum*) cultivated in Texas. The Texas pepper geminivirus (TPGV) was transmitted persistently by *B. tabaci* and was also transmitted mechanically to species of the Solanaceae. Electron microscopy of purified virions revealed typical geminate particles (Stenger *et al.*, 1990).

Pepper (*C. anuum*) plants affected with yellow leaf curl disease were collected from central Thailand and virus cultures were maintained in pepper plants. Transmission experiments were carried out with *B. tabaci*. Acquisition and inoculation threshold periods were 1 h and 30 min, respectively. The latent period was 10 h. Symptoms in cultured plants were the same as those observed in field plants (Samretwanich *et al.*, 2000).

Transmission of *Tomato leaf curl begomovirus* by two species of whitefly in chilli and tomato was done by Sri *et al.* (2007) in Indonesia. They found that the two species of whitefly, *B. tabaci* and *Trialeurodes vaporariorum*, were capable to transmit *Tomato leaf curl virus* (ToLCV) and evidenced that *B. tabaci* was more effective as insect vector of ToLCV in tomato and chilli pepper. A single *B. tabaci* was able to transmit ToLCV with a minimum acquisition and inoculation access period of 10 h. Transmission of ToLCV by *T. vaporariorum* required at least 10 insects per plant with a minimum acquisition and inoculation access period of 24 h. The transmission efficiency was increased with longer acquisition and inoculation access period of the insect and the higher number of insect per plant (Sri *et al.*, 2007).

A virus and its vector, pepper yellow leaf curl virus and *B. tabaci*, two new invaders of Indonesia have been studied (Barro *et al*, 2008). The invasion of pepper yellow leaf curl virus and *B. tabaci* was found associated with pepper yellow leaf curl disease, which was shown to have a probable origin in the central Thailand into the Indonesian islands of Sumatra then Java and Bali geographic region. The consequences of the invasion of a plant-infecting virus and its vector has been a massive increase in the scale and impact of begomoviruses in tomato and chilli production which has seen regional bans imposed on the planting of chilli, an important cash crop for many village farmers in Sumatra and Java (Barro *et al*, 2008).

Molecular Studies of Geminivirus Infecting Chilli Crops

Molecular Studies of Begomoviruses of Chilli Leaf Curl Disease Carried in India

Infection of two begomoviruses infecting chilli was investigated by Reddy *et al.* 2005. During a survey across the India, they collected samples from tomato and chilli plants. They detected presence of *Tomato leaf curl Bangalore virus* and *Tomato leaf curl New Delhi virus* infection in chilli crop. Pepper leaf curl Bangladesh virus was also found to be infecting tomato crop in India (Reddy *et al.*, (2005).

Further, *Tomato leaf curl New Delhi virus* (ToLCNDV) was shown to be associated with chilli leaf curl disease occurring in Lucknow (Khan *et al.*, 2006a). Natural incidence of leaf curl diseases was observed on various cultivars of chilli (Capsicum annuum) grown at Banthra research station of N.B.R.I., Lucknow, India. The virus isolate causing leaf curl diseases in chilli cultivars were transmitted by whiteflies. A begomovirus was detected from the samples showing leaf curl symptoms by dot blot hybridization tests (Figure 5a) using cloned probe of a begomovirus and confirmed by PCR using geminivirus detecting primers (Khan 2006). PCR amplification and nucleic acid hybridization technique appears to be sensitive enough to detect the presence of geminivirus infection in chilli and other commercially important plants.

In another study, two degenerate primers designed by Deng and Rojas were used during PCR for detection of geminivirus in chilli samples. Deng primers worked well and amplicon of expected size (~550 bp) was obtained which confirmed the association of geminivirus with chilli leaf curl disease. Since the Coat protein (CP) represents the most conserved gene in the family *Geminiviridae* which is informative as the entire sequence of the genome and is sufficient to classify a virus isolate. The CP nucleotide sequences/amino acid have been used to establish taxonomical relationship among WTGs. Therefore, the primers specific to CP and Replicase (Rep) genes of a well characterized geminivirus were employed. The PCR resulted in positive amplicons of ~800 bp, and ~1.1 kb respectively in all the chilli samples collected from Bahraich, Lucknow and Kanpur (Figure 5b &c). BLAST similarity search results of CP and Rep clones showed highest identity of chilli virus isolates with *Tomato leaf curl New Delhi virus* (ToLCNDV) strains, therefore, they were considered as strains of ToLCNDV infecting chilli (Khan 2006).

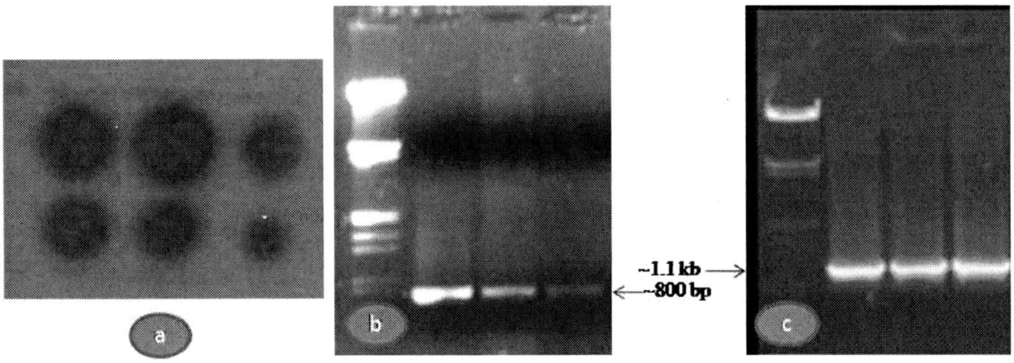

Figure 5. Detection of begomovirus from infected chilli leaf samples by dot blot hybridization tests and PCR analysis. Positive signal of hybridization with probe of a begomovirus (a) PCR products showing expected size amplicons ~800bp and 1.1kb with CP and Rep gene specific primers, respectively (b&c).

The complete DNA-A of ToLCNDV-chilli associated with the chilli leaf curl was cloned and sequenced (EU309045). The analysis of the sequence data revealed presence 2740 nt of complete DNA-A segment of *Tomato leaf curl New Delhi virus*-Chilli pepper isolate Bahraich which contained AV2, AV1, AC5, AC3, AC2, AC1 and AC4 ORFs (Figure 6a). BLAST search analysis of the virus isolate shared highest sequence identities with various isolates of ToLCNDV (AJ875157, AY428769, EO68246, Y16421, EF035482, AM286434, AF448058, DQ116880, DQ169056, GQ865546, AJ620187), therefore, it was considered as an isolate of ToLCNDV-Chilli.

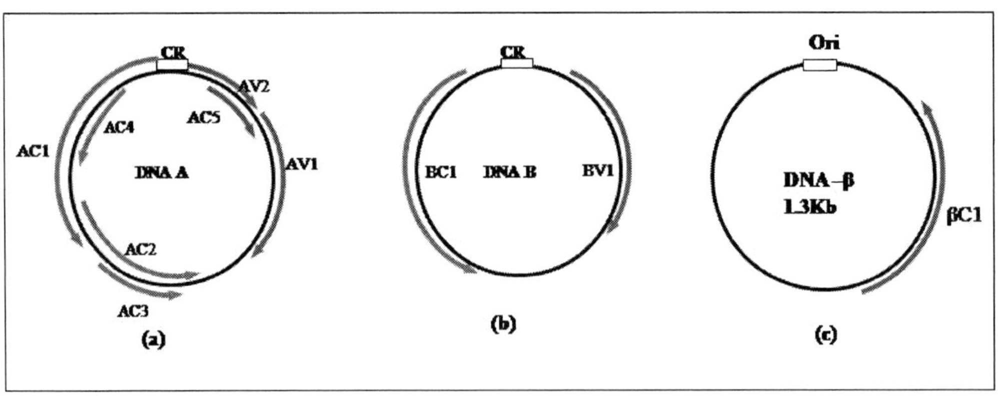

Figure 6. Genome organization of *Tomato leaf curl New Delhi virus*-Chilli pepper isolate showing different ORFs in the DNA-A (a), DNA-B (b) and associated beta DNA (c) components.

Movement protein (MP) specific primers employed with chilli samples were also capable to amplify the appropriate size (~875 bp) DNA fragments confirming the geminivirus contains the bipartite genome (Figure 6b). Blast search analysis of the complete nucleotide sequence and deduced amino acid of the MP ORF (under study) shared considerable sequence identity with most of the ToLCNDV strains originated from India, Pakistan and Bangladesh. It shared maximum 94% and 97% percent identity with Indian strain of ToLCNDV (U15017) at nucleotide and amino acid level respectively (Khan 2006).

DNA beta molecules are symptom modulating, single stranded circular DNA satellites associated with begomoviruses (family *Geminiviridae*). The molecules have been found associated with diverse plant species from India and neighboring countries. Thus, investigation of DNA beta satellite molecule was also carried out to confirm the association of satellite molecule with ToLCNDV-chilli isolates under study. The total DNA isolated from the chilli samples was employed initially for dot blot hybridization with DNA beta specific probe which showed strong signals of hybridization in all chilli samples (ToLCNDV). As expected to the primers, amplicons of approximately 1.35 kb were amplified in all three chilli samples. These findings confirmed the association of the satellite molecule with the leaf curl disease. Further the amplicon obtained from ToLCNDV-chilli Bahraich isolate was cloned and sequenced which reveled the full length nucleotide sequence of 1371 bp (DQ343289) (Figure 6c). BLAST search analysis of nucleotide sequence revealed high level sequence identity with a number of DNA beta components reported from a variety of plant species from different parts of the world. The analysis of the nucleotide sequence of the DNA beta associated with the ToLCNDV-chilli Bahraich isolate showed all the characteristics of DNA-β viz. satellite conserved region (SCR), adenine rich region (A-rich region) and an open reading frame beta CI (ORF beta CI) (Khan 2006).

Chilli leaf curl virus affecting chilli in the Jodhpur district of Rajasthan, India was studied by (Senanayake *et al.*, 2006). During December 2004, very high disease incidence (up to 100% of plants) was observed in farmer's fields in Narwa and Tinwari villages. The characteristic field symptoms were upward curling, puckering and reduced size of leaves (Figure 1). Severely affected plants were stunted and produced no fruit. The virus from field samples from Narwa village was transmitted by whitefly (*B. tabaci*) to 50-100% of chilli test plants, which produced vein clearing, curling and stunting symptoms. Electron microscopic examination of field samples revealed few, typical geminate particles. The presence of a begomovirus was confirmed by polymerase chain reactions (PCR) using the degenerate primers AVF28 5'-GCCCACATYGTC TTYCCNGT-3' and AV29R 5'-GGCTTYCT RT ACATRGG-3', which gave a *ca.* 1.0 kb product. Cloning and sequencing of the PCR product yielded a 995 bp sequence (Acc. No. DQ445255). A BLAST search of GenBank revealed close similarity of the sequence with the intergenic region and part of the replication initiator protein, AV1 and AV2 genes of Chilli leaf curl virus-[Pakistan: Multan] (ChiLCuV-[Pk:Mul]; AF336806).

Molecular Studies of Begomoviruses of Chilli Leaf Curl Disease Carried out Abroad

Genomic properties of a geminivirus isolated from peppers were carried out first time by Stenger *et al.* (1990) who isolated the replicatve form (RF) DNA of 2.6 kb that was double stranded, circular, and supercoiled. The RF DNA was linearized by digestion with the restriction enzymes *Eco*RI or *Hin*dIII, and cloned into *Escherichia coli* plasmid pUC 8. Analysis of cloned DNA by Southern hybridization and restriction endonuclease mapping indicated the cloning of two distinct from RF DNA. One TPGV DNA hybridized with DNA-A of tomato golden mosaic virus (TGMV); however, neither TPGV DNA hybridized with TGMV DNA-B. Infectivity assays using cloned TPGV DNAs demonstrated that both DNA species were required for systemic infection of test plants. These results indicated that TPGV is a typical whitefly-transmitted, bipartite geminivirus not previously known to occur in the United States (Stenger *et al.* 1990).

Complete nucleotide sequence of pepper huasteco virus and comparative analysis and with bipartite geminiviruses was done (Torres-Pacheco *et al.*, 1993). The complete nucleotide sequence of pepper huasteco geminivirus (PHV) isolated in Northern Mexico was determined. The PHV genome consisted two circular ssDNA molecules of 2631 bases (PHV A) and 2589 bases (PHV B). The genome organization of PHV showed to be a bipartite geminivirus with four open reading frames (ORFs) (AR1, ALl, AL2 and AL3) in component A and two (BR1 and BL1) in component B. An unexpected ORF was found in the complementary sense strand of PHV A. This ORF, termed AL5, was found entirely inside, but in the opposite orientation to AR1 (encoding the coat protein). AL5 showed some homology to equivalent but smaller ORFs predicted in other geminiviruses. Phylogeny trees based on pairwise comparisons of AR1, AL2, AL3, BL1 and BR1 predicted proteins placed PHV among the western hemisphere geminiviruses. A phylogeny tree based on ALl (replicase-encoding ORF), on the other hand, placed PHV with eastern hemisphere geminiviruses, i.e. *African cassava mosaic virus* and the Sardinia and Israel isolates of *Tomato yellow leaf curl virus* (Torres-Pacheco *et al.*, 1993).

In south and central Jamaica, leaf curling and severe dwarfing, with or without marginal chlorosis were observed on pepper and tomato plants since 1992. Genomic characterization using nonradioactive probes, PCR with degenerate primers and nucleotide sequencing were used to characterize the virus isolate. DNA sequence analysis showed that pepper and tomato samples have a similar geminivirus associated with them. Nucleotide sequence identity > 92% between the common regions of DNA-A and DNA-B confirmed the bipartite nature of the Jamaican geminivirus isolates. This new virus was given the name *Tomato dwarf leaf curl virus* (TDLCV) because of the dwarfing and leaf curling symptoms associated mostly with infected tomato plants (Roy *et al.*, 1999)

A new geminivirus associated with a yellow leaf curl disease of pepper in Thailand was reported by Samretwanich *et al.* (2000). They isolated the total DNA from pepper samples collected from central Thailand and amplified using PCR with geminivirus-specific degenerate primers. A PCR product of 2.7 kbp was amplified and cloned. Three independent clones were sequenced and analyzed, and an identical 32-base stem loop region and the unique sequence (TGGGGTC) of putative Rep binding site were found in the intergenic region (IR). The B component could not be detected. The nucleotide sequence of the coat protein gene was compared with 28 well-studied whitefly-transmitted geminiviruses which showed the highest sequence similarity (85%) with *Tomato leaf curl virus* from Taiwan (TwToLCV: accession number U88692), suggesting that it was a new geminivirus. Therefore, the virus isolate was designated *Pepper yellow leaf curl virus*.

Tomato yellow leaf curl virus infecting pepper plants in Cuba was reported by Quiñones *et al.* (2002). In Cuba, *Tomato yellow leaf curl virus* was for the first time reported in 1994 and later became the most serious constraint to tomato production. However, during a field survey in 2001, *C. annuum* showing mild interveinal yellowing and curling of leaves were observed in a greenhouse in Camagüey Province. PCR was carried out with CP and the intergenic region (IR) of the Cuban isolate of TYLCV-IS which resulted in positive amplification of expected sizes in all the symptomatic chilli samples. The CP fragment was digested with *Rsa*I, while the IR amplicon was digested with *Ava*II and *Eco*RI. In all cases the patterns obtained corresponded to digestion patterns for identical PCR fragments obtained from TYLCV-infected tomatoes. Furthermore, the IR amplicon sequence of one sample showed 99% identity with the corresponding region of the TYLCV-IS isolates of tomato in Cuba. This was the first report of TYLCV-IS infecting pepper in Cuba (Quiñones *et al.*, 2002).

Díaz-Plaza *et al.* (2002) detected geminiviruses infection in all chilli samples collected from 15 sites of the Yucatan Peninsula. The samples were analyzed by dot blot hybridization using general (gene of capsid protein of Pepper huasteco virus, PHV) and specific probes (common region of PHV, Pepper golden mosaic virus, PepGMV; Tomato mottle virus, ToMoV; Bean golden mosaic virus, BGMV; Tomato yellow leaf curl virus, TYLCV and Taino tomato mottle virus, TToMoV). Selected samples were then analyzed by PCR using several primer sets. Geminiviruses was detected in all of the sites sampled. The principal Geminiviruses detected were PHV, TToMV, PepGMV, TyLCV, ToMoV and BGMV. All of these viruses were detected in the 15 sites sampled with different incidence. The more common mixture was those where three different Geminiviruses were all at once, having 15% of the positive samples, the others mixtures were: with two Geminiviruses, 8%, with four 8%, with five 5% and with six 2%.

The cotton leaf curl disease affecting chili peppers in Pakistan has been investigated by (Hussain *et al.*, 2003). During a survey of cotton growing areas in 2002/2003 found typical Cotton leaf curl disease (CLCuD)symptoms, leaf curling (LC) and vein thickening (VT) on chili pepper (*C. annum*). To assess CLCuD in chili pepper, plants with LC and VT were collected from the Vehari and Khanewal districts of Punjab province. To identify the begomoviruses present, total DNA was isolated from leaf samples, resolved in agarose gels and blotted on to nylon membranes. Blots were probed with a non-radioactive probe derived from *Cotton leaf curl Multan virus* (Briddon *et al.*, 2001). A positive signal was detected in all symptomatic chili pepper samples. PCR using primers CLCV produced products of the expected size from all symptomatic samples. DNA-β was determined by PCR using a set of universal primers (Briddon *et al.*, 2003). An amplicon of the expected size (approx. 1350 nucleotides) was produced in assays of all symptomatic samples. Southern blots of extracts from symptomatic chili peppers, probed with CLCuD DNA-β yielded a strong signal in samples from Vehari and Khanewal districts. The results confirm the presence of CLCuD DNA-β in chili pepper (Hussain *et al.*, 2003).

Molecular Characterization of Chili leaf curl Begomoviruses from Pakistan was investigated by Shih *et al.* (2003). Leaf curl or yellowing symptoms, typical of those caused by begomovirus infection, are commonly observed in chili and tomato plantings in Pakistan. The chili sample with leaf curl symptoms was collected in 1998 in Multan (Punjab Province), in 2000. Virus DNA was first amplified by polymerase chain reaction using the degenerate primer pair PAL1v1978/PAR1c715. The expected 1.4-kb PCR products were obtained from the three samples. Based on the sequences of the 1.4-kb DNA products, specific primers were designed to complete each of the DNA-A sequences. Two primer pairs, DNABLC1/DNABLV2 and DNABLC2/DNABLV2, were used for the detection of DNA-B. The DNA-A of the virus associated with chili leaf curl from Pakistan (AF336806) consisted of 2,754 nucleotides, containing six predicted ORFs (AV1, AV2, AC1, AC2, AC3, and AC4).

Sequence comparisons showed highest sequence identity (87%) with *Tomato leaf curl Bangladesh virus* (AF188481). DNA beta of 1.3 kb was detected in the chili begomovirus isolate using Beta01/Beta02 primers. There was no evidence for the presence of a DNA-B in the chili begomovirus isolate when tested by the two DNA-B specific primer pairs. Based on DNA sequence comparisons, the chili leaf curl virus from Pakistan constituted a distinct, new monopartite begomovirus (Shih *et al*, 2003).

Tomato leaf curl New Delhi virus affecting chilli pepper was reported in Pakistan by Hussain *et al.* (2004). Chilli plants often show symptom similar to tomato leaf curl such as yellowing, leaf curling, reduction in leaf size and stunting of the plants. Generally, chilli and tomato crops overlap in the fields, it is likely that chilli may become infected with tomato begomoviruses. To ascertain the begomovirus infection in chilli plants, symptomatic plant samples were collected from several locations in the Punjab province of Pakistan. Total DNA was isolated from leaf samples, resolved in agarose gels and blotted onto nylon membranes and probed with the radioactively-labelled probes of DNA-A and DNA-B of ToLCNDV which gave strong signals of hybridization, confirmed the infection of bipartite begomovirus with the disease. PCR was also carried out with degenerate begomovirus DNA-A primers and a nearly full-length clone of this DNA-A product was partially sequenced and was found to share 95% sequence identity with DNA-A of ToLCNDV. These results confirmed the infection of ToLCNDV in chilli pepper in Pakistan. The presence of ToLCNDV was further confirmed by PCR using DNA-B specific primers (Padidam *et al.*, 1995), based on the movement protein gene of ToLCNDV. These primers produced products of the expected size from all samples positive in Southern hybridization analysis. These results prove the first confirmed finding of ToLCNDV in Pakistan (Hussain *et al.*, 2004).

The association of *Tomato severe leaf curl virus* (ToSLCV), Tomato leaf curl Sinaloa virus (ToLCSinV) and *Pepper golden mosaic virus* (PepGMV) with tomato and chilli was studied (Ala-Poikela *et al.*, 2005). During survey in 2003, samples from chilli, tomato and other plants showing geminivirus symptoms were collected from different parts of Nicaragua to diagnose and characterize the associated viruses. Chilli plants showed symptoms of mosaic, interveinal yellowing, curled leaves, leaf rolling and severe stunting of plants. During PCR assays with universal begomovirus primers, all tomato plants showing symptoms were infected with begomoviruses, while only 30–46% of chilli pepper and cushaw plants showing symptoms tested virus-positive. Sequence analysis of the virus isolated revealed the association of *Tomato severe leaf curl virus* (ToSLCV), Tomato leaf curl Sinaloa virus (ToLCSinV) and *Pepper golden mosaic virus* (PepGMV) with tomato and chilli (Ala-Poikela *et al.*, 2005).

During a survey in 2005, , in the pepper production areas of eastern Cuban, about 70% of pepper plants growing at an Experimental Station in Holguin province, showed a severe leaf yellow mosaic with crinkling and curling and plant stunting. Samples from symptomatic plants were collected and leaf samples analyzed by transmission electron microscopy. Total nucleic acid was extracted and assayed by non-radioactive nucleic acid hybridisation, using a TYLCV intergenic region probe and by polymerase chain reaction (PCR). During electron microscopic study, all symptomatic leaf samples showed the presence of typical geminivirus particles, whereas no any sample exhibited hybridization signal. PCR carried out with coat protein region amplifying degenerate primers (Wyatt & Brown, 1996), all samples produced amplicons of the expected size (576bp). PCR amplicons from three samples were cloned and BLAST analysis showed the highest identity (98%) within clones; between 93% and 94%

identity with *Cabbage leaf curl virus* (U65529 and AF065815, respectively). This report showed the occurrence of a new begomovirus in pepper plants in Cuba (Zubiaur *et al.*, 2006).

Tomato leaf curl Joydebpur virus infecting chilli in India was reported for the first time by Shih *et al.*, 2007. In the year 2004, chilli samples showing symptoms of begomovirus were collected from Ludhiyana, Punjab. PCR analysis of such 3 samples revealed presence of 1.4 kb DNA-A fragment and 1.3 kb DNA-beta, but no DNA-B was detected with their specific primers. Further, the genome organization (complete DNA-A, DQ6738 59) of the virus isolate shared highest sequence identity (90.3%) to that of *Tomato leaf curl Joydebpur virus* (ToLCJV, AJ875159) from Bangladesh (Maruthi *et al.*, 2006). On the basis of DNA-A sequence analysis, the ICTV species demarcation of 89% sequence identity and genome organization, the Punjab chilli begomovirus was considered to be a provisional strain of ToLCJV described from tomato in Bangladesh.

First report of a begomovirus associated with leaf curl disease of bell pepper was investigated by Tahir and Haider 2006. During a survey, C. *annuum* var. *grossum* showing leaf curl symptoms cultivated in vegetable growing areas around the Lahore city of Pakistan were collected. Overall the disease incidence was 60-70 % with severe leaf curl symptoms. PCR analysis with primers set designed from most conserved region (CP gene) of begomovirus, resulted positive amplification in all the symptomatic chilli samples, but no such amplicon was obtained with healthy samples taken as negative control. Sequence analysis of the CP gene of the virus isolate shared a maximum 84% sequence identity with *Tomato leaf curl New Delhi virus* [Pakistan:Solanum (syn. Solanum yellow leaf curl virus) segment A. Attempts to identify the presence of a DNA-β in the infected *C. annuum* var. *grossum* samples, using universal DNA-β primers (Briddon *et al.*, 2002) produced a *ca.* 1.4 kb product (Tahir and Haider 2006).

Molecular characterization of *Pepper yellow leaf curl Indonesia virus* in leaf curl and yellowing diseased tomato and pepper in Indonesia was investigated by Tsai *et al.* (2006). Tomato and chilli plant showing yellowing and leaf curl symptoms were collected from tomato and chilli fields near Bogor, Java, Indonesia in 2000. PCR was conducted with geminiviral DNA-A, DNA-B, and associated satellite DNA specific primers which resulted 1.4 kb DNA amplicon in DNA-A from both tomato and chilli samples but DNA-B and satellite DNA were not detected in any samples. Further, full length DNA-A of tomato (2,744 nt, DQ083765) and chilli (2,743nt, DQ083764) were sequenced by designing specific primers. The DNA-A of both begomoviruses contained six open reading frames, including two in the virion sense and four in the complementary sense, and the geminivirus conserved nanosequence-TAATATTAC in the loop of the hairpin structure of the intergenic region. Because of their high nucleotide sequence identities of 99%, the tomato- and chili-infecting begomovirus are considered the same virus. Blast search analysis of both the virus isolate also showed highest nucleotide sequence identity (95%) with *Pepper yellow leaf curl Indonesia virus* (AB189849). On the basis of the highest sequence identity, the virus isolates were considered the stain of *Pepper yellow leaf curl Indonesia virus* (Tsai *et al.*, 2006).

Sakata *et al*, (2008) determined the complete nucleotide sequences of begomoviruses from pepper with leaf curl and yellowing symptoms, tomato with leaf curl symptoms, and Ageratum with yellow vein in Indonesia. On the basis of genome organization and sequence homology, they were proposed to belong to a new species, *Pepper yellow leaf curl Indonesia virus* (PepYLCIV), which includes the new strains PepYLCIV-Tomato and PepYLCIV-Ageratum. These viruses had bipartite genomes. Pepper virus DNAs from Indonesia

(PepYLCIV, PepYLCIV-Tomato and PepYLCIV-Ageratum DNA-As) were noticeably distinct, forming a separate branch from the viruses infecting pepper. Considerable divergence was observed in the common region (CR) of the genomic components of PepYLCIV (77%), PepYLCIV-Tomato (82%) and PeYLCIV-Ageratum (75%). A stem-loop-forming region and a Rep-binding motif were identical in the CR of the three viruses. The CRs of PepYLCIV-Ageratum DNA-A was approximately 10 nucleotides longer than that of PepYLCIV DNA-A and PepYLCIV-Tomato DNA-A. A similar insertion was also found in the CR of PepYLCIV-Ageratum DNA-B. PepYLCIV DNA-A alone was infectious in pepper and *Nicotiana benthamiana* plants, and association with DNA-B increased symptom severity (Sakata *et al.*, 2008).

Tsai *et al*, (2009) found a new Begomovirus associated with yellow leaf curl diseases of tomato and pepper in Indonesia. In 2006, typical symptoms produced by begomoviruses, leaf curling, blistering, yellowing, and stunting, were observed in various tomato and pepper fields in North Sulawesi, Indonesia with incidence as high as 100 percent. DNA-A was detected in all the symptomatic tomato and chilli samples but no DNA-B and DNA-β could be detected by PCR. The full-length genomes of all 7 begomovirus isolates showed to have high nt identity (more than 95.1 %). Comparison with other sequences available in the GenBank database showed that the isolates have the highest nt sequence identity (66.5 %) with PepYLCIDV and less with other begomoviruses including those reported from Indonesia. On the basis of begomovirus species demarcation threshold of 89 % nt identity, the tomato and pepper begomovirus isolates from North Sulawesi constitute a distinct species in the genus *Begomovirus* for which the name Tomato leaf curl Sulawesi virus (ToLCSuV) was proposed. Phylogenetic analysis shows the ToLCSuV isolates form a cluster distinct from other Indonesian begomoviruses as well as begomoviruses from the neighboring Philippines (Tsai *et al.*, 2009).

Akhter *et al*, 2009 studied a severe leaf curl disease on chilies in Pakistan is associated with multiple Begomovirus components. Field visits of commercially grown chilies in 2007 and 2008 identified a very severe leaf curl disease with 100% incidence and severe yield losses at several locations in Faisalabad District, Punjab, Pakistan. Symptoms of the disease consisted severe leaf curl with cup-shaped, upward curling, yellowing, and stunted growth of plants. Total nucleic acid was isolated from the samples collected from 10 locations and PCR was performed with DNA-A amplifying begomovirus degenerate primers which resulted amplification of ~2.8 kb from all symptomatic plants. Southern hybridization analysis also gave strong signals of hybridization with a probe of *Cotton leaf curl Multan virus*. PCR products were cloned into a TA cloning vector and five clones were partially sequenced. Analysis of the CP sequence data resulted in identification of two begomovirus species; the first clone (FN179278) showed 94% sequence identity with a bipartite *Tomato leaf curl New Delhi virus* (ToLCNDV), while the second clone (FN252382) showed 97% sequence identity with a monopartite *Chili leaf curl Multan virus* (ChLCMV). DNA-B from the plants showing typical disease symptoms were also amplified and sequenced. Sequence analysis of the DNA B clone (FN179276) in the intergenic region and movement protein gene showed 94% identity with ToLCNDV DNA-B. DNA-β was also amplified with universal primes (β-01 and β-02) and its sequence analysis shared 90% identity with the previously cloned chili leaf curl beta satellite (Briddon *et al.*, 2003). The appearance of chili leaf curl disease at a higher incidence and symptom severity may be attributed to the synergistic action of geminivirus disease complex comprising a monopartite and a bipartite begomovirus along with DNA-beta satellite. High yield losses resulting from this severe disease threatens chili cultivation in the area and is forcing farmers to grow other crops (Akhter *et al.*, 2009).

MANAGEMENT STUDIES OF GEMINIVIRUS INFECTING CHILLI CROPS

Begomoviruses are transmitted by different whitefly vectors (which are a serious pest in themselves) and many can also be transmitted by mechanical means and grafting, but they are not seed transmitted. Begomovirus control is particularly difficult in open field crops due to the widespread presence and wide host range of whiteflies. Disease management may include pathogen exclusion, vector control, and elimination of possible pathogen and/or vector reservoir plants. In some cases, crop cultivars with increased virus resistance are available. The traditional host genetic resistance, chemical controls and cultural practices for successful control of viruses may be adopted.

There are several methods such as field sanitation, eradication of infected plants serving as primary source of virus inoculum, removal of weeds and alternate host plants from of fields, plantation practices, spraying of insecticides, use of virus free planting material and resistant varieties of vegetable crops are suggested to be practiced/ applied for control of begomovirus and its whitefly vector.

Use of Virus free Planting Material and Sanitation of Cropping Field

Use of virus free planting materials, regular surveillance for disease symptoms at early stages of plants, followed by cleaning and removal of commonly growing weeds viz. *Physalis minima, Nicandra physaloides, Solanum nigrum, Datura stramonium, D. metel* and other weeds (which act as potential reservoirs of begomovirus) from fields was found to be very effective.

Role of weeds/alternate hosts on the epidemiology of viral disease has been studied in case of many important viruses from India and abroad which are the major limiting factors to cultivation of chilli and other important crops. In India, 13 weeds species have been identified to harbour *Tomato leaf curl virus* in Karnataka state (Ramappa *et al.*, 1998). Singh *et al.* (1994) recorded, *Sida* sp. and *Abutilon indicum* as alternate hosts of *Cotton leaf curl virus* in Punjab. Valand and Muniyappa (1992) reported 18 weed hosts of *Tobacco leaf curl virus* occurring in southern India. *Nicotiana plumbaginifolia, Jatropha gossypifolia, Physalis minima, Coccinia grandis Solanum nigrum, Momordica charantia* (wild), *Sorghum vulgare* (widely growing) and *Luffa* spp. weeds have been found to be the hosts of geminivirus in north India (Khan *et al.*, 2003). Reddy *et al.* (2005) reported *Croton bonplandianum* to be the host of *Tomato leaf curl New Delhi virus* (ToLCNDV), and *Parthenium hysterophorus* was found to host of *Tomato leaf curl Karnataka virus* (ToLCKV).

Alternate sources of *Tomato leaf curl virus* (ToLCV) or *Tomato yellow leaf curl virus* (TYLCV) have been found in many weed hosts. Dalmon and Marchoux, 2000 reported that *Solanum nigrum, Datura stramonium* and *Malva* sp. could serve as reservoirs of TYLCV in France. In Spain numerous weed species viz. *Chenopodium murale, Datura stramonium,*

Malva parviflora, *Solanum nigrum*, *Convolvulus* sp. and *Cuscuta* sp. were recorded to be the reservoirs of TYLCV (Jorda *et al.*, 2001). Recently *Tomato leaf curl New Delhi virus* (ToLCNDV) has been found associated with a common weed *Eclipta prostrata* in Pakistan (Haider *et al.*, 2006), which is serious constraint of tomato, chilli and other crops on Indian subcontinent.

Therefore these weeds must be removed from and around the chilli fields to minimize the source of reservoirs of many begomoviruses which also provide the shelter to whitefly, the known vector for transmission of a variety of begomoviruses.

Chemical Management

The development of strategies for integrated disease management by spray of oils, viricides, pesticides and botanicals has been found to be significant in reducing the yield loss by many workers for various viral diseases (Verma *et al.*, 1980; 1985; 1996; Andrade *et al.*, 1995; Rezende and Costa, 1995; Singh and Awasthi, 2002; Kumar and Awasthi, 2003 and Kunkalikar *et al.*, 2006).

Natural incidence of ToLCNDV on chilli (*C. annuum*) cultivars growing at Banthra research station of N.B.R.I., Lucknow was found to be transmitted by whiteflies. Regular spraying of insecticide at every 21 days intervals was found to be effective for management of leaf curl disease of chilli. An improvement in growth and yield performance of treated chilli plants was also noticed as compared to the non treated plants (Khan *et al.*, 2006b).

To observe the impact of viral disease and their possible management by the use of insecticides, a field experiment was performed in randomized block design with four replications using 3 cultivars of Chilli namely Pusa Jawala, KDCS-810 and LCA-235 during 2003-04. Forty days old chilli seedlings were transplanted at Banthara farm site of NBRI in plot size of 2m X 3m with a spacing of 30cm X 30cm. After 15 days of transplantation infestation of whiteflies were noticed on chilli verities. Three rounds of spraying were undertaken at 21 days intervals during September-November. The first round of spraying was done when infestation of the insects was noticed on the plants. Then the regular spraying of 0.2% Malathian was continued at every 3 weeks intervals to controls the adverse effect of chilli leaf curl disease. A control bed without spray of each variety was also maintained to observe the effect of the Malathian insecticide on severity of the disease. The percentage of the disease infections was recorded in both the plots before spraying. Observations were also made on plant height, number of branches and red ripped fruits per plot in case of sprayed and unsprayed plot (Khan *et al.*, 2006b).

An improvement in plant growth of different chilli cultivars was observed as a result of insecticide (Malathian) spraying. The highest increase in plant height observed in LCA-235 followed by Pusa Jawala and KDCS-810 than the unsprayed plants. However, percent increase in height of sprayed plants was similar in LCA-235 (27%). It was minimum (18%) in KDCS-810 cultivars of chilli than the unsprayed plants. Malathian sprayed chilli also showed greater fruit yields at harvest. It resulted in to highest fruit yield of 933 gm^2 showed an increase of 31.4% over unsprayed plants. Appreciable enhancement in fruit yields was also recorded in case of KDCS-810 followed by Pusa Jawala. It is clear from the above results that the application/spraying of Malathian effectively reduced the infection of leaf curl virus disease as good crop growth reflected in ripped red fruit of chilli. These studies revealed that spraying of Malathian on different cultivars of chilli played a significance role for management of the insect borne leaf curl disease (Khan *et al.*, 2006b).

Since chilli fruits are used as raw and in food preparations and use of insecticides is known to cause adverse effect on human health therefore, spraying of insecticide on chilli should be avoided. Therefore an alternate method of disease management may be the best choice. Findings on management of viral disease by three sprayings of insecticide which resulted in increase of biomass and red fruit yield of chilli plants may be utilized for designing the efficient control measures of disease leading to better chilli production in India (Khan *et al.*, 2006b).

Singh and Dhakrani (2009) suggested that the chilli crop must be protected from seedlings to fruiting stage. It is necessary to apply systemic insecticide like furadan to nursery, and spray systemic insecticide like dimecron, dimethoate, metasystox, monocrotophos etc. to seedling/plants. Spraying of monocrotophos 36 EC or dimethoate 30 EC or phosalone 35 EC at 0.05 per cent concentration from weeks after transplanting at 10 days interval controls aphids and white flies and reduces disease incidence to a greater extent. For upward curling they suggested systemic insecticides clothianidin @ 0.06g/lit and for downward curling acaricides like Dicofol @2.5 ml/lit or ethion 50 EC @ 2.0 ml/lit or Millbeknock @0.5ml/lit or Fenazaquin @2.0 ml/lit may be applied. However, regular monitoring of disease is essential. Breeding resistant varieties appears to be the best approach to the problem. Puri red, Puri orange, G-2 and Jwala varieties are found to be resistant for leaf curl disease of chilli (Singh and Dhakrani, 2009).

Bio-management Strategy

The field efficacy of a bio-management strategy for leaf curl disease (LCD) of chilli a 2-year on farm experiment was conducted in farmers' fields in Jodhpur district in western Rajasthan (Arun Kumar, 2006). Chilli seeds were treated with raw cow's milk (RCM) for 24 hours in 1:1 ratio (i.e. RCM diluted to 50% by adding water) at the room temperature (30±2°C) and Trichoderma viride (6 g kg^{-1} seed) and T. viride (10 g m^{-2}) in nursery soil followed by dipping of nursery-raised saplings in RCM (15%) for 20 minutes before transplantation. After 20- days of transplanting the plants were sprayed with RCM (15%) for four times at 15 days interval. The farmers practice (FP) was treated as control. Treatment of bio-control agents was found superior over FP in all the replications providing about 17 to 65% protection over FP. Yield attributes like plant height, root length, number of branches plant-', number of fruits plant, fruit size, fruit weight and fruit yield plot showed an increase when compared to FP (Kumar, 2006).

Integrated Pest Management Strategy

Leaf curl disease transmitted by whiteflies (*B. tabaci)* and the incidence of disease becomes severe when the vector population increases in the field and in advanced stages the whole plant appears bushy, with stunted growth and fewer flowers. The infestations should be managed from seedling stage itself by using seeds from virus free plants. Mallikarjun *et al*, (2006) suggested some integrated pest management strategy for viral diseases on chilli crop.

They suggested that the nursery beds should be covered with nylon net or straw to protect the seedlings from viral infections. Spraying metasystox or rogar at 1ml per litre of water at 10 days' interval will control the disease incidence in the nursery. In the main field, infected plants should be removed as soon as they are noticed. Raising barrier crops such as maize, sorghum, pearl millet or snap bean around chillies in two to three rows will reduce the disease level. Applications of Diazinon, Metasystox, Melathion or Endosulphan 0.2 per cent at 15 days' interval till the harvest. Spraying of monocrotophos at 0.05 per cent at 10 days interval in the field but it should be stopped 15-20 days before harvest of the crop (Mallikarjun *et al.*, 2006).

Since the virus is generally transmitted by whitefly, so control measures of whitefly may be helpful. Whiteflies can be effectively attracted and controlled by yellow sticky traps, which are coated with grease and sticky oils. Spraying of Triazophos (2ml/litre) or Prophanophos or Spraying of any Neem product (5% Neem oil before egg laying) or 5 kg Neem Kernel extract per acre with any sticky material has been suggested as effective to manage the leaf curl disease of chilli (http://aaqua.persistent.co.in/aaqua/forum/viewthread?thread=8611).

Tesoriero and Azzopardi (2006) suggested strategy How to control TYLCV infection in chilli. Exclusion and early eradication are the best strategies for preventing the establishment of TYLCV and its vector, whitefy. Seedlings should be grown in facilities with insect screens (50 mesh) that exclude whitefies. Seedlings and crops should be regularly monitored for whitefies. Yellow sticky traps and plant inspections are necessary to detect early infestations and to identify the type of whitefies present. crops should be carefully monitored for virus symptoms if B. tabaci is detected. Use insecticides and biological control options for *B. tabaci* on seedlings and crops, depending on the level of virus risk. Infected plants should be sprayed with insecticide and carefully placed into plastic bags for disposal. Weed control is important if the virus has become established in the local area. When cropping is fnished, remove and dispose of crop residues immediately.

Use of Resistant Varieties

Leaf curl resistant genotypes under open field conditions have been reported, especially from India (e.g. Pusa Jwala and Pant C-1) (Tewari and Ramanujam, 1974; Mathai *et al.*, 1977). Different sources of resistance like: Punjab lal, Perennial, Jwala, Puri Orange and Puri Red have been identified against chilli leaf curl disease (Bal *et al.*, 1995). Further some chilli cultivars: Chilli Pragathi, DK-32, DK-6B, DKC-14, DKC-7B and DKC-8 were found highly resistant to the virus infection under field conditions (Promil, 2005).

Kumar *et al*, (2006) screened three hundred and seven genotypes belonging to four cultivated and one wild species of *Capsicum* against pepper leaf curl virus (PepLCV) causing devastating leaf curl disease of chilli (*C. annuum*). The selfed progenies of eight symptom-less and highly resistant lines were challenged by viruliferous white fly under glasshouse conditions, out of which only three genotypes, viz.GKC-29, BS-35 and EC-497636 showed no symptom. Using scion and root stalk of susceptible genotype (Pusa Jwala), these three putative symptom-less genotypes were further challenged by grafting and alternate grafting. The resistant reactions of GKC-29, BS-35, EC-497636 were confirmed because even after 50 days of successful grafting/alternate grafting, no viral symptom appeared on all the grafted plants of these genotypes. When subjected to PCR amplification with degenerate primers deigned to detect gemnivirus like PepLCV, the three symptomless genotypes did not show any amplification, suggesting that the resistant reaction in three identified symptomless resistant sources was because of the absence of viral genome and they are not symptom-less carrier (Kumar *et al.*, 2006).

Management through Virus Resistant Transgenic Chilli Plants

Genetically engineered resistance to viruses has been achieved in various plants by the use of viral genes is known as the pathogen derived resistance or PDR. Genetic engineering has been proved/is proving to be highly effective for controlling virus diseases in a wide range of crops grown worldwide (Wilson, 1993; Gonsalves and Slightom, 1993; Varma *et al.*, 2002).

The coat protein protection was reported to be successful against several RNA viruses however; few reports of engineered resistance against geminiviruses have also been reported. Kunik *et al.* in 1994 extended the concept of capsid protein mediated protection, which has been applied successfully with many RNA viruses, to the DNA geminivirus, TYLCV. Success has been reported with expression of other geminivirus transgene modified and unmodified like replication- associated protein (Rep), and movement protein gene. The use of antisense AC1 genes has been shown to confer resistance to *Tomato golden mosaic virus* by transgene expression and accumulation of the product (Day *et al.*, 1991) whereas in TYLCV resistance was conferred due to a dominant negative mutation (Bendhamane and Gronenborg, 1997).

Agrobacterium-mediated transformation of tomato was attempted for regeneration of transgenic tomato plants expressing *Tomato leaf curl virus* coat protein gene. The generated transgenic plants when challenged by TLCV through whiteflies, showed variable degrees of disease resistance/tolerance against TLCV infection (Raj *et al.*, 2005). Therefore, the coat protein of ToLCNDV characterized from the chilli may be utilized for the development of transgenic chilli for development of inbuilt resistance against ToLCNDV. Keeping this in mind the regeneration procedure of chilli was attempted which is a pre-requisite for development of transgenic chilli plants. *Capsicum annuum* cv. Pusa Jwala cotyledons (leaf disk) and hypocotyls (internodal segments) explants were cultured on MS basal medium supplemented with different combination and concentration of auxin and cytokinin. The medium supplemented with 5mg/l BAP (cytokinin) and 0.1mg/l NAA (auxin) was found optimum for the development of shoots from cotyledon explants while the supplementation of 10mg/l BAP and 0.1mg/l NAA was appropriate for hypocotyl explants.

CONCLUSION

Chilli (*C. annuum*) is an important and widely cultivated spice and vegetable crop on the Indian subcontinent. It has been regarded as one of the most permissive hosts for a number of viruses of various groups, such as *Cucumovirus*, *Nepovirus*, *Potyvirus*, Closterovirus and *Geminivirus* (Osaki and Inouye, 1981; Martelli and Quacquarelli, 1983). These viruses cause a number of diseases in chilli plants resulting in serious economic losses to chilli growers all over the world. In last few years, the leaf curl disease of chilli has become the major constraint for chilli production in most of the chilli growing areas (Reddy *et al.*, 2005; Senanayake *et al.*, 2006, Hussain *et al.*, 2004; Tsai *et al.*, 2006). Symptoms of chilli leaf curl disease consist of abaxial and adaxial curling of the leaves accompanied with puckering and blistering of inter-venial areas and thickening and swelling of the veins, distortion of leaves and reduced growth of the plants with fewer deformed fruits.

In recent years, the *Geminiviridae* family has received a great deal of attention and is becoming one of the most important and studied families of plant virus. Geminiviruses are of importance in the tropics and subtropics due to climatic factors favoring the multiplication and ability of vector i.e. whitefly (*B. tabaci*) for transmitting the geminivirus to other economic crops. The causal agent of leaf curl disease of chilli has been identified by various workers as a geminivirus from Bhutan, Bangladesh, China, India, Indonesia, Pakistan, Sri Lanka and Taiwan, (Shivanthan, 1982; Thinlay and Penjore, 1996; Tsai *et al.*, 2000; Shih *et al.*, 2003; Reddy *et al.*, 2005; Khan *et al.*, 2006a; Hussain *et al.*, 2004; Tsai *et al.*, 2006). The incidence and severity of the disease may be increasing due to the emergence of new geminiviruses through recombination or pseudo-recombination among strains and/or species in various crops (Varma and Malathi, 2003, Chakraborty *et al.*, 2003). The association of satellite molecule called DNA-β with both monopartite as well as bipartite begomoviruses are the other major factor which role in disease severity have been studied by many begomovirologist in various crops (Saunders *et al.*, 2000; Briddon *et al.*, 2001; Jose and Usha, 2003).

The number of begomoviruses causing disease in chilli is increasing very rapidly. In last few years, there has been dread increase of the begomovirus infection in chilli crop throughout the world. In the 2000s, about 10 begomoviruses had been reported to infect chilli from various part of the world but now the number has increased. So far, about 29 begomoviruses are known to infect chilli and amongst them 12 are reported to be tomato begomovirus (Table 1). This indicates that the number of chilli infecting begomoviruses could be significantly increased in the coming years. In India, although chilli leaf curl disease was reported as early as 1960 but the information was only available on host range and virus-vector relationship (Dhanraj and Seth 1968; Mishra *et al.*, 1963). However, there was no any report regarding the molecular characterization of the geminivirus (begomovirus) infecting chilli till 2005 (Reddy *et al.*, 2005; Khan *et al.*, 2006a; Senanayake *et al.*, 2006).

The leaf curl disease of chilli has moderate to high incidence and is widely distributed in India and other countries. In India, a very high disease incidence (up to 100% of plants) was recorded in 2004 from Jodhpur district, the major chilli growing area of Rajasthan state (Senanayake *et al.*, 2006). Beginning in the early 2000's, numerous epidemics of Begomoviruses have been reported in chilli pepper crops throughout Indonesia. These epidemics resulted in complete crop losses in some cases and causing farmers ceased planting the crops (Sri *et al.*, 2008). The severity of the disease depends upon the age of plant when gets it infected. During early stage of infection, the plants remained severely stunted and produced no fruit. Our recent observations of chilli fields in north India during 2003-2009 and the investigation from different parts of the world indicated that there has been a continuous increase in the incidence of leaf curl disease resulting in the severe economic losses (Hussain *et al.*, 2004; Reddy *et al.*, 2005; Senanayake *et al.*, 2006, Tsai *et al.*, 2006, Sri *et al.*, 2008). It has been reported in the literature that leaf curl disease of chilli is transmitted by whitefly *B.*

tabaci (Gennadius) biotypes (Khan, 2006; Senanayake *et al.*, 2006, Sri and Rahmayani, 2007, Barro *et al.*, 2008). The majority of known whitefly-transmitted viruses are transmitted by *B. tabaci* (Gennadius). The best known *B. tabaci* transmitted viruses are those of the genus *Begomovirus* of the family *Geminiviridae* causing serious losses in bean (Howarth *et al.*, 1985), cucurbits (Brown *et al.*, 2002; Singh *et al.*, 2001), tomato (Kheyr-Pour *et al.*, 19991; Chatchawankanphanich and Maxwell, 2002), chilli (Hussain *et al.*, 2004; Senanayake *et al.*, 2006) and cassava (Hong *et al.*, 1993; Pita *et al.*, 2001) in tropical and sub tropical regions.

Currently ToLCNDV is widely distributed throughout India and neighboring countries like Pakistan, Bangladesh and Thailand. The frequency with which new strains of *Tomato leaf curl New Delhi virus* (ToLCNDV) are appearing in several agricultural crops and non-crop species, indicate that the virus species has more virulent strains in field and pose a serious threat to vegetable cultivation particularly in South East Asia. ToLCNDV is known to infect tomato from more than a decade, but in last five years its number is increasing very fast. So far about 11 strains have been reported from a variety of plants like chilli, potato, bitter gourd, watermelon, pigeon pea, cowpea, tomato, *Eclipta* and *Croton*. It indicates that in coming years ToLCNDV will pose serious constraint to agricultural crops due to its more virulent strains with broad host range. Recombination and pseudorecombination due to mixed infection is known for the emergence of more virulent strains.

Therefore, strategies for management of chilli leaf curl disease/ its causal agent/ its transmitting vector must be taken care on priority basis. The proper surveillance beginning from raising of seedlings, their transplantation in field, every stage of growth till harvesting of crop must be considered as an important step of chilli cultivation. As soon as whiteflies, the leaf curl disease transmitting vector are noticed in the field or near by the chilli growing field, the spraying of an efficient insecticide must be done and it should be continued at every month till flowering. The vector control, elimination of possible pathogen and/or vector reservoir plants and weed plants may increase the chilli production. The traditional host genetic resistance, chemical controls and cultural practices may also be adopted for successful control of virus disease.

The regeneration protocol of cv. Pusa Jwala developed may be utilized for the transformation purpose for virus-derived resistance (VDR) against ToLCNDV. The coat protein, rep protein and movement protein genes of ToLCNDV isolate characterized during this study may be utilized for their mobilization into the genome of chilli for the development of inbuilt resistance against the leaf curl disease caused by ToLCNDV strains (Khan, 2006).

Acknowledgments

Authors are thankful to the Director, N.B.R.I., Lucknow for facilities and permission for summarizing the data generated by the researchers of plant virus laboratory.

References

Akhter, A., Qazi, J., Saeed, M. & Mansoor, S. (2009). A severe leaf curl disease on chilies in Pakistan is associated with multiple Begomovirus components. *Plant Disease, 93*, 962.

Ala-Poikela, M., Svensson, E., Rojas, A, Horko, T., Paulin, L., Valkonen, J.P.T. & Kvarnheden, A. (2005). Genetiv diversity and mixed infections of begomoviruses infecting tomato, pepper and cucurbit crops in Nicaragua. *Plant Pathology, 54*, 448-459.

Andrade, H., Garcia, E., Mora, A., Nieto, D. & Villawveva, A. (1995). Integrated *management of Papaya ringspot virus* in Vera Cruz, Mexico. *Tropical* Fruits *Newsletter, 14*, 7.

Anonymous. (2002). *Chillies of India, Publication of Spices Board*, Calicut, Kerala.

Bal, S.S., Singh, J. & Dhanju, K.C. (1995). Genetics of resistance of mosaic and leaf curl viruses in chilli (*Capsicum annuum*). *Indian J. Virology, 11*, 77-79.

Barro, P. J. , Sri, H.H., Don, F., Siti, S. & Shigenori, U. (2008). A virus and its vector, *pepper yellow leaf curl virus* and *Bemisia tabaci*, two new invaders of Indonesia. *Biological Invasions, 10*, 411-433.

Bendhamane, M. & Groneborn, B. (1997). Engineering resistance against *Tomato yellow leaf curl virus* using antisense RNA. *Plant Mol. Biol., 33*, 351-357.

Bezerra-Agasie, I.C., Ferreira, G.B., de Ávila, A.C. & Inoue-Nagata A.K. (2006). First Report of *Tomato severe rugose virus* in Chili Pepper in Brazil. *Plant Disease, 90*, 114.

Biacs, P.A., Daood, H.G. Pavisca, A. & Hajdu, F. (1989). Studies on the Careotenoid pigments of paprika (*Capsicum annuum* L.) *Am. Chem. Soic., 37* (2), 350-353.

Bidari, B.D. & Reddy, H.R. (1990). Identification of naturally occurring virus on commercial cultivars of chilli. *Mysore J. Agric Sci.*, 24, 42-51.

Bidari, B.D. & Reddy, H.R. (1994). A survey on mosaic viruses and their distribution in different chilli cultivars in Karnataka. *Mysore J. Agric Sci., 28*, 52-59.

Biswas, K.K., Pun, K.B., Pant, R.P. & Ahlawat, Y.S. (2005). Mosaic disease in chilli (*Capsicum annuum*) cv. Kalimpong Local in Darjeeling hills of West Bengal and its management. *Indian Phytopath., 58* (4), 456-461.

Bosland, P.W., Bailey, A.L. & Cotter, D.J. (1994). Growing chilli in New Mexico, *NM Coop. Ext. Serv. Circ.*, Las Cruces NM; H-230.

Bosland, P.W., Bailey, A.L., & Iglesias-Olivas, J. (1996). *Capsicum* pepper varieties and classification. *NM Coop. Ext. Serv. Circ.,* Las Cruces NM; 530.

Briddon, R.W., Bull, S.E., Amin, I., Idris, A.M., Mansoor, S., Bedford, I.D., Dhawan, P., Rishi, N., Siwatch, S.S., Abdel-Salam, A.M., Brown, J.K., Zafar, Y. & Markham, P.G. (2003). Diversity of DNA β; a satellite molecule associated with some monopartite begomoviruses. *Virology, 312*, 106-121.

Briddon, R.W., Mansoor, S., Bedford, I.D., Pinner, M.S., Saunders, K., Stanley, J., Zafar, Y., Malik, K.A. & Markahn, P.G. (2001). Identification of DNA β components required for induction of cotton leaf curl disease. *Virology, 285*, 234-243.

Brown, J.K. & Nelson, M.R. (1988). Transmission, host range, and virus-vector relationship of chino del tomato virus, a whitefly-transmitted geminivirus from Sinalo, Mexico. *Plant Disease, 72*, 866-869.

Brown, J.K. & Nelson, M.R. (1989). Two whitefly-transmitted geminivirus isolated from pepper affected with tigre disease. *Phytopathology, 79*, 908.

Brown, J.K. & Poulos, B.T. (1990). *Serrano golden mosaic virus*- a newly identified whitefly- transmitted geminivirus of pepper and tomato in the United States and Mexico. *Plant Disease, 74*, 720.

Brown, J.K. Idris, A.M. Alteri, C. & Stenger, D.C. (2002). Emergence of a new cucurbit infecting begomovirus species capable of forming viral reassortant with related viruses in the *Squash leaf curl virus* cluster. *Phytopathology, 92*, 734-742.

Chakraborty, S. Pandel, P.K., Banerjee, M.K. Kallo, G.W. & Faukuet, C.M. (2003). *Tomato leaf curl Gujarat virus*, a new begomovirus species causing a severe leaf curl disease of tomato in Vanarisi, India. *Phytopathology, 93*, 1485-1494.

Chatchawankanphanich, O. & Maxwell, D.P. (2002). *Tomato leaf curl Karnataka virus* from Bangalore, India, appears to be recombinant begomovirus. *Phytopathology, 92*, 637-645.

Chattopadhyay, B., Singh, A.K., Yadav, T., Fauquet, C.M., Sarin, N.B. & Chakraborty, S. (2008). Infectivity of the cloned components of a begomovirus: DNA beta complex causing chilli leaf curl disease in India. *Archives of Virology, 153*, 533-539.

Dalmon, A., & Marchoux, G. (2000). *Tomato yellow leaf curl virus* host plants. *Phytoma, 527*, 14-17.

Day, A.G., Bejarano, E.R., Buck, K.W., Burell, M. & Litchenstein, C.P. (1991). Expression of an antisense viral gene in transgenic tobacco confers resistance to the DNA virus, golden mosaic virus. *Proc. Natl. Acad. Sci. USA, 88*, 6721-6725.

Dhanraj, K.S. & Seth, M.L. (1968). Enation in *Capsicum annum* L (Chilli) caused by a new strain leaf curl virus. *Indian J. Hort, 25*, 70–71.

Díaz-Plaza, R., Avilés-Baeza, W.I., Peña-Ramirez R. & Ribera-Bustamante, R. (2002). *Proceedings of the 16th International Pepper Conference Tampico*, Tamaulipas, Mexico. November 10-12, 2002.

Fauquet, C.M. & Stanley, J. (2003). Geminivirus classification and nomenclature Progress and problems. *Annals of Applied Biology, 142*, 165-189.

Fauquet, C.M., Bisaro, D.M., Briddon, R.W., Brown, J.K., Harrison, B.D., Rybicki, E.P., Stenger, D.C. & Stanley, J. (2003). Revision of taxonomy criteria for species demarcation in the family Geminiviridae, and an updated list of begomovirus species. *Archives of Virology, 148*, 405-421.

Fauquet, C.M., Briddon, R.W., Brown, J.K., Moriones, E., Stanley, J., Zerbini, M. & Zhou, X. (2008). Geminivirus strain demarcation and nomenclature. *Archives of Virology, 153*, 783–821.

Francki, R.I.B. (1979). *Cucumber mosaic virus*. CMI/AAB No, 213.

Francki, R.I.B., Hatta, T., Boccardo, G. & Randles, J.W. (1980). The composition of chloris striate mosaic virus, a geminiviruses. *Virology, 101*, 233-241.

Fujisawa, I., Hanada, T. & Anang, S. (1986). Virus disease occurring on some vegetable crops in West Malaysia. *Tropical Agricultural Research Report.* Ministry of Agriculture, Forestry and Fisheries, Japan. 20, 78-84.

Garzon-Tiznado, J.A., Torres Pacheco, I., Ascencio Ibanez, J.T., Herrera Estrella, L. & Rivera Bustamante, R.F. (1993). Inoculation of peppers with infectious clones of a new geminivirus by a biolistic procedure. *Phytopathology, 83*, 514.

Goldberg, N.P. (1995). Chile pepper disease. NM Coop. Ext. Serv. Circ., Las Cruces NM, 549.

Gonsalves, D. & Slightom, J.L. (1993). Coat-protein mediated protection: Analysis of transgenic plants for resistance in a variety of crops. *Seminars Virol., 4*, 397-406.

Goodman, R.M. (1977). Single-stranded DNA genome in a whitefly transmitted plant virus. *Virology,* 83, 171-179.

Goodman, R.M., Bird, J. & Thongmeearkom, P. (1977). An unusual virus like particle associated with golden yellow mosaic of beans. *Phytopathology, 67*, 37-42.

Green, S.K. & Kim, J.S. (9191). Characteristics and control of viruses infecting peppers: a literature review. Asian vegetable and research development centre. *Technical Bull., 18*.

Green, S.K. (1992). Viruses in Asia Pacific Region. In: *Proceedings of the Conference on Chilli Pepper Production in the Tropics*, 98-129.

Haider, M.S., Tahir, M., Latif, S. & Briddon, R.W. (2006). First report of *Tomato leaf curl New Delhi virus* infecting *Eclipta prostrata* in Pakistan. *Pant Pathology, 55*, 285.

Harrison, B.D. & Robinson, D.J. (2002). Green shoot of geminivirology *Physiological and Molecular Plant Pathology, 60*, 215-218.

Harrison, B.D. (1985). Advances in geminivirus research. *Annu. Rev. Phytopath., 23*, 55-82.

Harrison, B.D., Barker, H., Bock, K.R., Gutheric, E.J., Meredith, G. & Atkinson, M. (1977). Plant viruses with circular single stranded DNA. *Nature, 270*, 760-762.

Hatta, T. & Francki, R.I.B. (1979). The fine structure of Chloris striate mosaic virus *Virology, 92*, 428-435.

Holguin-Pena, R.J. & Vazquez Juarez, R. (2004). *Pepper golden mosaic virus* affecting tomat crop in Baja, California Peninsula, Mexico. *Plant disease, 88*, 221.

Hong, Y.G., Robinson, D.J. & Harrison, B.D. (1993). Nucleotide sequence evidence for the occurrence of three distinct whitefly-transmitted geminiviruses in cassava. *Journal of General Virology, 74*, 2437-2443.

Howarth, A.J., Caton, J., Bossert, M. & Goodman, R.M. (1985). Nucleotide sequence of bean golden mosaic virus and a model for gene regulation in geminivirus. *Proc. Natl. Acad. Sci.* USA, *82*, 3572-3576.

Hussain, M., Mansoor, S., Amin, I., Iram, S., Zafar, Y., Malik, K.A. & Briddon, R.W. (2003). First report of cotton leaf curl disease affecting chili peppers. *Plant Pathology, 52*, 809.

Hussain, M., Mansoor, S., Iram, S., Zafar, Y. & Briddon, R.W. (2004). First report of *Tomato leaf curl New Delhi virus* affecting chilli pepper in Pakistan. *Plant Pathology, 53*, 794.

Jones, D.R. (2003). Plant viruses transmitted by whiteflies. *European Journal of Plant Pathology, 109*, 195-219.

Jorda, C., Font, I., Martinez, P., Juarez, N., Ortega, A. & Lacasa, A. (2001). Current status and new natural hosts of *Tomato yellow leaf curl virus* (TYLCV) in Spain. *Plant Dis., 85* (4), 445.

Jose, J. & Usha, R. (2003). Bhendi yellow vein mosaic disease in India is caused by association of a DNA β satellite with a begomovirus. *Virology, 305*, 310-317.

Joshi, R.D. & Dubey, L.N. (1973). Assement of losses due to CMV on Chilli. *Science and Culture, 39*, 521-522.

Joshi, R.D. & Suteri, B.D. (1976). *Calliopsis* mottle-a new virus disease. *Science and Culture, 42*, 616-618.

Kang, K.Y., Choi, J.I. & La, Y.J. (1973). Isolation and identification of virus affecting pepper in Korea. *J. of the Korean Society for Horticultural Science, 13*, 35-44.

Kapoor, P. (2005). Investigations on the viruses infecting bell pepper (*Capsicum annuum* L.) Ph. D. thesis (http://www.yspuniversity.ac.in/mpp/thesis/mpp-phd-abstracts.doc.).

Kaur, S. & Singh, J. (1985). Resistance to fruit rot disease under field conditions in pepper (*Capsicum annuum* L.). *Capsicum News Letter, 4*, 67.

Khan, M.S. (2006). Molecular characterization of geminivirus infecting chilli and weeds for development of diagnostics and effective management of the virus. Ph. D. Thesis submitted to University of Lucknow, Lucknow.

Khan, M.S., Raj, S.K. & Singh, B.P. (2003). Some weeds as new hosts of Geminivirus as evidenced by molecular probes. *Indian Journal of Plant Pathology, 21*, 82-85.

Khan, M.S., Raj, S.K. & Singh, R. (2006a). First report of *Tomato leaf curl New Delhi virus* infecting chilli (*Capsicum annuum*) in India. *Plant Pathology, 55*, 289.

Khan, M.S., Raj, S.K., Bano, T. & Garg, V.K. (2006b). Incidence and management of mosaic and leaf curl diseases in cultivars of chilli (*Capsicum annuum*). *J. Food Agriculture and Environment, 4* (1), 171-174.

Kheyr-Pour, A., Bendahmane, M., Matzeit, V., Accotto, G.P., Crespi, S. & Gronenborn, B. (1991). *Tomato yellow leaf curl virus* Sardinia is a whitefly-transmitted monopartite geminiviruses *Nucleic Acids Research, 19*, 6763-6779.

Kim, J. S., Kim, S. K., Lee, S. H. & Lee, M.W. (1990). A *Pepper vein chlorosis virus* causing stem necrosis and vein chlorosis on red pepper in Korea. *Korean J. Plant Pathol., 6*(3), 367-381.

Kim, J.S., Lee, K.H & Lee, S.H. (1991). Identification and characteristics of a *Pepper vein mosaic virus* from red pepper. *Res. Rept. RDA, 33* (3), 6-11.

Kiranmai, G., Sreenivasulu, P. & Naiudu, M.V. (1997). Characterization of cucumber mosaic cucumovirus isolates naturally infecting three Solanaceous vegetable crops in Andhra Pradesh. *Indian Phytopath, 50*, 416-425.

Kumar, S., Kumar, S., Singh, M., Singh, A.K. & Rai, M. (2006). Identification of host plant resistance to *Pepper leaf curl virus* in chilli (*Capsicum* species). *Scientia Horticulturae, 110*, 359-361.

Kumar, P. & Awasthi, L.P. (2003). Management of infection and spread of bottle gourd mosaic virus disease in bottle gourd through botanicals. *Indian Phyotopath., 56*, 361.

Kumar, A. (2006). On-farm management of leaf curl disease in chilli under arid farming system. *International Journal of Food, Agriculture and Environment, 4*, 180-182.

Kunik, T., Salamon, D., Zamir, D., Navot, N., Zeidan, M., Michelson, I., Gafni, Y. & Czosnek, H. (1994). Transgenic tomato plants expressing the tomato yellow curl virus capsid protein are resistant to the virus. *Bio/Technology, 12*, 500-504.

Kunkalikar, S., Byadgi, A.S., Kulkarni, V.R. & Reddy, M.K. (2006). Management of papaya ringspot virus disease. *Indian Journal of Virology, 17*, 39-43.

Lazarowitz, S.G. (1992). Geminivirus: genome structure and gene function. *Crit. Rev, Plant Sci., 11*, 327-349.

Mali, V.R., Nirmal, D.D., Patel, K.V. & Vyanjane, N.T. (1985). Studies on a virus causing distortion mosaic of *Datura fastuosa*. *Indian Phytopathol., 38*, 413- 417.

Mallikarjun, Kenganal, Y., Ranebennur, H., & Byadgi, A.S. (2006). Management of leaf curl in chilli. (http://www.hindu.com/seta/2006/07/27/stories/ 2006072700131600).

Martelli, G.P., & Quacquarelli, A. (1983). The present status of tomato and pepper viruses. *Acta Horticulturae, 127*, 39-64.

Maruthi, M.N., Rekha, A.R., Alam, S.N., Kader, K.A., Cork, A. & Colvin, J. (2006). A novel begomovirus with distinct genomic and phenotypic features infects tomato in Bangladesh. *Plant Pathology, 55*, 290.

Mathai, P.J., Dubey, G.S., Peter, K.V., Saklani,U.D. & Singh, N.P. (1977). Pant C-1 and Pant C-2-two new promising selections of chilli *Capsicum annuum* L. *South Indian Hort.*, *25*, 123-125.

Mishra, M.D, Raychaudhri, S.P. & Jha, A. (1963). Virus causing leaf curl of chilli (*Capsicum annum* L.). *Indian J. Microbiol*, *3*, 73-76.

Misra, A., Choudhary, D. P., Michra, P.K. & Jha, A. (1979). Electron microscopical and serological investigation on Indian chilli mosaic virus. *Zeitschrift fur Pflanzenkarankheiten und Pflanzenschutz*, *86*, 39-42.

Nagaraju & Reddy, H.R. (1980). Natural occurrence of pepper veinal mottle virus on bell pepper. *Indian Phytopathol, 33*, 282-284.

Nagaraju & Reddy, H.R. (1982). Occurrence of cucumber mosaic virus on bell pepper. *Indian J. Mycol. & Pl. Path.*, *12*, 217-219.

Nagaraju, & Reddy, H.R. (1981). Natural occurrence of pepper vein banding virus on bell pepper. *Mysore Journal of Agricultural Science*, *5*, 50-53.

Naqvi, Q.A. & Mahmood, K. (1975). Reaction of different varieties of brinjal, tomato and pepper to the new strain of brinjal mosaic virus. *Geobios, 2*, 189-190.

Naqvi, Q.A., Haseeb, A., Saxena, S.K. & Mahmood, K. (1978). Influence of Marigold mottle virus, the renifoem nematode, Rotylenchulus renifomis and the stunt nematode on the growth of chilli and on nematode multiplication. *Acta Botanica, 6*, 174-175.

Naqvi, Q.A., Mahmood, K. & Saxena, S.K. (1976). Varietal response of vegetable against Launaea mosaic virus. *Geobios, 3*, 203-205.

Narasimhan, V. & Alagianagalingam, M.N. (1986). Potassium in the management of chilli mosaic disease. *J. Potassium Res.*, *2*, 59-64.

Osaki, T. & Inouye, T. (1981). Tobacco leaf curl virus. *CMI/AAB, Bulletin*, 232.

Padidam, M., Fauquet, C.M. & Beachy, R.N. (1995). Tomato leaf curl geminivirus from India has a bipartite genome and coat protein is not essential for infectivity. *Journal of General Virology, 76*, 25-35.

Persley, D.M., O'Brien, R. & Syme, J.S. (1989). Vegetables crops: a disease management guide. Qeensland Department of Primary Industries, Information Series QI 88019, Brisbane.

Pita, J.S., Fondong, V.N., Sangare, A., Otim-Nape, G.W., Ogwal, S. & Fauquet, C.M. (2001). Recombination, pseudorecombination and synergism of geminiviruses are determinant keys to the epidemic of severe cassava mosaic disease in Uganda. *J. Gen. Virol.*, *82*, 655-665.

Proctor, C.H. (1975). Studies on tomato aspermy virus from Chrysnathemum indicum l. in New Zealand. *New Zealand Journal of Agricultural Research, 18*, 387-390.

Purseglove, J.W., Brown, E.G., Green, C.L., & Robbins, S.R.J. (1981). *Spices* 1: Longman Inc, NewYork.

Quiñones, M., Fonseca, D. & Martínez, Y. (2002). First Report of *Tomato yellow leaf curl virus* Infecting Pepper Plants in Cuba. *Plant Disease, 86*, 73.

Raj, S.K., Singh, R., Pandey, S.K. & Singh, B.P. (2005). *Agrobacterium*-mediated tomato transformation and regeneration of transgenic lines expressing *Tomato leaf curl virus* coat protein gene for resistance against TLCV infection. *Current Science, 88*, 1675-1679.

Ramappa, H.K., Muniyappa, V. & Colvin, J. (1998). The contribution of tomato and alternate hosts plants to tomato leaf curl virus inoculation pressure in different areas of south India. *Ann. Appl. Biol.*, *133*, 187-198.

Reddy, C.R.V., Colvin, J., Muniyappa, V. & Seal, S. (2005). Diversity and distribution of begomoviruses infecting tomato in India. *Archives of Virology, 150*, 845-867.

Reisman, D., Riccardi, R.P. & Goodman, R.M. (1979). Size and topology of single stranded DNA from bean golden mosaic virus. *Virology, 97*, 388-395.

Rezende, J.A.M. & Costa, A.S. (1995). Alternatives for integrated control of papaya ringspot virus. *Acta Horticult., 370*, 129-132.

Rishi, N. & Dhawan, P. (1989). Characterization and identification of viruses causing mosaic diseases of *Capsicum* in nature. *Indian Phytopath., 55*, 284-298.

Roye, M.E., Smith, K., McLaughlin, W.A., Nakhla, M. & Maxwell, D.P. (1999). Tomato dwarf leaf curl virus, a new bipartite geminivirus associated with tomatoes and peppers in Jamaica and mixed infection with tomato yellow leaf curl virus. *Plant Pathology, 48*, 370-378.

Sakata, J., Yutaka Shibuya, Y., Sharma, P. & Ikegami, M. (2008). Strains of a new bipartite begomovirus, pepper yellow leaf curl Indonesia virus, in leaf-curl-diseased tomato and yellow-vein-diseased ageratum in Indonesia. *Archives of Virology, 12*, 2307-2313.

Samretwanich, K., Chiemsombat, P., Kittipakorn, K. & Lkegami, M. (2000). A new geminivirus associated with a yellow leaf curl disease of pepper in Thailand. *Plant disease, 84*, 1047.

Sanogo, S. (2003). Chile pepper and the threat of the wilt diseases. Plant Health Progress. PHP-2003-0430-01-RV.

Satya Prakash, Singh, S.J., Singh, R.K. & Upadhyaya, P.P. (2002). Distribution, incidence and detection of potyvirus on chilli from eastern Uttar Pradesh. *Indian Phytopathology, 55*, 284-298.

Saunders, K., Bedford, I.D., Briddon, R.W., Markham, P.G., Wong, S.M. & Stanley, J. (2000). A unique virus complex causes Ageratum yellow vein disease. *Proceeding of National Academy of Science of the United Stated of America, 97*, 6890-6895.

Senanayake, D.M.J.B., Mandal, B., Lodha, S. & Varma, A. (2006). First report of Chilli leaf curl virus affecting chilli in India. *Plant Pathology, 56*, 343.

Shih, S.L, Tsai W.S, Green, S.K., Khalid, S., Ahmad, I. & Rezaian, M.A. (2003). Molecular characterization of tomato and chilli leaf curl *begomoviruses* from Pakistan. *Plant Disease, 87*, 200.

Shih, S.L., Tsai, W.S. Green, S.K. & Singh, D. (2007). First report of *Tomato leaf curl Jodebpur virus* infecting chilli in India. *Plant Pathology, 56*, 341.

Shivanthan, P. (1982). Epidemiology of chilli leaf curl disease in Sri Lanka. *Tropical Agriculturist, 138*, 99-109.

Simons, J.N. (1957). Three strains of Cucumber mosaic virus affection bell peppers in the Everglades area of south Florida. *Phytopathology, 47*, 145-150.

Singh, A., Cheema, S.S. & Hundal, J.S. (1990). Identification of virus(es) associated with chilli mosaic syndrome. *J. Research of PAU, 27*, 75-83.

Singh, D. & Singh, S.J. (2000). Chilli little leaf-a new phytoplasma disease in India. *Indian Phytopathology, 53*, 309-310.

Singh, J., Sohi, A.S., Mann, H.S. & Kapur, S.P. (1994). Studies on whitefly, *Bemisia tabaci* (Genn.) transmitted cotton leaf curl disease in Punjab. *Journal of Insect Science, 7*, 194-198.

Singh, R., Raj, S.K. & Chandra, G. (2001). Association of a monopartite begomovirus with yellow mosaic disease of Pumpkin (*Cucurbita maxima*) in India. *Plant Disease, 85*, 9.

Singh, S. &Awasthi, L.P. (2002). Prevention of infection and spread of bean common mosaic virus disease of mungbean and urdbean through botanicals. *Indian J. Mycol. Pl. Pathol., 32*, 141.

Singh, S.S. & Dhakrani, K.V.K. (2009). Chilli leaf curl disease. (http://aaqua.persistent.co.in/aaqua/forum/viewthread? thread=11170).

Smith, P.G., Villalon, B. & Villa, P.L. (1987). Horticultural classification of peppers grown in the United States. *Hort Science, 22*, 11-12.

Somos, A. (1984). The Paprika. Akademiai Kiado, Budapest. South Pacific Seeds. Hot Chilli Growing for Mechanical Harvest. South Pacific Seeds, Griffith, NSW.

Sri, H.H. & Rahmayani, E. (2007). Transmission of Tomato leaf curl begomovirus by Two Different Species of Whitefly (Hemiptera: Aleyrodidae). *Plant Pathol. J., 23*, 57-61.

Sri, H.H., Aidawati, N. & Sri, S. (2008). Current studies on divergence and management of whitefly-transmitted geminiviruses in Indonesia. *Indian J. Virology, 19*, S-77.

Stanley, J., Bisaro, D.M., Briddon, R.W., Brown, J.K., Fauquet, C.M., Harrison, B.D., Rybicki, E.P. & Stenger, D.C. (2005). Geminiviridae. In: Fauquet, C.M., Mayo, M.A. Maniloff, J. Desselberger, U. and Ball, L.A. (eds), Virus Taxonomy, VIIIth Report of the ICTV. Elsevier/Academic Press, London, pp 301-326.

Stenger, D.C., Duffus, J.E. & Villalon, B. (1990). Biological and genomic properties of a geminivirus isolated from peppers. *Phytopathology, 80*, 704-709.

Sugiura, M., Bandaranayake, C.M. & Hemachandra, G.H. (1975). Chilli viruses disease in Sri Lanka. Technical Bulletin, Tropical Agricultural Research Report Ministry of Agriculture, Forestry and Fisheries, Japan, *8*, 62.

Tahir, M. & Haider, M.S. (2006). First report of a begomovirus associated with leaf curl disease of bell pepper. *Plant Pathology, 55*, 570.

Tesoriero, L. & Azzopardi, S. (2006). *Tomato yellow leaf curl virus* in Australia. *Prime facts* (http://www.dpi.nsw.gov.au/_data/assets/pdf_file/0019/77113/Tomato-yellow-leaf-curl-virus-in-Australia-Primefact-220.pdf).

Tewari, V.P. & Ramanujam, S. (1974). Grow Pusa Jwala, a disease resistant high yielding chilli. *Indian Farming, 24*, 20.

Thinlay, & Penjore, U. (1996). Survey of leaf curl virus of chilli and tomato in Bhutan. *Proceeding of the phase I Final Workshop of the South Asian Vegetable Research Network*, Kathmandu, Nepal, p. 216.

Tobias, I. (1983). *Cucumber mosaic virus* isolates on pepper in Hungary. *Capsicum Newsletter, 2*, 129-129.

Tochihara, H. (1970). Some properties of chrysanthemum mild mottle virus and comparison of this virus with cucumber mosaic virus. *Ann. Phytopath Soc. Japan, 36*, 1-10.

Tomlinson, J.A. (1987). Epidemiology and control of virus disease of vegetable. *Ann. Appl. Biol., 110*, 661-681.

Torres-Pacheco, I., Garz6n-Tiznado, J.A., Herrera-Estrella, L. & Rivera-Bustamante R.F. (1993). Complete nucleotide sequence of pepper huasteco virus: analysis and comparison with bipartite geminiviruses. *Journal of General Virology, 74*, 2225-2231.

Tsai, W.S., Green, S.K. & Rashid, M.H. (2000). *Chilli leaf curl virus*. A component, complete sequence. *EMBL/ Gen bank/ DDBJ Database*. Accession no. AF 314531.

Tsai, W.S., Shih, S.L., Green, S.K., Rauf, A., Hidayat, S.H. & Jan, F.J. (2006). Molecular characterization of *Pepper yellow leaf curl Indonesia virus* in leaf curl and yellowing diseased tomato and pepper in Indonesia. *Plant Disease, 90*, 247.

Tsai, W.S., Shih, S.L., Green, S.K., Lee, L.M. & Luther, G.C. (2009). Identification of a New Begomovirus Associated with Yellow Leaf Curl Diseases of Tomato and Pepper in Sulawesi, Indonesia. *Plant Disease, 93*, 321.

Umaharan, P., Padidam, M., Phelps, R.H., Beachy, R.N. & Fauquet, C.M. (1998). Distribution and diversity of geminiviruses in Trinidad and Tobago. *Phytopathology, 88*, 1262-1268.

UmaUngs, W.D., Woodbridge, C.G. & Csizinszky A.A. (1977). Screening peppers for resistance to *Curly top virus*. *Hort. Science, 12*, 161-162.

Valand, G.B. & Myniyappa, V. (1992). Epidemiology of tobacco leaf curl virus in India. *Ann. Appl. Biol., 120*, 257-267.

van Regenmortel, M.H.V., Fauquet, C.M., Bishop, D.H.L., Carstens, E.B., Estes, M.K., Lemon, S.M., Maniloff, J., Mayo, M.A., McGeoch, D.J., Pringle, C.R. & Wickner, R.B. (2000). Virus Taxonomy: *Classification and Nomenclature of Viruses*. San Diego, California, USA, Academic Press. pp. 1162.

Varma, A. & Malathi, V.G. (2003). Emerging geminivirus problems: A serious threat to crop production. *Annals of Applied Biology, 142*, 145-146.

Varma, A., Jain, R.K. & Bhat, A.I. (2002). Virus resistant transgenic plants for environmentally safe management of viral diseases. *Indian J. Biotechnol., 1*, 73-86.

Verma, H.N., Srivastava, S., Varsha & Kumar, D. (1996). Induction of systemic resistance in plants against viruses by a basic protein from *Clerodendron aculeatum* leaves. *Phytopathology, 86*, 485-492.

Verma, H.N., Awasthi, L.P., Kumar, V., Chaudhary, B., Rastogi, P. & Diwedi, S.D. (1980). Control of plant virus diseases by extracts from higher plants. *Ind. Bot. Soc., 59*, 30.

Verma, H.N., Rastogi, P. & Prasad, V. (1985). Possible control of natural virus infection on *Vigna radiata* and *Vigna mungo* by plant extracts. *Indian J. Pl. Pathol., 3*, 21-24.

Villalon, B. (1975). Virus disease of bell peppers in south Texas. *Plant Dis. Reptr., 59*, 858-862.

Villalon, B. (1981). Breeding peppers to resist virus diseases. *Plant Disease, 65*, 557-561.

Wilson, T.M.A. (1993). Strategies to protect crop plants against viruses:Pathogen-derived resistance blossoms. *Proc. Natl. Acad. Science USA., 90*, 3134-3141.

Wyatt, S.D. & Brown, J.K. (1996). Detection of subgroup III geminivirus isolates in leaf extracts by degenerate primers and polymerase chain reaction. *Phytopathology, 86*, 1288-1293.

Xi, Z., Xu, S. & Mang, K. (1982). A small ssDNA-containing virus-bean distortion dwarf virus. *Acta Microbiologica Sinica, 22*, 293-299.

Zubiaur Y.M., Martín, Y. M. & Pantoja, M.Q. (2006). A new *begomovirus* infecting pepper plants in Cuba. *Plant Pathology, 55*, 817.

In: Emerging Geminiviral Diseases and their Management ISBN 978-1-61668-620-8
Editors: P. Sharma, R. K. Gaur, et al. pp. 105-128

Chapter 5

EMERGING BEGOMOVIRUS PROBLEMS IN INDIA

Yogesh Kumar, Vipin Hallan* and A.A. Zaidi
Institute of Himalayan Bioresource Technology (CSIR), Palampur, Himachal Pradesh, India.

ABSTRACT

Begomoviruses are one of the most important plant pathogens all over the world, especially in tropical and sub-tropical parts. The viruses from this group infect a range of crops viz. vegetables, cash crops, ornamentals, tropical fruits and even weeds which sometimes can act as latent hosts of the virus. Among these, solanaceous plants are important as food crops and are grown in a large part of the world and also cultivated throughout India. Hot and humid climate in some parts of India is suitable for spread of begomoviruses, especially affecting communities dependent on produce from these crops as sole source of very valuable protein component of diet. Many begomoviruses have been detected and characterized from important solanceous crops like tomato, chili, potato and tobacco. Due to global warming, green house cultivation, appearance and increase in population of whitefly biotype 'B' and its intrusion into new areas, begomovirus diseases are emerging as serious problem in more temperate areas of India. Human activities like introduction of new crops, exchange of germplasm and movement of infected plant materials from one place to another have also contributed significantly in the spread of these diseases. Moreover, begomoviruses are evolving very fast due to mutations, pseudo-recombination and interspecific recombination in mixed infections and new species are emerging frequently. Association of satellite DNAs with many begomoviruses has further increased the severity of the problem. One of the primary steps in managing these diseases is to create a clear picture of diversity and geographical distribution of the causal pathogens. This chapter summarizes symptomatology, virus transmission and molecular characterization of various emerging begomoviruses in the last ten years, in different parts of India.

* Correspondence concerning this article should be addressed to: Vipin Hallan, E-mail: rnaivi@gmail.com.

INTRODUCTION

The genus *Begomovirus* belongs to the family *Geminiviridae,* members of which have unique twin particle morphology and single stranded circular DNA genomes of 2.5 to 3.0 kb in size. Geminiviruses constitute the second largest group of plant viruses, which infect a wide range of economically important plants throughout the world, mainly in tropical and sub-tropical parts. The genomic DNA replicates through double-stranded DNA intermediates by rolling circle and recombination-dependent replication in the nuclei of infected cells (Jeske et al., 2001; Preiss & Jeske, 2003). Based on host range and mode of transmission, geminiviruses are divided into four genera namely *Mastrevirus, Curtovirus, Topocuvirus* and *Begomovirus* (Fauquet et al., 2003). Compared to other three genera, begomoviruses have emerged as more serious threat to agriculture worldwide. Begomoviruses infect dicots and are transmitted by whitefly vector *Bemisia tabaci* (Gennadius). Begomoviruses are either monopartite having single genomic component referred to as DNA-A or bipartite having two genomic components almost equal in size referred to as DNA-A and DNA-B (Hanley-Bowdoin et al., 1999). Additional components DNA-β and/or DNA1 have recently been found associated with some monopartite begomoviruses (Briddon et al., 2003, 2004). The DNA-A and DNA-B of a bipartite begomovirus shows no sequence similarity except for a highly conserved (>90%) non-coding region of approximately 200 nucleotides (nt) known as common region (CR). CR contains sequences essential for transcription and replication, which are replication associated protein (Rep) binding sites (iterons) and a conserved nonanucleotide (TAATATTAC) stem-loop structure (Moffat 1997; Fauquet et al., 2003). Typically, DNA-A encodes six ORFs, two in virion sense (AV1 and AV2) and four (AC1, AC2, AC3, AC4) in complementary sense. DNA-B encodes two ORFs, one in virion sense (BV1) and the other in complementary sense (BC1). Most of the New World begomoviruses lack the AV2 ORF (Harrison and Robinson, 1999).

Bemisia tabaci, begomovirus vector is very prevalent in hot and humid climate, in such conditions dual affect of virus infection and vector feeding/multiplication is highly damaging to the crops. The vectors not only suck precious juice from the plants, they spread viruses effecting quality and productivity. Hence, due to presence of begomoviruses and presence of their vector in ample quantity, these infections/diseases are a serious constraint to crop cultivation. Many of the begomoviral diseases are devastating and very frequent emergence of new begomoviruses shows that these are still evolving. Two aspects are important from the point of view of recent trends of begomovirus spread (Seal et al., 2006), which are: emergence of new begomoviruses through recombination and mutation and invasion by the vector *B. tabaci* to newer areas with introduction of greenhouse culture and changing climatic conditions (greenhouse affect). This chapter will describe incidences of emerging begomoviruses in last ten years in India, with a focus on emerging begomoviruses in sub-temperate areas where cultivation of crops is mainly possible in greenhouses due to harsh climate in field. Further, the virulent vector whitefly biotype B is becoming established into newer areas. It is no longer limited to its native habitat/geographical boundary and this demonstrates its adaptive capability. This may be attributed to existence of variants in whitefly population, which are continuously evolving. Several biotypes of *B. tabaci* are known viz. An, B, D, K, L, M and Q. Biotype B is the only known vector of begomoviruses and recently *B. argentifolii* has been identified as a race of *B. tabaci* (Barro et al. 2005).

However several races of *B. tabaci* have been identified from around the world through biological, molecular and phylogenetic characterization of ITS1 and mitochondrial cytochrome oxidase gene (Barro et al., 2005).

There are scattered records wherein efforts have been made to document threatening plant diseases of fungal, bacterial and viral nature. Geminiviruses are important components in this list (http://www.ces.ncsu.edu/depts/ent/clinic/Emerging/vdis.htm). Their intrusion into temperate regions and appearance of diseases caused by a complex of begomovirus and satellite DNAs are serious concerns (Kumar et al., 2008). Natural mutations and recombination among diverse begomoviruses contribute to emergence of most of the agriculturally devastating begomovirus complexes. Association of different satellite DNAs further complicates the diversity and classification of begomoviruses (Umaharan et al., 1998; Padidam et al., 1999; Monci et al., 2002; Garcia-Andres et al., 2006; Briddon et al., 2003, 2004). The biochemical processes involved in recombination and emergence of more virulent strains/species and the evolutionary processes that determine which of the recombinants will survive, are still unknown. However, it has been speculated that recombination hot-spots and cold-spots exist within the genome of begomoviruses (Chatchawankanphanich and Maxwell 2002; Ndunguru et al., 2005; Garcia-Andres et al., 2007). The frequency of emergence of new species could be exemplified by the fact that more than 800 complete DNA-A sequences are available in EMBL database, which represent $\geq$ 200 different species based on the 89 % identity for species demarcation (Fauquet and Stanley, 2005; Fauquet et al., 2008).

EMERGING BEGOMOVIRUSES IN TROPICAL AREAS

Begomoviruses cause numerous severe diseases of cultivated crops of the families *Solanaceae*, *Cucurbitaceae*, *Fabaceae*, and *Malvaceae* and pose a threat to their cultivation (Varma and Malathi, 2003) in different parts of India (Figure 1). This section will summarize the emerging begomovirus problems in the last ten years in India.

Tomato

Tomatoes rank fourth among cultivated vegetable crops worldwide with a production of more than 100 million metric tons a year (Peralta et al., 2005). It is one of the most important vegetables in India and its cultivation is constrained by tomato leaf curl disease (ToLCD), which was first reported in Northern India in 1948 (Vasudeva and Sam Raj, 1948). The symptoms induced by begomovirus infection include leaf curling, mosaic, bright yellow spots on leaves, reduction in leaf size, vein clearing, crinkling, chlolosis, stunting, reduced or no fruiting and occasional formation of enations (Varma and Malathi, 2003). The disease causes considerable crop losses. In a study, tomato cv. Pusa Ruby, Pusa Early Dwarf and Maitri showed 66.40, 67.00 and 88.00% reduction in yield respectively due to ToLCD. Maitri showed the highest disease incidence of 62.16%, while Pusa Ruby showed the lowest, 58.27% (Shahnaz and Kumar, 2004).

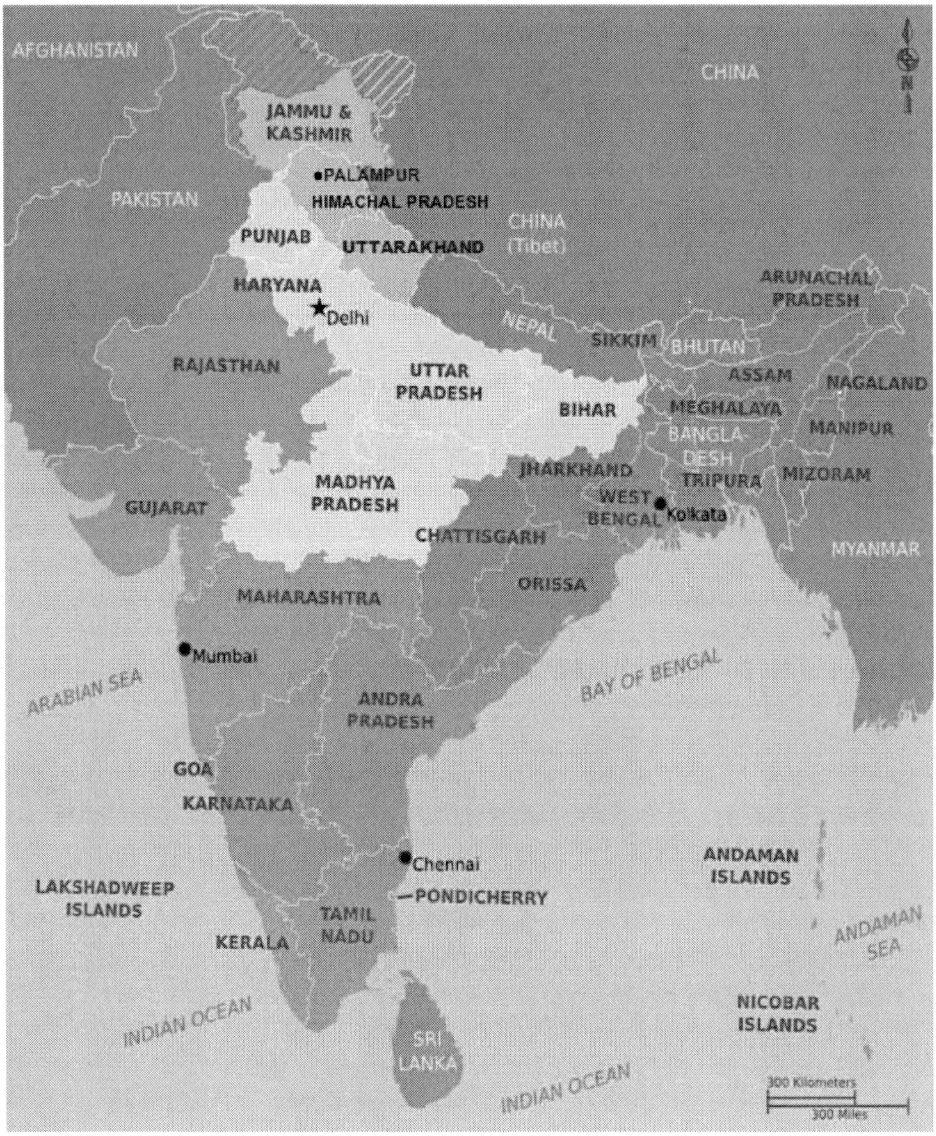

Figure 1. Geographical map of India, showing different states and the sub-temperate region (Palampur), where begomoviruses are spreading.

In November 2001, tomato plants showing yellowing of leaf lamina, upward leaf curling, leaf distortion, shrinking of leaf surface, and stunted growth were observed in tomato-growing areas in Varanasi and Mirzapur districts of eastern Uttar Pradesh, India. Based on DNA sequences obtained, the causal pathogen was identified as a begomovirus (Chakraborty et al., 2003b).

In 2002, a new monopartite begomovirus, tomato leaf curl Karnataka virus (U38239) was characterized from Bangalore, South India, infecting tomatoes. The virus was transferred to tomato by whitefly biotype B, sap inoculation and particle bombardment and it produced typical yellowing and leaf curl symptoms. The isolate appeared to be a recombinant virus (Chatchawankanphanich and Maxwell, 2002).

In 2003, a new bipartite species was found causing a severe leaf curl disease of tomato in Varanasi (Chakraborty et al., 2003a), Uttar Pradesh, India. The virus was named as Varanasi isolate of *Tomato leaf curl Gujarat virus* (ToLCGV-[Var]; AY190290, AY190291). The virus could be transmitted by grafting and through whitefly transmission in a persistent manner. ToLCGV-[Var] could infect and cause symptoms in tomato, pepper (*Capsicum annuum* cv. California Wonder), *Nicotiana benthamiana* and *N. tabacum* (cv. Xanthi) when partial tandem dimeric constructs of DNA-A and DNA-B were co-inoculated by particle bombardment. DNA-A alone was also infectious, but symptoms were milder and took longer to develop (Chakraborty et al., 2003a).

In order to assess variability of tomato leaf curl begomovirus (TLCB) in India, Reddy et al. tested 69 tomato plants and 34 weeds or neighbouring crops for begomoviruses, collected from all the major tomato producing areas of India. Eighty-one of the 103 samples were found positive in PCR using begomovirus genus-specific primers. Coat protein (CP) genes from 29 samples were PCR amplified and characterized. Phylogenetic analyses of the CP sequences revealed five different TLCB clusters each having <88% identity to the others. Four clusters represented known Indian begomoviruses, whereas one cluster contained sequences originating from Haryana State with highest identity (89%) with *Croton yellow vein mosaic virus*. Sixty-five begomovirus positive samples were characterized further by PCR for DNA-β and DNA-B for four Indian TLCB species, using universal begomovirus primers. The majority of samples represented monopartite TLCBs associated with DNA-β components. All four known TLCBs appeared to be present throughout India. TLCBs were also found to be present in chilli, cowpea, okra and tobacco crops and in some common weeds. *Papaya leaf curl virus* and *Pepper leaf curl Bangladesh virus* were also detected in tomato. Mixed begomovirus infections, were evident in 13 samples (Reddy et al., 2005a).

In 2008, isolates of two distinct begomovirus species, the bipartite severe strain of the species *Tomato leaf curl New Delhi virus* (ToLCNDV-[IN:ND:Svr:92]) and the Varanasi strain of the species *Tomato leaf curl Gujarat virus* (ToLCGV-[IN:Var:01], mono/bipartite) infecting tomato, causing severe yield losses in northern India were characterized. This study also investigated the infectivity properties of genomic components of these two species. Both were infectious in *N. benthamiana, N. tabacum* and *L. esculentum*. Pathogenicity was enhanced when DNA-A of ToLCNDV-[IN:ND:Svr:92] was trans-complemented with ToLCGV-[IN:Var:01] DNA-B, and increase in accumulation of ToLCGV-[IN:Var:01] DNA-B was consistently observed with it. A synergistic interaction between these two viruses was also evident in the study. This was the first report demonstrating a virulent pseudorecombinant between two distinct species of begomoviruses that infect tomato, and was the second report on synergism between begomoviruses (Chakraborty et al., 2008).

Potato

Tomato leaf curl New Delhi virus has emerged as a serious virus disease in India. Delayed setting in winter in subtropical countries due to global warming and changed cropping pattern has helped this virus to establish on potato (Garg, 2005). Potato leaf curl disease (PoLCD) was observed for the first time in northern India in 1999 (Usharani et al., 2004a). A distinct strain of tomato leaf curl New Delhi virus (ToLCNDV-Pot) was found to be associated the disease. The virus was sap-transmissible and bipartite in nature. A defective DNA-β was also detected, associated with an old world bipartite begomovirus for the first time (Usharani et al., 2004b).

Chilli

A leaf curl disease in the chilli (*Capsicum annuum*) cultivars Surajmukhi, Anhra Jyoti, X 235, Chandramukhi, Pusa Jawala, Longi-Jpani, Chanchal, Chaman and Selection 54 was observed in Uttar Pradesh, India. The causal virus could be transmitted from infected to healthy chilli plants through whiteflies. The virus was detected by dot-blot hybridization, using the total DNA extracted from infected chilli leaf samples and an Indian tomato leaf curl virus-specific probe. This study was the first evidence of association of a begomovirus with leaf curl disease of chilli in India (Raj et al., 2005a).

Capsicum annuum plants were observed to exhibit symptoms consisting of leaf curling, shortening of internodes and petioles, crowding of leaves and stunting of whole plants in fields around Lucknow, Uttar Pradesh, India. The disease was experimentally transmitted from infected to healthy chilli and tomato seedlings by the whitefly. Inoculated chilli plants developed typical symptoms of the disease. Tomato plants developed severe leaf curl symptoms similar to those of leaf curl disease of tomato caused by ToLCNDV. PCR amplification and sequence analyses confirmed the identity of the virus as ToLCNDV. This was the first report of this begomovirus affecting chilli crops in India (Khan et al., 2006).

In December 2004, up to 100% disease incidence was observed in *Capsicum annuum* plants in Jodhpur district, Rajasthan, India. Infected plants exhibited symptoms consisting of upward curling, puckering and reduced size of leaves. Severely affected plants were stunted and produced no fruit. The virus was transmitted to healthy chilli plants by whiteflies. PCR amplification and sequence analysis (DQ445255) showed close similarity of the isolate with the intergenic region and part of the replication initiator protein, AV1 and AV2 genes of *Chilli leaf curl virus*-[Pakistan: Multan] (ChilCuv-[Pk:Mul]; AF336806), reported from Pakistan. This was the first report of ChilCuV affecting chilli in India (Senanayake et al., 2007).

Symptoms of mild yellowing, severe leaf curling, leaf distortion, stunting and blistering were observed in June 2004 in chilli fields in Ludhiana, Punjab, India. The pathogen was identified as *Tomato leaf curl Joydebpur virus* (ToLCJV), based on the sequence data. This was the first report of ToLCJV infecting chilli naturally in India (Shih et al., 2007).

Another incidence of chilli leaf curl disease was observed in 2008 in Varanasi, India. Sequence analysis revealed that the monopartite genome (EF190217) of causal virus had the highest sequence identity (95%) with that of chilli leaf curl virus-PK[PK:Mul:98]. The DNA-β (EF190215) shared maximum sequence identity (88%) with a DNA-β satellite associated with tomato leaf curl disease from Rajasthan (ToLCBDB-[IN:Raj:03]). Infectious clones of the viral genome and the DNA-β produced typical leaf curl symptoms in chilli and *N. benthamiana* (Chattopadhyay et al., 2008).

Cassava

Cassava mosaic disease (CMD) causes considerable crop losses in Africa and Asia. Recombination between their genomic components has generated new pathotypes with enhanced virulence in Africa. Two bipartite begomoviruses, *Indian cassava mosaic virus* (ICMV) and *Sri Lankan cassava mosaic virus* (SLCMV), were characterized from cassava plants exhibiting mosaic symptoms from central India and Sri Lanka, respectively. ICMV could be transmitted by sap-inoculation to *Nicotiana benthamiana* and it produced leaf curl symptoms. SLCMV was found to be much more virulent in *N. benthamiana*, producing severe stunting, leaf curl and chlorosis. These symptoms were also reproduced by agroinoculation and mechanical inoculation of cloned genomic components (DNAs A and B) into *N. benthamiana.* SLCMV DNA B was found to be originated from ICMV DNA-B by a recombination event involving the SLCMV DNA-A intergenic region. SLCMV DNA-A alone was able to produce a severe upward leaf roll symptom, when introduced into *N. benthamiana* by agroinoculation reminiscent of the phenotype associated with some monopartite begomoviruses (Saunders et al., 2002). Later in 2005, an isolate of SLCMV was reported for the first time in India to cause cassava mosaic disease in Kerala (Dutt et al., 2005). It has been observed that in majority, SLCMV is more widespread than ICMV in India. The diversity of the Indian SLCMV-like sequences was rather low compared to the variability of African viruses associated with cassava mosaic disease. Phylogenetic analysis indicated evidence of recombination events between ancestors of SLCMV and ICMV (Rothenstein et al., 2006).

Cucurbits

Pumpkin yellow mosaic disease of *C. maxima* was observed in and around Lucknow, Uttar Pradesh, India in 2001. The isolate was transmitted by whiteflies to *C. maxima* and it reproduced the yellow mosaic symptoms. The virus was sequenced (AF381977) and identified as a strain of monopartite tomato leaf curl virus (TLCV). The study revealed association of a monopartite geminivirus with yellow mosaic disease of pumpkin for the first time in India (Singh et al., 2001).

In 2001, severe symptoms consisting of disease of upward curling, shortening, and distortion of leaves were observed in bitter melon (*Momordica charantia*) in Lucknow, Uttar Pradesh, India. The infected fruits were stunted and deformed. Disease incidence was as high as 100%. The virus could be transmitted from infected bitter melon to *Nicotiana tabacum* by *Bemisia tabaci*. It produced leaf curl symptoms in the inoculated plants indicating that the pathogen could be a begomovirus. This was the first report of association of a begomovirus infecting bitter melon (Khan et al., 2002a).

Pumpkin yellow vein mosaic virus (PYVMV) causes yellow vein mosaic disease, which occurs frequently in pumpkins (*Cucurbita moschata*) in India. It was detected in India for the first time in 2001, from Coimbatore, Tamil Nadu, India (Maruthi et al., 2003). The virus has a limited host range. Infected plants show vein yellowing, chlorotic patches, stunting symptoms and flowers drop prematurely, greatly reducing yields. PYVMV is transmitted readily and in a persistent manner by the whiteflies. It was detected serologically in pumpkin plants and in viruliferous *B. tabaci* using polyclonal antibodies in a double-antibody sandwich enzyme-linked immunosorbent assay. PCR based characterization and coat protein sequence analysis revealed PYVMV to be most similar to ToLCNDV and appeared to be a new strain of this virus (Muniyappa et al., 2003).

Sechium edule (Chayote) plants were observed to be affected by a yellow mosaic disease in West Bengal, India. The affected leaves developed yellow spots, mosaic and upward curling with occasional enations. Virus particles could not be detected in electron microscopy of leaf extracts. However, PCR and dot-blot hybridization showed an association of ToLCNDV. This was the first report of the occurrence of yellow mosaic disease of chayote in India, associated with a begomovirus (Mandal et al., 2004). In 2006, emergence of begomovirus infection in bottle gourd (*Lagenaria siceraria*), ivy gourd (*Coccinia cordifolia*), pumpkin (*Cucurbita moschata*), ridge gourd (*Luffa acutangula*) and watermelon (*Citrullus lanatus*) was documented for the first time in northern India (Sohrab et al., 2006). The association of ICMV with yellow mosaic disease of bittergourd (*Momordica charantia*) was detected for the first time in Tamil Nadu, South India (Rajinimala and Rabindran, 2007). In 2008, a begomovirus causing yellow vein mosaic disease of pumpkin was characterized at molecular level and based on the highest (96%) sequence identities and close phylogenetic relationships with *Squash leaf curl China virus* species, it was identified as strain of *Squash leaf curl China virus*. DNA-B was also detected by Southern blotting (Singh et al., 2008). In 2008, yellow mosaic disease of *L. cylindrica* was reported to be associated with ToLCNDV, for the first time in India (Sohrab et al., 2008).

Cotton

Cotton leaf curl disease (CLCuD) is a serious problem and it has emerged as an epidemic disease in Western India. During 1997-98, a sudden increase in cotton leaf curl disease incidence was observed in Rajasthan, Punjab and Haryana. Upto 60% of crop losses were observed (Sharma, 2003). The diseased plants show leaf curling, stunted growth, a poor yield of cotton fibre and the plants may develop leaf-like outgrowths from the veins on the undersides of leaves. This disease is caused by a monopartite begomovirus cotton leaf curl virus (CLCuV) and satellite DNA complex. This disease has spread into almost all the cotton growing belts in the country. The aetiology of CLCuD is uncertain and complex (Sharma & Rishi, 2004). An outbreak of cotton leaf curl disease was observed, caused by cotton leaf curl Rajasthan virus (CLCuRV). The disease has become a major threat to cotton production in the northwest India (Radhakrishnan et al., 2004). A number of other distinct begomoviruses have been characterized, associated with this disease, all of which require a satellite DNA-β component to induce symptoms in cotton. An additional, satellite-like component, DNA 1, is also found in diseased plants, although it is not required for disease development (Briddon, 2003). Complete genomic sequence of DNA-A component of two CLCuD-causing begomoviruses, cotton leaf curl Kokhran virus-Dabawali (CLCuKV-Dab), tomato leaf curl Bangalore virus-Cotton-[Fatehabad] (ToLCBV-Cotton-[Fat]) and partial sequences of two other isolates cotton leaf curl Rajasthan virus-Bangalore (CLCuRV-Ban) and cotton leaf curl Kokhran virus-Ganganagar (CLCuKV-Gang) were reported in 2004, from infected cotton plants. Extensive variability was observed in this group of viruses. The AC4 ORF and the region from 2411-424 (771 nt) was least conserved among these viruses (Kirthi et al., 2004).

CLCuV could be transmitted from infected cotton plants to tomato (*Lycopersicon esculentum*), tobacco (*Nicotiana tabacum*), cotton (*Gossypium hirsutum*), okra (*Abelmoschus esculentus*), chilli (*Capsicum annuum*), China rose (*Hibiscus rosa-sinensis*), *Zinnia elegans*, *Ageratum conyzoides* and *Sida alba* plants by whiteflies, giving an acquisition and inoculation access feeding period of 24 h. Chilli and China rose did not show any visual symptoms, while all other plants took a longer time for symptom development. Vein thickening and leaf curling were the main symptoms produced after inoculation. Enations were observed in cotton, tomato and tobacco. These results revealed a wide host-range of CLCuV (Kang et al., 2004). *Gossypium barbadense* cv. MLC leaves exhibiting mild leaf curl, vein thickening and enation symptoms were observed in Bangalore, Karnataka, India. The disease could be maintained by successive *B. tabaci* inoculations on cotton. The DNA-A sequence data of the virus isolate suggested it to be a distinct Begomovirus species, which was named as cotton leaf curl Bangalore virus (Reddy et al., 2005b).

Legumes

An unknown begomovirus was found associated with guar plants [*Cyamopsis tetragonoloba* (L.) Taub] showing leaf curling, chlorosis and stunting symptoms in Lucknow, India. The isolate could be transmitted by whitefly. In Southern hybridization, the isolate reacted positively with α-P^{32} labeled DNA probes prepared from *Tomato leaf curl virus* and *Cotton leaf curl virus,* confirming the occurrence of a begomovirus on diseased guar. This was the first report demonstrating the association of an unknown begomovirus with guar in India (Khan et al., 2002b). Later on in 2003, the virus causing the disease was identified as tomato leaf curl virus. The disease was transmitted from infected to healthy plants by the whitefly (Khan et al., 2003).

Yellow mosaic disease (YMD) of soybean was observed in 2004, in Delhi, India. Genomic DNA of the causing begomovirus were cloned, sequenced and evaluated for infectivity. Based on sequence identities, it was designated as a soybean isolate of mungbean yellow mosaic India virus (MYMIV-Sb). Phylogenetic analysis revealed dichotomy between two distinct species, namely, MYMIV and mungbean yellow mosaic virus (Usharani et al., 2004a).

In 2004, severe leaf distortion disease was observed in cowpea plantations in northern India. Symptoms included yellow mosaic with downward leaf curling, vein swelling and vein enations. The causal virus could be easily transmitted to cowpea seedlings by *B. tabaci*, resulting in different symptoms. The total nucleic acid extracted from plants with symptoms hybridized to radiolabelled probes specific to DNA A and DNA B of MYMIV. The samples were also probed for the presence of satellite DNA-β by Southern hybridization. DNA-β was detected consistently in samples showing enation symptoms. However, no hybridization was observed with an extract from leaves showing only yellow mosaic symptoms. Although MYMIV alone could produce yellow mosaic symptoms, but the presence of DNA-β enhanced the pathogenicity of the virus. This study reported the association of a satellite DNA-β with a bipartite begomovirus infecting a leguminous host for the first time in India (Rouhibakhsh and Malathi, 2005).

In 2005, complete nucleotide sequences of MYMIV and MYMV were determined, infecting soybean in India. Sequence analyses showed that MYMIV was present in central India, and MYMV in southern India. Multiple DNA-B components could be detected with the soybean strain of MYMV species. Analyses of the common region of the genomic components of these two virus isolates indicated considerable variability in the origin of replication, which did not affect their infectivity. Recombination events were detected among legume-infecting begomoviruses from South-East Asia through phylogenetic analyses (Girish and Usha, 2005).

Cultivation of dolichos (*Lablab purpureus*) has been affected by dolichos yellow mosaic disease (DYMD), which is associated with a begomovirus tentatively named *Dolichos yellow mosaic virus* (DoYMV). The symptoms of the disease consist of bright yellow mosaic patches on leaves, with significant yield reductions. In February 2004, a sample from a DYMD-affected dolichos plant was found positive in PCR for a distinct begomovirus of dolichos collected from Mysore, Karnataka, India. The putative virus was transmitted by the whiteflies from dolichos to dolichos cv. Mani (Maruthi et al., 2006a, 2006b). An 80-90% yellow mosaic disease incidence was observed on soybean during August 2005 in Lucknow, Uttar Pradesh, India. The disease agent was transmitted experimentally by *B. tabaci,* and symptoms developed after 23 days. Fifty-one soybean leaf samples were tested for begomovirus infection and the nucleotide sequences revealed 95% identity with *Cotton leaf curl Kokhran virus* (CLCKV). This was the first report of soybean as a new host of CLCKV (Raj et al., 2006a).

In 2005, pigeon pea plants were found to exhibit symptoms consisting of yellow mosaic, mottling, shortening of leaves and stunting in most pigeon pea-growing areas in India at low incidence. Based on the high nucleotide sequence identities, the virus associated with the disease was identified as an isolate of ToLCNDV. This was the first report of the occurrence of a begomovirus on pigeon pea from India (Raj et al., 2005b). Later in September 2005, a leaf crumple disease of soybean was observed in Lucknow, Uttar Pradesh, India. The disease incidence was as high as 80%. The diseased plants showed severe yellowing, crumpling and distortion of leaves. The causal begomovirus (DQ343284) was identified as *Tomato leaf curl Karnataka virus* (ToLCKV) based on high sequence identity (96-98%) with this virus. This was the first report of ToLCKV infection of soybean in India (Raj et al., 2006b). In 2008, MYMIV, with a distinct DNA-B component having higher similarity to MYMV was characterized and identified as the causal pathogen of cowpea golden mosaic disease. The infectious clones of the virus could infect cowpea, mungbean, blackgram and French bean by agroinoculation (John et al., 2008).

Bhendi

Production of bhendi or okra (*Abelmoschus esculentus*), an important vegetable crop of India is hindered to great extent by yellow vein mosaic disease. The disease is caused by a complex of monopartite begomovirus bhendi yellow vein mosaic virus (BYVMV) and a satellite DNA-β component. BYVMV alone could infect bhendi systemically but produces only mild leaf curling in this host. DNA-β is required for induction of typical symptoms of the disease. The DNA-β had a number of features in common with DNA-β reported for ageratum yellow vein disease and cotton leaf curl disease (Jose and Usha, 2003).

Other Crops

During July/August 2001, a severe virus-like disease affected 80-90% of sunn hemp plants in experimental plots at the National Botanical Research Institute, Lucknow, Uttar Pradesh, India. Symptomatic plants exhibited interveinal chlorosis, yellowing, curling and shortening of leaves, and stunting of plants. Whiteflies could transmit the associated virus from diseased to healthy sunn hemp plants, which developed symptoms similar to those occurring in naturally infected plants. Based on PCR amplification of its DNA with begomovirus specific primers and its strong hybridization with a DNA-A probe of Indian tomato leaf curl virus, the causal pathogen was identified as a begomovirus (Khan et al., 2002c).

A yellow vein net disease was observed on several *Calendula officinalis* plants in Uttar Pradesh, India. Disease symptoms consisted of vein yellowing, shortening of leaves and petioles and stunting. This was the first report of a natural infection of *C. officinalis* by a Begomovirus (Khan et al., 2005). In 2006, electron microscopy, Southern hybridization and PCR-mediated amplification indicated the association of a begomovirus associated with yellow mosaic disease of *Clerodendron inerme*. Typical symptoms were bright yellow spots along the midrib, which coalesce to give a mosaic, reduction in leaf size, and stunting of the plant (John et al., 2006). A begomovirus and a satellite DNA-β were found to be associated with yellow vein mosaic disease of mesta (*Hibiscus cannabinus* and *Hibiscus sabdariffa*), for the first time in mesta growing regions of India (Paul et al., 2006). Full-length DNA-A of a begomovirus infecting mesta was cloned and sequenced. The genome was identical to the DNA-A of monopartite begomoviruses originating from the Old World and complete nucleotide sequence consisted of 2728 nucleotides in length, having the highest sequence identity (83.5%) with an Indian begomovirus causing cotton leaf curl disease. The distinct species was given a name *Mesta yellow vein mosaic virus* (Chatterjee et al., 2005; Chatterjee and Ghosh, 2007a).

During September 2004, in Kolar and Bangalore districts of Karnataka, mosaic symptoms on naturally grown Jatropha plants were observed. Disease incidence ranged from 12.50 to 46.66%. Symptoms consisted of mosaic, reduced leaf size, leaf distortion, blistering and stunting. The disease could be transmitted to healthy plants through whiteflies. Inoculated plants developed typical symptoms such as veinal netting, chlorotic specks, leaf distortion and stunting (Narayana et al., 2006). Leaf curl disease of radish was observed in 2007, and the virus was successfully transmitted to radish cv. Minu Early by *B. tabaci*. Sequence analysis of the viral isolate revealed the association of a distinct begomovirus, DNA-α and a DNA-β complex with the disease. The name radish leaf curl virus was given to the new species (Singh et al., 2007).

Croton leaf curl disease was observed on ornamental croton (*Codieum variegatum*) in and around Bengaluru, South India. The diseased plants showed thickening of leaf veins, severe leaf curling, decrease in leaf size and stunting. The pathogen could be transmitted to healthy croton plants by grafting, and through whiteflies. A begomovirus was found associated with the disease and based on nucleotide sequence identities (94.96%), the virus was most closely related to *Ageratum yellow vein virus* reported from Pakistan (Mahesh et al., 2008).

In 2007, typical begomovirus-like symptoms characterized by chlorotic specks on leaves, curling and malformation of leaves, severe reduction in leaf size, partial or complete sterility were seen on infected jatropha plants in Karnataka state, South India. Jatropha mosaic disease (JMD) caused significant yield losses and disease incidences were up to 47 %. The virus could be transmitted by whiteflies and grafting to the dodder plants *Cuscuta subinclusa*. Based on PCR amplification of core CP sequences of ~575 and phylogenetic analysis, the virus isolate clustered in a separate group, close to *Indian cassava mosaic virus* and *Sri Lankan cassava mosaic virus* and shared highest nucleotide identities (90-95%) with them. These results confirmed that JMD in India was caused by begomoviruses and they were most closely related to cassava mosaic viruses from the Indian sub-continent (Narayana et al., 2007; Raj et al., 2008a). Association of a whitefly-transmitted Begomovirus with yellow net disease of *Dimorphotheca sinuata* was reported in 2007 for the first time from India based on PCR amplification and sequence analysis of the coat protein gene region (Raj et al., 2007a). A strain of ToLCNDV and a DNA-β was found associated with yellow vein net disease of Bimili jute (*Hibiscus cannabinus* L.) in India. The virus was identified by dot blot hybridization, PCR and sequence analysis (Raj et al., 2007b).

In 2008, ToLCNDV was found to be associated with leaf curl disease of papaya (*Carica papaya* L.) by PCR and confirmed by highest sequence similarities and close phylogenetic relationships (Raj et al., 2008b). The association of a distinct begomovirus with yellow vein net disease in grain amaranths (*Amaranthus cruentus* L.) in India was detected by PCR and the begomovirus possessed highest identities and closest relationship with *Ageratum enation virus* (Raj et al., 2008c). Yellow mosaic disease of jute (*Corchorus capsularis*) was observed and found to to be associated with a New World begomovirus for the first time in India, based on DNA-A sequence (Ghosh et al., 2008).

Weeds

Tomato leaf curl virus was found to naturally infect *Parthenium hysterophorus and Sonchus asper* plants in Uttar Pradesh, India, during 2002-03. *B. tabaci* could transmit the virus from these symptomless infected weeds to tomato plants, which exhibited leaf curling and twisting symptoms at 15-25 days after transmission (Ansari et al., 2005).

In 2007, disease incidence of yellow vein mosaic (YVM) disease of kenaf (*Hibiscus cannabinus*) was observed in northern India. The disease was found in both commercial and experimental farms in the Bahraich district, Uttar Pradesh, India and caused significant reduction in plant height and crop yield. Southern hybridization with begomovirus-specific probe and PCR amplification with DNA-β and coat protein primers confirmed the association of begomovirus with the disease (Ghosh et al., 2007). Later in 2007, full-length DNA-A of a begomovirus infecting mesta (*Hibiscus cannabinus* L.) was cloned and sequenced. The complete nucleotide sequence of the DNA-A molecule shared highest sequence identity (83.5%) with an Indian begomovirus causing cotton leaf curl disease. Thus it was considered as a novel begomovirus species, and the name *Mesta yellow vein mosaic virus* was given (Chatterjee and Ghosh, 2007a). A DNA-β component distinct from other known geminiviruses or nanovirus components was also identified associated with mesta yellow vein mosaic disease (Chatterjee and Ghosh, 2007b). In 2008, six betasatellite isolates associated with YVM disease in mesta crops from three different geographical locations of India were characterized. These six isolates and the one previously reported from eastern India could be divided into two distinct types based on the sequence identities. Another inceidence of YVM disease of mesta in northern India was recorded and the causal begomovirus was identified as a distinct monopartite species associated with a DNA-β satellite. This begomovirus species shared low sequence identity with the *Mesta yellow vein mosaic virus* (Das et al., 2008, 2008a).

A begomovirus with DNA-A and a DNA-β molecule was detected for the first time in the eastern part of India in *Urena lobata* showing yellow vein mosaic disease symptoms (Chatterjee et al., 2007).

EMERGING BEGOMOVIRUS PROBLEMS IN SUB-TEMPERATE REGIONS

Tomato Leaf Curl Disease

Sambyal et al. collected tomato samples showing leaf curl symptoms from different areas of Solan district of Himachal Pradesh at different altitudes ranging from 600-1650 mts. The samples showed positive reaction with antisera of tomato leaf curl virus. PCR also confirmed the presence of tomato leaf curl virus in the samples. Begomovirus infection could be detected from samples collected from locations with 1650 m altitude. This was the first molecular evidence of a natural begomovirus infection in a sub-temperate area in India (Sambyal et al., 2006).

A distinct bipartite begomovirus was found to associated with tomato plants showing yellowing, curling, and crumpling of the leaves (Figure 2), in Palampur region in Northern India. The region has a sub-temperate climate with altitude around 1300m and was previously unaffected by these viruses and whiteflies. Whiteflies were first observed in earlier years of this decade, mainly in the greenhouses where crops like tomato, cucurbits, gerbera, and chrysanthemum were cultivated. Begomovirus symptoms were not prevalent in this area earlier, and leaf curl symptoms on tomato were first observed in 2004-2005. The complete DNA-A and DNA-B components were amplified from the total DNA isolated from infected samples, through rolling circle amplification (RCA) using ϕ-29 DNA polymerase. DNA components were cloned and sequenced by primer walking. The DNA-A of the isolate was comprised of 2,756 nucleotides, encoding six open reading frames (ORFs) and DNA-B that of 2,725 nucleotides, encoding two ORFs. Genome organization of the DNA components was similar to an old world bipartite begomovirus. Sequence comparisons showed that DNA-A and its IR had the close identity (86% and 84%, respectively) with the *Tomato leaf curl New Delhi virus* (ToLCNDV; DQ116885) and some other begomoviruses (<84%) reported from cucurbits and tomato. The data indicated that the isolate to be a distinct begomovirus species, which was given a name tomato leaf curl Palampur virus (ToLCPMV). DNA-B had the maximum sequence identity (73%) with ToLCNDV (AY150305) reported from Pakistan. The common region of DNA-A and DNA-B showed 94% sequence similarity with each other. In a region like Palampur (and other regions) once established in greenhouses, whitefly populations can easily flourish. In sub-temperate areas, plant seedlings are often transported from tropical and sub-tropical regions and cultivated in the greenhouses, which could account for emergence of the disease in this region (Kumar et al., 2008).

Figure 2. Tomato plant showing yellowing, curling, and crumpling of the leaves

Ageratum Yellow Mosaic Disease

Ageratum is an obnoxious weed, commonly called as "Neela Phulnu" is widespread in different parts of Himachal Pradesh and two species viz., *Ageratum conyzoides* Linn. and *A. houstonianum* Mill are found widely. *A. houstonianum* grows best in moist and shady conditions. It is spread in abundance along the road sides, field bunds, water channels, tea gardens, orchards, pastures and grasslands. The occurrence of a virus like disease showing yellowing of veins, crinkling of young leaves, non-flowering and marginal necrosis in severely infected plants has been reported on *A. houstonianum* (Sharma *et al.*, 2001). Yellow mosaic disease of *Ageratum houstonianum* occurs widely in different parts of Himachal Pradesh and the leaves exhibit vein yellowing followed by yellow vein mosaic pattern (Figure 3). A number of other weeds have also been found to show symptoms typical of geminiviruses (Figure 4), which shows the emergence of probably new begomoviruses, as these were positive for universal geminiviral PCR primers described earlier (Wyatt et al. 1996).

Severely infected plants show necrosis, poor flowering and poor seed setting. In a study, the virus did not show seed transmission but it could be transmitted through whitefly and grafting only to *A. houstonianum, A. conyzoides* and *Abelmoschus esculentus.* Immuno-diagnostic studies (ISEM and ELISA) revealed the association of geminate particles measuring 17-28 nm in diameter with the disease. The infected samples reacted positively with the antiserum of ICMV, suggesting association of begomovirus. Based on the particle morphology, serological relationship, ultrastructural changes and transmission studies it was evident that a begomovirus is associated with yellow mosaic disease of *A. houstonianum* in

Himachal Pradesh. The virus seemed to be similar to okra yellow vein mosaic virus and thus, these weed species (*A. houstonianum* and *A. conyzoides)* could be an alternative host of okra yellow vein mosaic virus in the region (Sharma et al., 2007).

Figure 3. Ageratum leaves showing yellow vein mosaic.

Figure 4. Some unknown weeds showing typical geminiviral symptoms.

Chilli Leaf Curl Disease

In November 2006 and June 2008 leaf samples of chilli plants showing leaf curl symptoms (Figure 5) were collected from Palampur. The samples were screened for begomoviruses by Southern hybridization and PCR, using degenerate primers AV494 and AC1048 (Wyatt et al., 1996) at standard conditions. Six samples gave positive amplifications. The amplicons were cloned into pGEM-T Easy vector and sequenced. The sequences were submitted to EMBL (accession numbers AM773441, FM210473, FM210474, FM210475, FM210476, FM210477). The virus was identified as *Chilli leaf curl virus,* based on the sequence identities in BLAST search (Unpublished).

Figure 5. Chilli plant showing leaf curl symptoms.

Potato Leaf Curl Disease

In March 2007, apical leaf curl and mottling disease of potato (Figure 6) was observed in parts of Kangra district of Himachal Pradesh. Twenty samples showing mild to severe symptoms were collected and total DNA was isolated. Initial screening by slot-blot hybridization, using α-P^{32} dATP labeled probe specific to coat protein gene of ToLCNDV showed positive results in ten samples. From the positive samples DNA was amplified by RCA with the help of TempliPhi^{TM} DNA Amplification Kit (GE Healthcare, USA) as per the standardized protocol (Kumar et al., 2008). Amplified samples were cloned and sequencing of the three recombinant clones from of each of the two DNA components was accomplished by primer walking, sequencing both strands. Infecting isolate was identified as strain of ToLCNDV in a BLAST search (99% identity). DNA-A and DNA-B of the isolate comprised of 2740 and 2693 nucleotides respectively and had a typical Old World begomovirus like genome organization with six and two ORFs encoded by DNA-A and DNA-B respectively. The study is the first evidence of association of ToLCNDV infecting potato plants in sub-temperate zone of Himachal Pradesh (Unpublished).

Figure 6. Potato plant showing leaf curl and mottling symptoms.

Conclusion

The increasing incidence of new begomovirus-related problems, and their damaging outbreaks are serious threat to sustainable crop production throughout the world. Many new begomovirus problems are emerging due to evolution of virulent variants through natural recombination in mixed infections. These viruses are expanding rapidly in terms of their geographical distribution and genetic variability. Global warming, increase in the whitefly populations and emergence of its new biotypes are grave concerns to world agriculture. Human interventions have also played a significant role in spreading these viruses to newer areas. There can not be a single control measure for these epidemics, and impact of human activities on begomovirus evolution can only be minimized by managing and changing the practices that have led to the emergence of these viruses and their vector populations.

Understanding of vector-virus relationship needs to be improved. It would be important to understand the influence of *B. tabaci* populations on the distribution and spread of begomoviruses. Due to lack of molecular methods, progress on *B. tabaci* studies have been slow (Brown, 2000). Studying the correlations between crops grown, the feeding preferences of whiteflies, and the emergence of new viruses in mixed infections would be an important aspect. Climate change could also be the underlying factor for evolution of this vector. Due to global warming, a shift in begomovirus problems from summer crops to winter crops can also be expected.

There are many other factors for emergence of new begomovirus problems including changing cropping systems, increased green-house agricultural practices, movement of infected planting material, introduction of susceptible plant varieties and changing weather events. Introductions of new crops or varieties in a country must first be assessed for their susceptibility on a small scale for a few growing seasons. The exchange of vegetative plant material from one area to the other should be highly cautious. Quarantine practices and testing of imported plating materials should be extremely thorough and accurate. In future crop management, focus should be made on reducing vector populations and factors leading to evolution of more virulent begomovirus variants need to be well understood. Research focus is also needed on interactions between various satellite molecules and begomoviruses, since there has been a clear affect of such interactions on disease severity. Availability of natural host resistance should be explored more and cropping practices leading to reduce the evolution of resistance-breaking virus variants should be adopted.

To conclude, multidisciplinary research is required to understand the complex virus/vector, host/pathogen, host/vector and virus/satellite interactions. Virologists, entomologists and plant breeders should come together and give a combined effort in understanding and managing these serious begomovirus disease problems. However, the current picture of begomovirus distribution and evolution indicates that these problems are going to become more serious in future.

Acknowledgments

The authors are thankful to the Director, IHBT (Council of Scientific and Industrial Research), for providing necessary facilities to carry out the research work and an inhouse project on plant viral diagnostics. Yogesh Kumar is grateful to CSIR for awarding him Senior Research Fellowship. Useful support from Guru Nanak Dev University, Amritsar, India is highly acknowledged. This is IHBT publication number 1085.

References

Ansari, N. A. & Tewari, J. P. (2005). Hitherto unrecorded weed reservoirs of tomato leaf curl virus (TLCV) in Eastern U.P. and its role in epidemiology of the disease. *Journal of Living World,12*(2), 22-26.

Barro, P. J. D., Trueman J. W. H. & Frohlich, D. R. (2005). *Bemisia argentifolii* is a race of *B. tabaci* (Hemiptera: Aleyrodidae): the molecular genetic differentiation of *B. tabaci* populations around the world. *Bulletin of Entomological Research, 95*, 193–203.

Briddon, R. W. (2003). Cotton leaf curl disease, a multicomponent begomovirus complex. *Molecular Plant Pathology, 4*(6), 427-434.

Briddon, R. W., Bull, S. E., Amin, I et al. (2003) Diversity of DNA beta, a satellite molecule associated with some monopartite begomoviruses. *Virology, 312*, 106-121.

Briddon, R. W., Bull, S. E., Amin, I., Mansoor, S., Bedford, I. D., Rishi, N., Siwatch, S. S., Zafar, Y., Abdel-Salam, A. M. & Markham, P. G. (2004). Diversity of DNA 1: a satellite-like molecule associated with monopartite begomovirus-DNA-β complexes. *Virology*, *324*, 462-474.

Brown, J. K. (2000). Molecular markers for the identification and global tracking of whitefly vector-*Begomovirus* complexes. *Virus Research*, *71*, 233-260.

Chakraborty, S., Pandey, P. K., Banerjee, M. K., Kalloo, G. & Fauquet, C. M. (2003a). *Tomato leaf curl Gujarat virus*, a new Begomovirus species causing a severe leaf curl disease of tomato in Varanasi, India. *Phytopathology*, *93*(12), 1485-1495.

Chakraborty, S., Pandey, P. K., Banerjee, M. K., Kalloo, G. & Fauquet, C. M. (2003b). A new begomovirus species causing tomato leaf curl disease in Varanasi, India. *Plant Disease*, *87*(3), 313.

Chakraborty, S., Vanitharani, R., Chattopadhyay, B. & Fauquet, C. M. (2008). Supervirulent pseudorecombination and asymmetric synergism between genomic components of two distinct species of begomovirus associated with severe tomato leaf curl disease in India. *Journal of General Virology*, *89*(3), 818-828.

Chatchawankanphanich, O. & Maxwell, D.P. (2002). *Tomato leaf curl Karnataka virus* from Bangalore, India, appears to be a recombinant begomovirus. *Phytopathology*, *92*(6), 637-645.

Chatterjee, A. & Ghosh, S. K. (2007a). A new monopartite Begomovirus isolated from *Hibiscus cannabinus* L. in India. *Archives of Virology*, *152*(11), 2113-2118.

Chatterjee, A. & Ghosh, S. K. (2007b). Association of a satellite DNA beta molecule with mesta yellow vein mosaic disease. *Virus Genes*, *35*(3), 835-844.

Chatterjee, A., Roy, A. & Ghosh, S. K. (2007). First record of a begomovirus associated with yellow vein mosaic disease of *Urena lobata* in India. *Australasian Plant Disease Notes*, *2*(1), 27-28.

Chatterjee, A., Roy, A., Padmalatha, K. V., Malathi, V. G. & Ghosh, S. K. (2005). Occurrence of a Begomovirus with yellow vein mosaic disease of mesta (*Hibiscus cannabinus* and *Hibiscus sabdariffa*). *Australasian Plant Pathology*, *34*(4), 609-610.

Chattopadhyay, B., Singh, A. K., Yadav, T., Fauquet, C. M., Sarin, N. B. & Chakraborty, S. (2008). Infectivity of the cloned components of a begomovirus: DNA beta complex causing chilli leaf curl disease in India. *Archives of Virology*, *153*(3), 533-539.

Das, S., Ghosh, R., Paul, S., Roy, A. & Ghosh, S. K. (2008). Complete nucleotide sequence of a monopartite begomovirus associated with yellow vein mosaic disease of mesta from north India. *Archives of Virology*, *153*(9), 1791-1796.

Das, S., Roy, A., Ghosh, R., Paul, S., Acharyya, S. & Ghosh, S. K. (2008). Sequence variability and phylogenetic relationship of betasatellite isolates associated with yellow vein mosaic disease of mesta in India. *Virus Genes*, *37*(3), 414-424.

Dutt, N., Briddon, R. W. & Dasgupta, I. (2005). Identification of a second begomovirus, Sri Lankan cassava mosaic virus, causing cassava mosaic disease in India. Archives-of-*Virology*,*150*(10), 2101-2108.

Fauquet, C. M. & Stanley, J. (2005). Revising the way we conceive and name viruses below the species level: a review of geminivirus taxonomy calls for new standardized isolate descriptors. *Archives of Virology*, *150*, 2151-2179.

Fauquet, C. M., Bisaro, D. M., Briddon, R. W., Brown, J. K., Harrison, B. D., Rybicki, E. P., Stenger, D. C. & Stanley, J. (2003). Revision of taxonomic criteria for species demarcation in the family Geminiviridae, and an updated list of begomovirus species. *Archives of Virology*,*148*, 405-421.

Fauquet, C. M., Briddon, R. W., Brown, J. K., Moriones, E., Stanley, J., Zerbini, M. & Zhou, X. (2008). Geminivirus strain demarcation and nomenclature. *Archives of Virology. 153*, 783-821.

Garcia-Andres, S., Monci, F., Navas-Castillo, J. & Moriones, E. (2006). Begomovirus genetic diversity in the native plant reservoir *Solanum nigrum*: evidence for the presence of a new virus species of recombinant nature. *Virology, 350*, 433-442.

Garcia-Andres, S., Tomas, D. M., Sanchez-Campos, S., Navas-Castillo, J. & Moriones, E. (2007). Frequent occurrence of recombinants in mixed infections of tomato yellow leaf curl disease-associated begomoviruses. *Virology, 359*, 302-312.

Garg, I. D. (2005). Virus and virus-like diseases of potato and their management. In: R. C. Sharma. & J. N. Sharma (Eds.), Challenging problems in horticultural and forest pathology: pp. 200-222. Indus Publishing Company, New Delhi, India.

Ghosh, R., Paul, S., Das, S., Palit, P., Acharyya, S., Das, A., Mir, J. I., Ghosh, S. K. & Roy, A. (2008). Molecular evidence for existence of a New World begomovirus associated with yellow mosaic disease of *Corchorus capsularis* in India. *Australasian Plant Disease Notes, 3*(1), 59-62.

Ghosh, R., Paul, S., Roy, A., Mir, J. I., Ghosh, S. K., Srivastava, R. K. & Yadav, U. S. (2007). Occurrence of begomovirus associated with yellow vein mosaic disease of kenaf (*Hibiscus cannabinus*) in Northern India. *Plant Health Progress*, May, 1-3.

Girish, K. R. & Usha, R. (2005). Molecular characterization of two soybean-infeceting begomoviruses from India and evidence for recombination among legume-infecting begomoviruses from South-East Asia. *Virus Research, 108*(1/2), 167-176.

Hanley-Bowdoin, L., Settlage, S. B., Orozco, B. M., Nagar, S. & Robertson, D. (1999). Geminiviruses: models for plant DNA replication, transcription, and cell cycle regulation. *Critical Reviews in Plant Sciences,* 18, 1-106.

Harrison, B. D. & Robinson, D. J. (1999). Natural genomic and antigenic variation in whitefly-transmitted geminiviruses. *Annual Review of Phytopathology, 37*, 369-398.

Jeske, H., Luetgemeier, M. & Preiss, W. (2001). DNA forms indicate rolling circle and recombinationdependent replication of *Abutilon mosaic virus*. *EMBO Journal, 20*, 6158-6167.

John, P., Sivalingam, P. N., Haq, Q. M. I., Kumar, N., Mishra, A., Briddon, R. W. & Malathi, V. G. (2008). Cowpea golden mosaic disease in Gujarat is caused by a *Mungbean yellow mosaic India virus* isolate with a DNA B variant. *Archives of Virology, 153*(7), 1359-1365.

John, P., Sivalingam, P. N., Kumar, N., Mishra, A., Ahlawat, Y. S. & Malathi, V. G. (2006). A new begomovirus associated with yellow mosaic disease of *Clerodendron inerme. Plant Pathology, 55*(2), 291.

Jose, J. & Usha, R. (2003). Bhendi yellow vein mosaic disease in India is caused by association of a DNA beta satellite with a begomovirus. *Virology, 305*(2), 310-317.

Kang, S. S., Athar, M. & Cheema, S. S. (2004). Additional hosts for cotton leaf curl geminivirus. *Journal of Research, 41*(1), 85-88.

Khan, A. A., Naqvi, Q. A., Khan, M.S., Singh, R. & Raj, S. K. (2005). First report of a begomovirus infecting *Calendula* in India. *Plant Pathology, 54*(4), 569.

Khan, J. A., Siddiqui, M. K. & Singh, B. P. (2002a). The association of begomovirus with bitter melon in India. *Plant Disease, 86*(3), 328.

Khan, J. A., Siddiqui, M. K. & Singh, B. P. (2002c). The natural occurrence of a begomovirus in sunn hemp (*Crotalaria juncea*) in India. *Plant Pathology, 51*(3), 398.

Khan, J. A., Sohrab, S. S. & Aminuddin. (2003). Guar leaf curl disease from India is caused by *Tomato leaf curl virus*. *Plant Pathology, 52*(6), 796.

Khan, J. A., Sohrab, S. S., Aminuddin, Gupta, R. K. (2002b). Detection of a Begomovirus affecting guar (*Cyamopsis tetragonoloba* (L.) Taub.) in India. *Zeitschrift-fur-Pflanzenkrankheiten-und-Pflanzenschutz, 109*(1), 68-73.

Khan, M. S., Raj, S. K. & Singh, R. (2006). First report of *Tomato leaf curl New Delhi virus* infecting chilli in India. *Plant Pathology, 55*(2), 289.

Kirthi, N., Priyadarshini, C. G. P., Sharma, P., Maiya, S. P., Hemalatha, V., Sivaraman, P., Dhawan, P., Rishi, N. & Savithri, H. S. (2004). Genetic variability of begomoviruses associated with cotton leaf curl disease originating from India. *Archives of Virology, 149*(10), 2047-2057.

Kumar, Y., Hallan, V. & Zaidi, A. A. (2008). Molecular characterization of a distinct bipartite begomovirus species infecting tomato in India. *Virus Genes, 37*, 425-431.

Mahesh, Y. S., Shankarappa, K. S., Rangaswamy, K. T., Prameela, H. A., Aswathanarayana, D. S. & Maruthi, M. N. (2008). Detection of a begomovirus associated with leaf curl disease of ornamental croton (*Codieum variegatum*) in Southern India. *Journal of Plant Pathology, 90*(1), 143.

Mandal, B., Mandal, S., Sohrab, S. S., Pun, K. B. & Varma, A. (2004). A new yellow mosaic disease of chayote in India. *Plant-Pathology, 53*(6), 797.

Maruthi, M. N., Colvin, J., Briddon, R. W., Bull, S. E. & Muniyappa, V. (2003). *Pumpkin yellow vein mosaic virus*: a novel begomovirus infecting cucurbits. *Journal of Plant Pathology, 85*(1), 64-65.

Maruthi, M. N., Manjunatha, B., Rekha, A. R., Govindappa, M. R., Colvin, J. & Muniyappa, V. (2006a). *Dolichos yellow mosaic virus* belongs to a distinct lineage of Old World begomoviruses; its biological and molecular properties. *Annals of Applied Biology, 149*(2), 187-195.

Maruthi, M. N., Rekha, A. R., Govindappa, M. R., Colvin, J. & Muniyappa, V. (2006b). A distinct begomovirus causes Indian dolichos yellow mosaic disease. *Plant Pathology. 55*(2), 290.

Moffat, A. S. (1997). Geminiviruses emerge as serious crop threat. *Science, 286*, 1835.

Monci, F., Sanchez-Campos, S., Navas-Castillo, J. & Moriones, E. (2002). A natural recombinant between the geminiviruses *Tomato yellow leaf curl Sardinia virus* and *Tomato yellow leaf curl virus* exhibits a novel pathogenic phenotype and is becoming prevalent in Spanish populations. *Virology, 303*, 317-326.

Muniyappa, V., Maruthi, M. N., Babitha, C. R., Colvin, J., Briddon, R. W. & Rangaswamy, K. T. (2003). Characterisation of pumpkin yellow vein mosaic virus from India. *Annals of Applied Biology, 142*(3), 323-331.

Narayana, D. S. A., Rangaswamy, K. T., Shankarappa, K. S., Maruthi, M. N., Reddy, C. N. L., Rekha, A. R. & Murthy, K. V. K. (2007). Distinct begomoviruses closely related to cassava mosaic viruses cause Indian Jatropha mosaic disease. *International Journal of Virology, 3*(1), 1-11.

Narayana, D. S. A., Shankarappa, K. S., Govindappa, M. R., Prameela, H. A., Rao, M. R. G. & Rangaswamy, K. T. (2006). Natural occurrence of Jatropha mosaic virus disease in India. *Current Science, 91*(5), 584-586.

Ndunguru, J., Legg, J. P., Aveling, T. A., Thompson, G. & Fauquet, C. M. (2005). Molecular biodiversity of cassava begomoviruses in Tanzania: evolution of cassava geminiviruses in Africa and evidence for East Africa being a center of diversity of cassava geminiviruses. *Virology Journal, 2*, 21.

Padidam, M., Sawyer, S. & Fauquet, C.M. (1999). Possible emergence of new geminiviruses by frequent recombination. *Virology, 265*, 218-225.

Paul, S., Ghosh, R., Roy, A., Mir, J. I. & Ghosh, S. K. (2006). Occurrence of a DNA beta-containing begomovirus associated with leaf curl disease of kenaf (*Hibiscus cannabinus* L.) in India. *Australasian Plant Disease Notes, 1*(1), 29-30.

Peralta, I. E., & Spooner, D. M. (2005). History, origin and early cultivation of tomato (Solanaceae). In: M. K. Razdan & A. K. Mattoo (Eds.), *Genetic improvement of solanaceous crops: Vol 2*, pp. 1-24. Science Publishers, Enfield, NH.

Preiss, W. & Jeske, H. (2003). Multitasking in replication is common among geminiviruses. *Journal of Virology, 77*, 2972-2980.

Radhakrishnan, G., Malathi, V. G. & Varma, A. (2004). Detection of DNA A and DNA beta associated with cotton leaf curl and some other plant diseases caused by whitefly transmitted geminiviruses. *Indian Phytopathology, 57*(1), 53-60.

Raj, S. K., Khan, M. S., Pandey, S. K., & Singh, B. P. (2005a). Association of a begomovirus with leaf-curl disease of chilli (*Capsicum annuum*). *Indian Journal of Agricultural Sciences, 75*(6), 364-366.

Raj, S. K., Khan, M. S. & Singh, R. (2005b). Natural occurrence of a begomovirus on pigeonpea in India. *Plant Pathology, 54*(6), 809.

Raj, S. K., Khan, M. S., Snehi, S. K., Kumar, S. & Khan, A. A. (2007a). Natural occurrence of a Begomovirus on *Dimorphotheca sinuata* in India. *Australasian Plant Disease Notes, 2*(1), 25-26.

Raj, S. K., Khan, M. S., Snehi, S. K. & Roy, R. K. (2007b). Yellow vein netting of Bimili jute (*Hibiscus cannabinus* L.) in India caused by a strain of *Tomato leaf curl New Delhi virus* containing DNA beta. *Australasian Plant Disease Notes, 2*(1), 45-47.

Raj, S. K., Khan, M. S., Snehi, S. K., Srivastava, S. & Singh, H. B. (2006a). A yellow mosaic disease of soybean in northern India is caused by *Cotton leaf curl Kokhran virus*. *Plant Disease, 90*(7), 975.

Raj, S. K., Khan, M. S., Snehi, S. K., Srivastava, S. & Singh, H. B. (2006b) First report of *Tomato leaf curl Karnataka virus* infecting soybean in India. *Plant Pathology, 55*(6), 817.

Raj, S. K., Snehi, S. K., Kumar, S., Khan, M. S. & Pathre, U. (2008a). First molecular identification of a begomovirus in India that is closely related to *Cassava mosaic virus* and causes mosaic and stunting of *Jatropha curcas* L. *Australasian Plant Disease Notes, 3*(1), 69-72.

Raj, S. K., Snehi, S. K., Khan, M. S., Singh, R. & Khan, A. A. (2008b). Molecular evidence for association of *Tomato leaf curl New Delhi virus* with leaf curl disease of papaya (*Carica papaya* L.) in India. *Australasian Plant Disease Notes, 3*(1), 152-155.

Raj, S. K., Snehi, S. K., Khan, M. S., Chandra, P. & Pandey, R. M. (2008c). First molecular identification of a begomovirus associated with yellow vein net disease in grain amaranth (*Amaranthus cruentus* L.) in India. *Australasian Plant Disease Notes, 3*(1), 129-131.

Rajinimala, N. & Rabindran, R. (2007). First report of Indian cassava mosaic virus on bittergourd (*Momordica charantia*) in Tamil Nadu, India. *Australasian Plant Disease Notes, 2*(1), 81-82.

Reddy, R. V. C., Colvin, J., Muniyappa, V. & Seal, S. (2005a). Diversity and distribution of begomoviruses infecting tomato in India. *Archives of Virology, 150*(5), 845-867.

Reddy, R. V. C., Muniyappa, V., Colvin, J. & Seal, S. (2005b). A new begomovirus isolated from *Gossypium barbadense* in southern India. *Plant Pathology, 54*(4), 570.

Rothenstein, D., Haible, D., Dasgupta, I., Dutt, N., Patil, B. L. & Jeske, H. (2006). Biodiversity and recombination of cassava-infecting begomoviruses from southern India. *Archives of Virology, 151*(1), 55-69.

Rouhibakhsh, A. & Malathi, V. G. (2005). Severe leaf curl disease of cowpea-a new disease of cowpea in northern India caused by *Mungbean yellow mosaic India virus* and a satellite DNA beta. *Plant Pathology, 54*(2), 259.

Sambyal, P. S., Bhardwaj, S. V., Malathi, V. G., Thakur, P. D. & Chakrabarti SK (2006). A Geminivirus Infecting Tomato (*Lycopersicon esculentum* Mill.) in Himachal Pradesh. In: D. K. Kishore, S. K. Sharma & K. K. Pramanick (Eds.), *Temperate Horticulture: Current Scenario: xxii, pp 467-470.* New India Publishing Agency, New Delhi, India.

Saunders, K., Salim, N.,Mali, V. R., Malathi, V. G., Briddon, R. W., Markham, P. G. & Stanley, J. (2002). Characterisation of *Sri Lankan cassava mosaic virus* and *Indian cassava mosaic virus*: evidence for acquisition of a DNA B component by a monopartite begomovirus. *Virology*, 2002, 293(1), 63-74.

Seal, S. E., VandenBosch, F. & Jeger, M. J. (2006). Factors Influencing Begomovirus Evolution and Their Increasing Global Significance: Implications for Sustainable Control. *Critical Reviews in Plant Sciences, 25*, 23-46.

Senanayake, D. M. J. B., Mandal, B., Lodha, S. & Varma, A. (2007). First report of Chilli leaf curl virus affecting chilli in India. *Plant Pathology, 56*(2), 343.

Shahnaz, E. & Kumar, K. (2004). Effect of leaf curl disease on yield of tomato. *Progressive Horticulture, 36*(1), 155-156.

Sharma, O. P., Sharma, P. N. & Rana, R. (2001). Prevalence and distribution of Ageratum yellow mosaic disease in Himachal Pradesh. *Himachal Journal of Agricultural Research, 27*, 46-50.

Sharma, O. P., Sharma, P. N., Rana, R. & Garg, I. D. (2007). Identification and characterization of virus associated with yellow mosaic disease of *Ageratum* in Himachal Pradesh. *Indian Journal of Virology, 18*(2), 89-92.

Sharma, P. & Rishi, N. (2004). Influence of weather variables on the incidence of cotton leaf curl virus disease. *Indian Journal of Plant Protection. 32*(1), 160-161.

Shih, S. L., Tsai, W. S., Green, S. K. & Singh, D. (2007). First report of *Tomato leaf curl Joydebpur virus* infecting chilli in India. *Plant Pathology, 56*(2), 341.

Singh, A. K., Chattopadhyay, B., Pandey, P. K., Singh, A. K., Chakraborty, S. (2007). A new *Begomovirus* species causing leaf curl disease of radish in India. *Plant Disease, 91*(8), 1053.

Singh, R., Raj, S. K. & Chandra, G. (2001). Association of a monopartite begomovirus with yellow mosaic disease of pumpkin (*Cucurbita maxima*) in India. *Plant Disease, 85*(9), 1029.

Singh, R., Raj, S. K. & Prasad, V. (2008). Molecular characterization of a strain of *Squash leaf curl China virus* from North India. *Journal of Phytopathology,156*(4), 222-228.

Sohrab, S. S., Mandal, B., Ali, A. & Varma, A. (2006). Molecular diagnosis of emerging begomovirus diseases in cucurbits occurring in northern India. *Indian Journal of Virology*, 2006, 17(2), 88-95.

Sohrab, S. S., Mandal, B., Pant, R. P. & Varma, A. (2008). First report of association of *Tomato leaf curl virus*-New Delhi with Yellow Mosaic Disease of *Luffa cylindrica* in India. *Plant Disease, 87*(9), 1148.

Umaharan, P., Padidam, M., Phelps, R. H., Beachy, R. N. & Fauquet, C. (1998). Distribution and diversity of geminiviruses in Trinidad and Tobago. *Phytopathology, 88*, 1262-1268.

Usharani, K. S., Surendranath, B., Haq, Q. M. R. & Malathi, V. G. (2004a). Yellow mosaic virus infecting soybean in northern India is distinct from the species infecting soybean in southern and western India. *Current Science, 86*(6), 845-850.

Usharani, K. S., Srivastava, A., Padmalatha, K. V. & Malathi, V. G. (2004b). First report of the association of a defective satellite DNA beta molecule with a bipartite genome begomovirus causing potato leaf curl disease in India. *Journal of Plant Pathology, 86*(2), 178.

Varma, A. & Malathi, V. G. (2003). Emerging geminivirus problems: a serious threat to crop production. *Annals of Applied Biology, 142*(2), 145-164.

Vasudeva, R. S. & Sam Raj, J. (1948). A leaf curl disease of tomato. *Phytopathology, 38*, 364-369.

Wyatt, S. D., & Brown ,J. K. (1996). Detection of subgroup III geminivirus isolates in leaf extracts by degenerate primers and polymerase chain reaction. *Phytopathology, 86*, 1288-1293.

In: Emerging Geminiviral Diseases and their Management ISBN 978-1-61668-620-8
Editors: P. Sharma, R. K. Gaur, et al. pp. 129-146

Chapter 6

YELLOW DWARF DISEASE COMPLEX, A MAJOR PROBLEM FOR TOMATO CROP IN EAST ASIA

***M. Ikegami*[1,*#] *and P. Sharma*[2]**

[1]Graduate School of Agricultural Science Tohoku University, Miyagi 981-8555, Japan
[2]Division of Crop Improvement, Directorate of Wheat Research, Karnal 132001, India.

ABSTRACT

Yellow dwarf disease is a very acute problem for tomato cultivation in central part of Japan. The yellow dwarf disease of tomato plants has been reported to be caused by *Tobacco leaf curl virus* (TbLCV) in 1978. After complete nucleotide sequence analysis, TbLCV was named as *Tobacco leaf curl Japan virus* (TbLCJV). Recentry *Honeysuckle yellow vein mosaic virus* (HYVMV) and TbLCJV, which are associated with DNAβ satellites were isolated from tomato plants showing yellow dwarf disease as well as diseased honeysuckle plants. In this chapter we explain the recent information of the etiology of yellow dwarf disease complex affecting tomato cultivation.

INTRODUCTION

Tomato yellow dwarf disease is a very acute problem for tomato (*Solanum lycopersicum*) cultivation in Nara Prefecture, Osaka Prefecture and Wakayama Prefecture, Japan. Since the first report of the occurrence of a yellow dwarf disease of tomato plants, caused by *Tobacco leaf curl virus* (TbLCV) in Japan by Osaki and Inouye (1978), attempts were made to characterize the causal agent based on biological and molecular properties (Ikegami et al., 1987). After complete nucleotide sequence analysis, TbLCV was named as *Tobacco leaf curl*

* Correspondence concerning this article should be addressed to: M. Ikegami, E-mail: m3ikegami@nodai.ac.jp
#Present address: NODAI Research Institute, Tokyo University of Agriculture, Tokyo 156-8502, Japan.

Japan virus (TbLCJV) (Shimizu and Ikegami, 1999). However, several begomoviruses are reported to infect tomato, honeysuckle, tobacco and *Eupatorium makinoi* in Japan (Sharma et al., 1997, 1998a; Shimizu and Ikegami, 1999). Based on vector transmission and electronic microscopic studies Osaki and associate (1979) have suggested that the causal agent of yellow vein mosaic disease of honeysuckle, a common perennial wild and ornamental plant, is TbLCJV. Ogawa et al (2009) have suggested that two begomoviruses are associated with tomato yellow dwarf disease. *Honeysuckle yellow vein mosaic virus* (HYVMV) exhibits yellowing of main and small veins followed by mosaic pattern, while *Honeysuckle yellow vein virus* (HYVV) showing yellow net symptoms accompanied by small elliptical shaped enation are typical symptoms evoked by these viruses. However, it has not been documented earlier whether HYVMV causes yellow dwarf disease on tomato plants and there were no molecular data available on the virus(es) causing yellow dwarf disease of tomato, a major vegetable crop grown in Japan. Recently, HYVMV has been isolated from tomato plants with yellow dwarf disease in the field (Kitamura et al., 2008). Knowledge of the viruses infecting economical crops and identification of weed hosts are crucial steps for management of the diseases by developing resistant cultivars. Therefore, in this chapter we explain the recent information of the etiology of yellow dwarf disease complex affecting tomato cultivation in Japan (Kitamura et al., 2008; Ogawa et al., 2008; Ito et al., 2009); (1) HYVMV and HYVMV / Honeysuckle yellow vein mosaic betasatellite [Japan:Nara:2006] (HYVMB [JR:Nar:06]) as well as TbLCJV and TbLCJV/ Tobacco leaf curl Japan betasatellite [Japan:Ibaraki:2006] (TbLCB [JR:Iba:06]) are causal agents of yellow dwarf disease on tomato plants, (2) honeysuckle is a potential reservoir weed host of these viruses, (3) three defective molecules of approximately half size in addition to the genomic DNA are found. (4) TbLCB [JR:Iba:06] and HYVMB [JR:Nar:06] are *trans*-replicated by tomato yellow leaf curl virus (TYLCV-IL[JR:Toc:07]), inducing more severe tomato yellow leaf curl disease-related symptoms than TYLCV-IL[JR:Toc:07] alone.

ETIOLOGY OF DISEASE:

Association of HYVMV-Nara1, HYVMV-Nara2, and TbLCJV-Hs[Iba] Begomoviruses with Yellow Disease of Tomato

HYVMV-Nara1 and HYVMV-Nara2, a satellite DNA (Honeysuckle yellow vein mosaic betasatellite [Japan:Nara:2006]; HYVMB-[JR:Nar:06] and defective DNAs were obtained from honeysuckle (*Lonicera japonica*) showing characteristic yellow vein mosaic symptoms in Nara Prefecture, Japan. Tobacco leaf curl Japan virus-Honeysuckle[Ibaraki] (TbLCJV-Hs[Iba]) and a satellite DNA (Tobacco leaf curl Japan betasatellite [Japan:Ibaraki:2006]; TbLCJB-[JR:Iba:06]) were also isolated from honeysuckle plants exhibited typical yellowing of veins and small elliptical shaped enations along veins on the under side of the leaves in Ibaraki Prefecture, Japan.

The complete DNA sequences of HYVMV-Nara1 and HYVMV-Nara2 were 2762 nucleotides (nts), and TbLCJV-Hs[Iba] consisted of 2763 nts, with a typical genome organization of Old World begomoviruses. A 33-base potential stem loop-forming region (5'-GCGGCCATCCGTATAATATTACCGGATGGCCGC-3') of HYMV-Nara1 was identical to those of HYVMV-Nara2 and TbLCJV-Hs[Iba]. This nucleotide sequence was also identical to those of TbLCJV and *Tobacco leaf curl Kochi virus*-[KK] (TbLCKoV-[KK]), but not to those of HYVMV. The iteron or high affinity Rep binding sites differed in HYVMV-Nara1, HYVMV-Nara2 and TbLCJV-Hs[Iba]. The HYVMV-Nara1 iteron was GGTGT (nucleotides 2667-2671, 2675-2679, and 2634-2630). The HYVMV-Nara2 iteron was GGTACC (nucleotides 2625-2630, 2661-2666, 2679-2684, 2630-2625, 2666-2661, and 2684-2679). The TbLCJV-Hs[Iba] iteron was GGTGTA (nucleotides 2669-2674, 2677-2682, and 2636-2631). Efforts to detect DNA-B components associated with HYVMV-Nara1, HYVMV-Nara2, and TbLCJV-Hs[Iba] were unsuccessful. From these results, it was concluded that HYVMV-Nara1, HYVMV-Nara2, and TbLCJV-Hs[Iba] are monopartite begomoviruses.

Association of Betasatellites with Yellow Dwarf Disease

HYVMB [JR:Nar:06] was isolated from the same diseased honeysuckle plant as HYVMV-Nara1 and HYVMV-Nara2 and TbLCJB [JR:Iba:06] from the same honeysuckle plant as TbLCJV-Hs[Iba]. HYVMB [JR:Nar:06] and TbLCJB [JR:Iba:06] consist of 1346 and 1337 nucleotides, respectively. Each DNAβ contained one complementary sense ORF (βC1). The βC1 of HYVMB [JR:Nar:06] and TbLCJB [JR:Iba:06] consisted of 351 nucleotides. HYVMB [JR:Nar:06] and TbLCJB [JR:Iba:06] have 83% nucleotide sequence identity. The sequences of HYVMB [JR:Nar:06] and TbLCJB [JR:Iba:06] contain an A-rich region between nucleotides 708 to 952 and 704 to 967 (61% and 64% identities within A-rich region, respectively). The TbLCJB [JR:Iba:06] iteron (GGTGTA) was occurring at nucleotides 1094-1099. The HYVMB [JR:Nar:06] iteron was GGTACC (nucleotides 1104-1110, and 1110-1104) and GGTGTA (1094-1099, and 1076-1072). The HYVMV-Nara1 iteron (GGTGT) was found in HYVMB [JR:Nar:06] and TbLCJB [JR:Iba:06]. The HYVMV-Nara2 iteron (GGTACC) was found in HYVMB [JR:Nar:06] but not found in TbLCJB [JR:Iba:06]. The TbLCJV-Hs[Iba] iteron (GGTGTA) was found in both HYVMB [JR:Nar:06] and TbLCJB [JR:Iba:06] .

Phylogenetic Analysis Accordingly to Begomoviruses

The complete nucleotide sequences of HYVMV-Nara1, HYVMV-Nara2, and TbLCJV-Hs[Iba] were compared with those of other begomoviruses in Japan (Table 1). HYVMV-Nara1 shared 94% nucleotide identity with HYVMV-Nara2 and also showed a high identity to isolates of HYVMV and TbLCJV, *Honeysuckle yellow vein Kobe virus* (HYVKoV) at 90-92%. The highest identity was found to HYVMV-[Mae] and HYVMV-[Uki] at 92%. HYVMV-Nara2 had total nucleotide sequence identities with TbLCJV of 93%, as compared with HYVMV of 92%, HYVMV-[Uki] and HYVMV-[Yam] of 90%. Thus, HYVMV-Nara1 and HYVMV-Nara2 have high nucleotide sequence identities with TbLCJV and HYVMV. Amino acid sequence of ORF V1 (coat protein) of HYVMV-Nara1 had amino acid sequence identities with HYVMV-Nara2 of 100%, HYVMV of 99% and TbLCJV of 88%.

Table 1. Total nucleotide sequence identities (%) among begomoviruses infecting tomato, honeysuckle, eupatorium and tobacco from Japan (Ogawa et al., 2008)

virus	TbLCJV	TbLCJV-[JP2]	TbLCJV-[JP3]	TbLCJV-Hs[Iba]	HYVMV-Nara1	HYVMV-Nara2	HYVKoV-[Hy12]	TbLCKoV-[KK]	HYVMV	HYVMV-[Mae]	HYVMV-[Uki]	HYVMV-[Yam]	HYVV-[SP1]	HYVV-[UK1]	HYVV-[UK2]	HYVKgV-[Tob:KG5]	EpYVV	EpYVV-[Tb]	EpYVV-[Yam]	EpYVV-[MNS2]	EpYVMV-[SOJ3]
TbLCJV	*																				
TbLCJV-[JP2]	**94**	*																			
TbLCJV-[JP3]	**94**	**98**	*																		
TbLCJV-Hs[Iba]	**90**	**92**	**92**	*																	
HYVMV-Nara1	**90**	**90**	**90**	**90**	*																
HYVMV-Nara2	**93**	88	89	86	**94**	*															
HYVKoV-[Hy12]	89	89	**90**	88	**90**	89	*														
TbLCKoV-[KK]	88	89	89	**90**	89	86	**90**	*													
HYVMV	88	86	86	86	**90**	**92**	88	87	*												
HYVMV-[Mae]	86	87	87	87	**92**	89	89	89	**94**	*											
HYVMV-[Uki]	87	88	88	87	**92**	**90**	89	89	**94**	**98**	*										
HYVMV-[Yam]	86	85	86	86	**90**	**90**	88	89	**92**	**91**	**91**	*									
HYVV-[SP1]	87	86	86	86	**91**	**92**	86	86	89	88	89	89	*								
HYVV-[UK1]	86	86	85	86	85	85	**90**	88	86	85	86	87	**90**	*							
HYVV-[UK2]	87	86	86	86	85	86	**91**	89	87	86	86	88	**91**	**98**	*						
HYVKgV-[Tob:KG5]	86	87	88	88	89	87	88	89	88	89	**90**	88	86	85	86	*					
EpYVV	81	81	82	81	81	81	81	82	81	80	81	82	81	82	82	81	*				
EpYVV-[Tb]	81	81	82	81	81	81	81	82	81	80	81	82	81	82	82	81	**99**	*			
EpYVV-[Yam]	81	80	81	82	81	78	81	81	81	80	81	83	81	81	82	81	**90**	**90**	*		
EpYVV-[MNS2]	81	80	81	81	81	78	81	81	81	80	80	82	81	81	82	81	**92**	**92**	**93**	*	
EpYVMV-[SOJ3]	79	79	80	80	79	78	80	80	78	78	79	78	78	78	79	79	87	87	87	88	*

The shaded part indicates the same species.

TbLCJV-Hs[Iba] had total nucleotide sequence identities with TbLCJV of 90-92%, as compared with HYVMV of 86-88%. Similarly V1 (coat protein) of TbLCJV-Hs[Iba] also showed higher amino acid sequence identities with TbLCJV isolates of 93-94% compared to those of HYVV, HYVMV and HYVKoV.

The phylogenetic tree based analysis of the complete nucleotide sequences showed that TbLCJV-Hs[Iba] clustered with TbLCJV and TbLCJV-[JP3]. HYVMV-Nara1 and HYVMV-Nara2 clustered with HYVMV, HYVMV-[Mae] and HYVMV-[Yam] clade. The phylogenetic tree based on relationship of the complete amino acid sequence of C1 showed that HYVMV-Nara1, HYVMV-Nara2 and TbLCJV-Hs[Iba] clustered with TbLCJV and TbLCJV-[JP3] and HYVKoV-[Honeysuckle12] (HYVKoV-[Hy12]) while C4 showed that HYVMV-Nara1 is clustered within TbLCJV-Hs[Iba] and HYVMV-Nara2 clustered with TbLCJV. These results suggested HYVMV-Nara1 and HYVMV-Nara2 have hybrid genomes.

AFFINITIES TO WITH OTHER BEGOMOVIRUSES AND DNAβ MOLECULES:

Phylogenetic Analysis Accordingly to Betasatellites

The phylogenetic analysis based on alignments of the corresponding nucleotide sequences of HYVMB [JR:Nar:06], and TbLCJB [JR:Iba:06] and selected DNAβmolecules associated with other begomoviruses was carried out. HYVMB [JR:Nar:06] and TbLCJB [JR:Iba:06] were closely related to DNAβ-Hy12, DNAβ-U92-2, and DNAβ-Norfolk, which were isolated from naturally infected honeysuckle plants. HYVMB [JR:Nar:06] and TbLCJB [JR:Iba:06] had high nucleotide sequence identities (83% and 79%) with tobacco leaf curl Japan betasatellite [Japan:Kochi:2006] (TbLCJB [JR:Koc:06]), which was isolated from naturally infected tomato plants in Japan. HYVMB [JR:Nar:06] and TbLCJB [JR:Iba:06] had lower nucleotide sequence identities with DNAβ (EpYVV DNAβ-m) isolated from naturally infected *Eupatorium* plants of 69 and 68%, respectively. The phylogenetic tree based on relationship of the complete amino acid sequence of βC1 product showed that HYVMB [JR:Nar:06] and TbLCJB [JR:Iba:06] clustered with DNAβ-Hy12, DNAβ-U92-2, DNAβ-Norfolk and TbLCJB [JR:Koc:06].

Association of Defective DNAs with Yellow Dwarf Disease

The 1.3 kbp HYVMV defective DNAs (Hyd-1, Hyd-3 and Hyd-4) isolated from the diseased honeysuckle plant containing HYVMV-Nara1 and HYVMV-Nara2 were sequenced and compared with those of other begomoviruses isolated from Japan. Each defective DNA molecule contained sequences from TbLCJV, TbLCJV-[JP3], HYVMV-Nara1 and HYVMV-Nara2, but large parts had been deleted and the remaining sequences had been duplicated and inverted in various ways (Table 2). The Hyd-1 and Hyd-4 iterons (GGTACC) were the same as those of HYVMV-Nara2 and TbLCJV. The Hyd-3 iteron (GGTG) was the same as that HYVMV-Nara1.

Previously, it have been reported the occurrence of defective molecules of DNA of TbLCJV in the field in Japan (Sharma et al., 1998b). This defective DNA indicated a large deletion near the coat protein gene and several rearrangements in reprication-associated protein (C1) and the intergenic region.

The defective DNAs were detected in a HYVMD infected honeysuckle plant in the field, and investigation of the sequence confirmed it to be a truncated TbLCJV, HYVMV-Nara1 and HYVMV-Nara2 DNAs. Subgenomic ss-DNA molecules of about half the size of the genomic DNA have also been detected in plants infected with other begomoviruses, and have high genome sequence identity with the respective helpers from which they derived. Sequence analysis of DNAs showed its close relationship to the TbLCJV isolates for which the overall nucleotide sequence identity is 97-100% (Table 2). The partial sequences of HYVMV-Nara1 and HYVMV-Nara2, the helper viruses of defective DNA molecules, showed high nucleotide sequence identity to the defective molecules and with other TbLCJV isolates suggesting that these defective DNA molecules are indeed derived from TbLCJV. In contrast, the defective form of CLCuV-PK DNA-A contains insertion of non-viral sequences (Liu et al., 1998).

The defective DNA molecules may prove useful in identifying the replication origins and partial encapsidation that contribute to the production of defective DNAs in monopartite begomoviruses. Besides their biological significance, these defective DNA molecules provide yet another example of recombination in begomovirus DNA.

Table 2. Homorogous regions in defective DNAs (Ogawa et al., 2008)

		Maximum matching virus		
Defective DNAs	region (nt.)	virus	region (nt.)	homology
Hyd-1	1-107	TbLCJV-[JP3]	1-107	97%
	108-333	HYVMV-Nara2	1213-988*	98%
	334-892	TbLCJV-[JP3]	1754-2312	98%
	893-985	TbLCJV-[JP3]/ HYVMV-Nara1	1214-1122/ 1212-1120*	96%
	986-1165	TbLCJV/ HYVMV-Nara2	2420-2599/ 2420-2599	99%
	1166-1208	TbLCJV/ HYVMV-Nara2	2600-2642/ 2600-2642**	100%
	1209-1370	TbLCJV	2600-2762	100%
Hyd-3	1-107	TbLCJV-[JP3]	1-107	97%
	108-333	HYVMV-Nara1	1212-987*	95%
	334-892	TbLCJV-[JP3]	1754-2312	98%
	893-985	TbLCJV-[JP3]/ HYVMV-Nara1	1214-1122/ 1212-1120*	96%
	986-1377	HYVMV-Nara1	2372-2762	99%
Hyd-4	1-107	TbLCJV-[JP3]	1-107	97%
	108-333	HYVMV-Nara2	1213-988*	98%
	334-892	TbLCJV-[JP3]	1754-2312	98%
	893-985	TbLCJV-[JP3]/ HYVMV-Nara1	1214-1122/ 1212-1120*	96%
	986-1327	TbLCJV/ HYVMV-Nara2	2420-2762/ 2420-2762	99%

*; inversion, **;duplication

Recombination

Kitamura et al. (1994) reported that TLCJV has a chimeric genome which may have arisen by recombination between TbLCJV-[Jp2] or TbLCJV-[Jp3]-like and HYVMV-like ancestor. It has been examined the possibility that HYVMV-Nara1 and HYVMV-Nara2 have chimeric genomes by using a recombination detection program (RDP) (Martin et al., 2005). The RDP analysis provided evidence of recombination between HYVMV and TbLCJV are shown in Figurc 1. Statically significant recombination events were detected between HYVMV and TbLCJV in the parts of IR, Rep gene and V2, C2, C3 ORFs. There was no significant recombination detected in TbLCJV-Hs[Iba] when similarly analyzed. Hence, from these observations, we conclude that HYVMV-Naral has a hybrid genome, with a major parent HYVMV, and the minor parents TbLCJV and HYVKoV while HYVMV-Nara2 has a hybrid genome, with a major parent HYVMV and a minor parent TbLCJV.

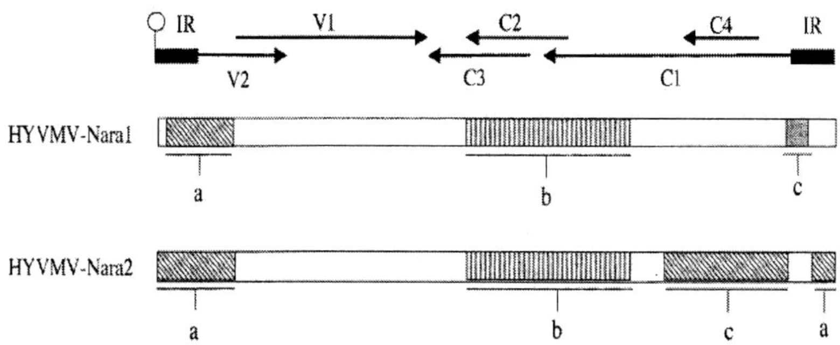

Virus	Event	region (nt.)	Origin (nt.)	*P* -value
HYVMV-Nara1	a	31-279	TbLCJV (37-312)	1.052×10^{-13}
	b	1268-1938	TbLCJV-[JP3] (1312-1991)	3.227×10^{-16}
	c	2610-2686	HYVKoV-[Hy12] (2666-2751)	1.116×10^{-17}
HYVMV-Nara2	a	2708-208	TbLCJV (2776-312)	1.903×10^{-14}
	b	1269-1939	TbLCJV-[JP3] (1312-1991)	4.660×10^{-16}
	c	2090-2605	TbLCJV (2145-2660)	9.060×10^{-23}

□ ; HYVMV-[Yam] ▨ ; TbLCJV ▥ ; TbLCJV-[JP3] ▩ ; HYVKoV-[Hy12]

Figure 1. A schematic representation of the recombinant regions in HYVMV-Naral and HYVMV-Nara2 (Ogawa *et al.*, 2008). Potential recombination break points and the probability that the indicated regions do not have a recombination origin are presented. The shadowing patterns indicate the viral genomic DNA origins of the recombinant region. The open circles indicate the stem loop-forming region. The arrows represent predicted open reading frames in both orientations ("C" for complementary sense and "V" for viral sense).

More than 133 officially recognized geminivirus species are known to infect plants (Fauquet and Stanley, 2005), and many more remain to be characterized, particularly from weed hosts, which acts as reservoirs of virus populations. It is widely accepted that recombination plays an important role in the evolution and genetic diversification of emerging begomoviral populations (Galvao et al., 2003; Kitamura et al., 2004; Kon et al., 2007; Padidam et al., 1999). HYVMV-Naral, HYVMV-Nara2 and TbLCJV-Hs[Iba] were isolated from naturally infected honeysuckle plants. When two distinct begomoviruses infect a single plant in the field, the viruses may replicate simultaneously in the same cell, which is a prerequisite for recombination (Morilla et al., 2004). Mixed infections leading to new begomovirus species in nature via recombination has been demonstrated (Chatchachawankanphanich and Maxwell, 2002; Galvao et al., 2003; Padidam et al., 1999). Here, recombination analysis using RDP program showed that the genome of HYVMV-Nara1 has a hybrid genome, with a major parent HYVMV, and the minor parents TbLCJV and HYVKoV while HYVMV-Nara2 has a hybrid genome, with a major parent HYVMV and a minor parent TbLCJV. Frequent recombination events may have increased sequence identity between HYVMV, TbLCJV and other begomoviruses in Japan. However, they probably also show high identity because they recently split from a common ancestor. Several begomovirus species that have previously been isolated and identified have arisen by recombination (Kitamura et al., 2004; Kon et al, 2007; Kirthi et al., 2002; Zhou et al., 2003). These results strongly suggest that recombination is a significant contributor to begomoviruses evolution in south East Asia.

Infectivity and Symptoms Induced by HYVMV-Nara2/HYVMB-[JR:Nar:06] and TbLCJV-Hs[Iba]/TbLCJB-[JR:Iba:06]

The clones containing partial and tandem repeats of genomic DNA and DNAβin pBI121 was constructed (Figure 2): 1.2-mer of TbLCJV-Hs[Iba] (pBHIG1.2), 2.0-mer of TbLCJB-[JR:Iba:06] (pBHIβ2.0), and 1.8-mer of HYVMV-Nara2 (pBHN2G1.8).

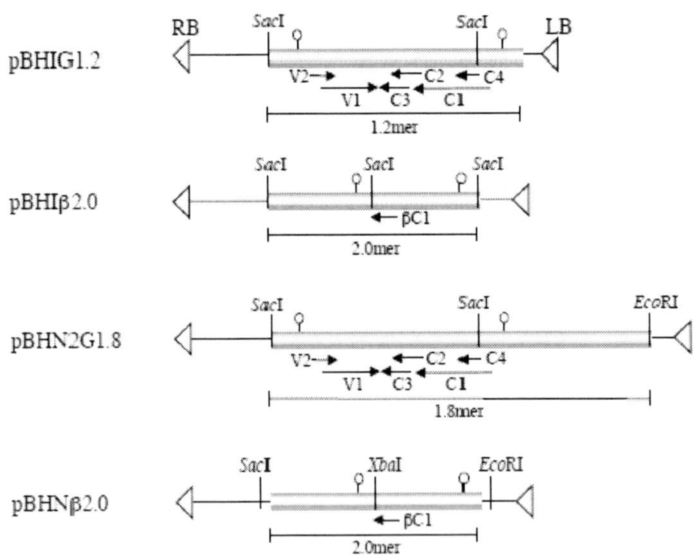

Figure 2. Construction of infectious clones of TbLCJV-Hs[Iba] (pBHIG1.2), HYVMV-Nara2 (pBHN2G1.8), TbLCJB [JR:Iba:06] (pBHIβ2.0) and HYVMB [JR:Nar:06] (pBHNβ2.0) (Ogawa *et al.*, 2008). The open circles indicate the stem- loop-forming region. The arrows represent predicted open reading frames in both orientations ("C" for complementary sense and "V" for viral sense). The restriction sites used for cloning are as follows: *Sac* I (nt 2327) on TbLCJV-Hs[Ibar] genome DNA, *Sac* I (nt 369) on TbLCJB [JR:Iba:06], *Sac* I (nt 2326) on HYVMV-Nara2 genome DNA and *Xba* I (nt 220) on HYVMB [JR:Nar:06].

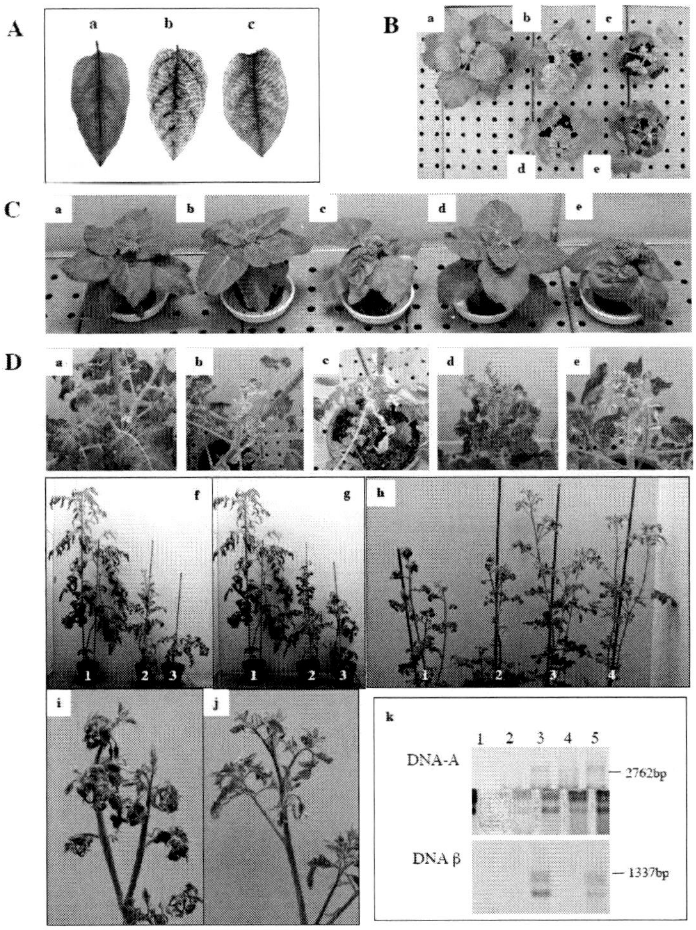

Figure 3. (A) The leaves of honeysuckle plants inoculated with (a) mock-agroinoculation; (b) TbLCJV-Hs[Iba] and TbLCJB [JR:Iba:06]; (c) HYVMV-Nara2 and HYVMB [JR:Nar:06]. (B) *N. benthamiana* plants inoculated with (a) mock-agroinoculation (21dpi);(b) TbLCJV-Hs[Iba] (21dpi); (c) TbLCJV-Hs[Iba] and TbLCJB [JR:Iba:06] (21dpi); (d) HYVMV-Nara2 (21dpi); (e) HYVMV-Nara2 and HYVMB [JR:Nar:06] (21dpi). (C) *N. glutinosa* plants inoculated with (a) mock agroinoculation (35 dpi); (b) TBLCJV-Hs[Iba] (35 dpi); (e) TbLCJV-Hs[Iba] and TbLCJB [JR:iba:06] (35 dpi); (d) HYVMV-Nara2 (35 dpi); (e) YVMV-Nara2 and HYVMB [JR:Nara;06]. (D) Tomato plants inoculated with (a) mock-agroinoculation (35dpi); (b) TbLCJV-Hs[Iba] (35dpi); (c) TbLCJV-Hs[Iba] and TbLCJB [JR:Iba:06] (35dpi); (d) HYVMV-Nara2 (35dpi); (e) HYVMV-Nara2 and HYVMB [JR:Nar:06] (35dpi); (f-1) mock-agroinoculation (63dpi.); (f-2) TbLCJV-Hs[Iba] (63dpi); (f-3) TbLCJV-Hs[Iba] and TbLCJB [JR:Iba:06] (63dpi); (g-1) mock-agroinoculation (63dpi); (g-2) HYVMV-Nara2 (63dpi); (g-3) HYVMV-Nara2 and HYVMB [JR:Nar:06] (63dpi); (h) Tomato plants inoculated with 1,2 TbLCJV-Hs[Iba) (63dpi); (3,4) HYVMV Nara2 (63dpi), (i) Tomato plants inoculated with TBLCJV-Hs[Iba] (63 dpi), (j) Tomato plants inoculated with HYVMV-Nara2;(k) Southern blot analysis of tomato plants inoculated with (1) mock-agroinoculation (35dpi); (2) TbLCJV-Hs[Iba] (35dpi); (3) TbLCJV-Hs[Iba] and TbLCJB [JR:Iba:06] (35dpi); (4) HYVMV-Nara2 (35dpi); (5) HYVMV-Nara2 and HYVMB [JR:Nar:06] (35dpi).HYVMV-Nara2 (upper gel) and HYVMB [JR:Nar:06](lower gel)were used as probes (Ogawa *et al.*, 2008).

The infectivity of genomic DNA and DNAβ (either alone or together in combination) was tested onto honeysuckle, tomato and *N. benthamiana* plants using these clones. Honeysuckle plants co-agroinoculated with HYVMV-Nara2 and HYVMB-[JR:Nar:06] showed yellow vein mosaic symptoms (Figure 3Ac, Table 3), whereas honeysuckle plants co-agroinoculated with TbLCJV-Hs[Iba] and TbLCJB-[JR:Iba:06] exhibited yellow vein and enations like symptoms on under side of leaves (Figure 3Ab, Table 3). The symptoms observed under experimental studies on honeysuckle plants were very similar to those of plants infected under field conditions.

Table 3. Infectivity and symptoms induced by HYVMV-Nara2 (pBHN2G1.8) and TbLCJV-Hs[Iba] (pBHIG1.2) with and without HYVMB-[JR:Nar:06] (pBHNβ2.0) and TbLCJB-[JR:Iba:06] (pBHIβ2.0) (Ogawa et al., 2008)

Test plant and inoculum	Infected/ inoculated[a]	Symptoms
Lonicera japonica		
TbLCJV-Hs[Iba] and TbLCJB-[JR:Iba:06]	1/2 (2)[b]	Transparent vein yellowing and small enations
HYVMV-Nara2 and HYVMB-[JR:Nar:06]	1/2 (2)	Systemic mosaic and vein yellowing
Solanum lycopersicum		
TbLCJV-Hs[Iba]	5/16 (4)	Yellowing, leaf crinking and mild stunting
TbLCJV-Hs[Iba] and TbLCJB-[JR:Iba:06]	1/34 (5)	Yellowing, leaf crinking and severe stunting
HYVMV-Nara2	15/15 (3)	Yellowing, leaf crinking and mild stunting
HYVMV-Nara2 and HYVMB-[JR:Nar:06]	5/23 (3)	Yellowing, leaf crinking and severe stunting
Nicotiana benthamiana		
TbLCJV-Hs[Iba]	16/18 (6)	Mild leaf curling, mild yellowing and mild stunting
TbLCJV-Hs[Iba] and TbLCJB-[JR:Iba:06]	15/15 (4)	Severe leaf curling, severe leaf crinking, mild yellowing and severe stunting
HYVMV-Nara2	13/13 (4)	Mild leaf curling, mild yellowing and mild stunting
HYVMV-Nara2 and HYVMB-[JR:Nar:06]	17/20 (4)	Severe leaf curling, severe leaf crinking, mild yellowing and severe stunting
TbLCJV-Hs[Iba] andHYVMB-[JR:Nar:06]	6/10 (2)	Severe leaf curling, severe leaf crinking, mild yellowing and severe stunting
HYVMV-Nara2 and TbLCJB-[JR:Iba:06]	2/12 (2)	Severe leaf curling, severe leaf crinking, mild yellowing and severe stunting

a; Number of plants infected/inoculated.
b;Numbers in parenthesis represent replication of individual experiments.

Tomato plants mock-agroinoculated with the empty binary vector mobilized into *Agrobacterium* cells did not develop any symptom (Figure 3Ca), whereas those agroinoculated with HYVMV-Nara2 and HYVMB-[JR:Nar:06] developed more severe stunting symptoms than those developed by HYVMV-Nara2 alone (Figure 3Ce and 3Cg3). Plants agroinoculated with TbLCJV-Hs[Iba] were infected systemically and produced yellowing and leaf crinkling symptoms on upper leaves and mild stunting symptoms later stages (Figure 3Cb and 3Cf2), but when co-inoculated with TbLCJB-[JR:Iba:06] produced more severe stunting symptoms, and failure of new leaves (Figure 3Cc and 4Cf3). TbLCJV-Hs[Iba]/TbLCJB-[JR:Iba:06] produced more severe dwarf symptoms than that of HYVMV-Nara2/HYVMB-[JR:Nar:06] in tomato plants (Figures 3Cf3 and 3Cg3).

Agro-inoculation of infectious clones of HYVMV-Nara2 and TbLCJV-Hs[Iba] to *N. benthamiana* resulted in mild leaf curling, mild yellowing and mild stunting of the plants (Figure 3Bd and b) followed by typical severe leaf curling, crinkling, yellowing and severe stunting when co-inoculated with HYVMB-[JR:Nar:06] and TbLCJB-[JR:Iba:06], respectively (Figure 3Bc and e). The presence of viral genomic DNA and DNAβs in infected tomato (Figure 3Ch), *N. benthamiana* and honeysuckle plants were confirmed by Southern blot and PCR analysis.

When each of virus/DNAβ complex agroinoculated onto tomato, showed typical yellow dwarf symptoms resembling those observed in tomato infested plants under natural conditions. Although genomic DNA alone is infectious and produces mild and delayed symptoms on *N. benthamiana* and tomato (Figure 3Bb, d, 3Cb, d), the association of DNAβ increases symptoms severity (Figure 3Bc, e, 3Cc, e). In both HYVMV-Nara2 and TbLCJV-Hs[Iba], genomic DNA alone was shown to be infectious, but DNAβ increased symptoms severity. We have also cloned and sequenced HYVMV (Kitamura et al., 2008) and TbLCJV with or without DNAβ from yellow dwarf diseased tomato plants in the fields (Shimizu and Ikegami, 1999; Ogawa et al., unpublished results). Using agroinoculation studies, Koch's postulates were fulfilled and the etiology of this begomovirus plant disease was resolved. Our studies demonstrate that HYVMV/DNAβ, and TbLCJV/DNAβ cause yellow dwarf disease and that honeysuckle act as a reservoir host for causal agent of tomato yellow dwarf disease.

Studies to determine whether TbLCJB-[JR:Iba:06] is *trans*-replicated by HYVMV-Nara2 and HYVMB-[JR:Nar:06] by TbLCJV-Hs[Iba] were performed using agroinculation. When TbLCJB-[JR:Iba:06] was inoculated with HYVMV-Nara2 into *N. benthamiana, it trans*-replicated and produced severe leaf curling, severe leaf crinkling, mild yellowing, and severe stunting (Table 3). When HYVMB-[JR:Nar:06] was inoculated with TbLCJV-Hs[Iba] into *N. benthamiana,* it was capable of being *trans*-replicated by the helper virus. Therefore, both DNAβ were replicated in the presence of either virus.

Geminivirus DNA satellites are dependent on respective helper begomovirus for replication, movement within plants and transmission between plants (Cui et al., 2004; Kon et al., 2006, 2007; Saunders et al, 2000, 2003, 2004). Different begomovirus genomic DNA sequences possess different iteron sequences in their IR region, which facilitates sequence-specific Rep binding to initiate rolling circle replication (Arguello-Astorga et al., 1994; Chatterji et al., 2000; Hanley-Bowdoin et al., 2000). Available evidence indicates that DNAβs does not contain iteron sequences of their helper viruses, yet they still depend on helper viral DNA for their replication (Kon et al., 2006, 2007; Jose and Usha, 2003; Saunders et al., 2000; Briddon et al., 2001). In the present case, DNAβ satellites contain the iteron of their helper viruses, but that both DNAβs were *trans*-replicated in the presence of either virus. Interestingly, in HYVMB-[JR:Nar:06], one iteron (GGTACC) is identical to that of its helper virus, HYVMV-Nara2 while second iteron (GGTGTA) is similar to the TbLCJV-Hs[Iba]. HYVMV-Nara2 contains a *Rep* gene binding motif repeat of GGTACC different from that found in TbLCJB-[JR:Iba:06]. Thus, *trans*-replication is unlikely low (Table 3) because of probable sequence incompatibility, suggesting a more relaxed specificity for begomovirus Rep binding during their replication. This is the first evidence that DNAβ satellites contain iterons in direct sequence-specific binding sites and therefore depends on helper viral DNA for *trans*-replication and are able to functionally interact with helper viral DNA component to produce severe systemic infection. However, DNAβ also appears to be capable of being replicated by the Rep proteins from a diverse range of begomoviruses (Alberter et al., 2005; Briddon et al., 2003; Mansoor et al., 2003; Saeed et al., 2005).

INTERACTION OF BEGOMOVIRUS WITH DIVERSE BETASATELLITE-CASE STUDY

Tomato yellow leaf curl diseases are a very acute problem for tomato cultivation throughout the world (Czosnek and Laterrot, 1997). Tomato yellow leaf curl virus (TYLCV), identified as the causative agent (Navot et al., 1991), is known to severely stunt tomato plants in early growth stages, in addition to causing downward cupping, discoloring, and deformation of leaves. Although TYLCV from Oman is associated with betasatellites (Khan et al., 2008), many other TYLCV strains have shown no association. On the other hand, betasatellites are associated with many monopartite begomoviruses in South East Asia (Kon et al., 2006, 2007; Matsuda et al., 2008; Ogawa et al., 2008). Therefore co-agroinoculation with TYLCV-IL [JR:Toc:07] and two betasatellites associated with tomato yellow dwarf disease in Japan (TbLCJB-[JR:Iba:06] and HYVMB-[JR:Nar:06]), as well as with Tomato leaf curl Philippines betasatellite [Philippines:Laguna1:2008] (ToLCPHB-[PH:Lag1:08]) has been carried out (Ito et al., 2009).

The infectivity and symptoms induced by 1.4-mer of DNA of TYLCV-IL [JR:Toc:07] either alone or with diverse betasatellites, 2.0-mer of HYVMB-[JR:Nar:06], 2.0-mer of TbLCJB-[JR:Iba:06] and 2.0-mer of ToLCPHB-[PH:Lag1:08] has been examined. Tomato and *N. benthamiana* plants systemically infected with TYLCV-IL[JR:Toc:07] and heterogeneous betasatellites induced more severe symptoms than those induced by TYLCV-IL[JR:Toc:07] alone. These results are summarized in Figure 4 and Table 4. Plants mock-agroinoculated with *A. tumefaciens* cells harboring the empty binary vector developed no symptoms (Figure 4Aa and 4Ba). Inoculation of tomato plants with TYLCV-IL[JR:Toc:07] alone produced symptoms of bright yellowing, small leaves and mild stunting (Figure 4Ab). In contrast, combined inoculation with TYLCV-IL[JR:Toc:07] and TbLCJB-[JR:Iba:06], HYVMB-[JR:Nar:06] or ToLCPHB-[PH:Lag1:08] produced symptoms of yellowing, severely deformed small leaves and severe stunting (Figure 4Ac - 5Ae); the combination with HYVMB-[JR:Nar:06] and ToLCPHB-[PH:Lag1:08] induced more severe symptoms than that with TbLCJB-[JR:Iba:06].

Table 4. Infectivity and symptoms induced by TYLCV-IL[JR:Toc:07] with different DNAβ components

Table 4. Infectivity and symptoms induced by TYLCV-IL[JR:Toc:07] with different DNAβ components.

Test plant and inoculum	Infectivity[a]	Symptoms
Solanum lycopersicum		
TYLCV-IL[JR:Toc:07]	29/33(3)[b]	Bright yellowing, small leaves and mild stunting
TYLCV-IL[JR:Toc:07] and TbLCJB-[JR:Iba:06]	1/22(2)	Yellowing, severe deformed small leaves and severe stunting
TYLCV-IL[JR:Toc:07] and HYVMB-[JR:Nar:06]	3/22(2)	Yellowing, severe deformed small leaves and severe stunting
TYLCV-IL[JR:Toc:07] and ToLCPHB-[PH:Lag1:08]	2/22(2)	Yellowing, severe deformed small leaves and severe stunting
Nicotiana benthamiana		
TYLCV-IL[JR:Toc:07]	7/7(2)	Mild downward leaf curling, leaf crinkling, yellowing and stunting
TYLCV-IL[JR:Toc:07] and TbLCJB-[JR:Iba:06]	11/11(2)	Downward leaf curling, severe leaf crinkling and crumpling, bright yellowing, and severe stunting
TYLCV-IL[JR:Toc:07] and HYVMB-[JR:Nar:06]	11/11(2)	Severe downward leaf curling, severe leaf crinkling and crumpling, yellowing, severe stunting, and vein swelling
TYLCV-IL[JR:Toc:07] and ToLCPHB-[PH:Lag1:08]	12/12(2)	Severe downward leaf curling, severe leaf crinkling and crumpling, yellowing, severe stunting, and vein swelling

a: Number of plants infected/number of plants inoculated.
b: Numbers in parenthesis replication of individual experiments.

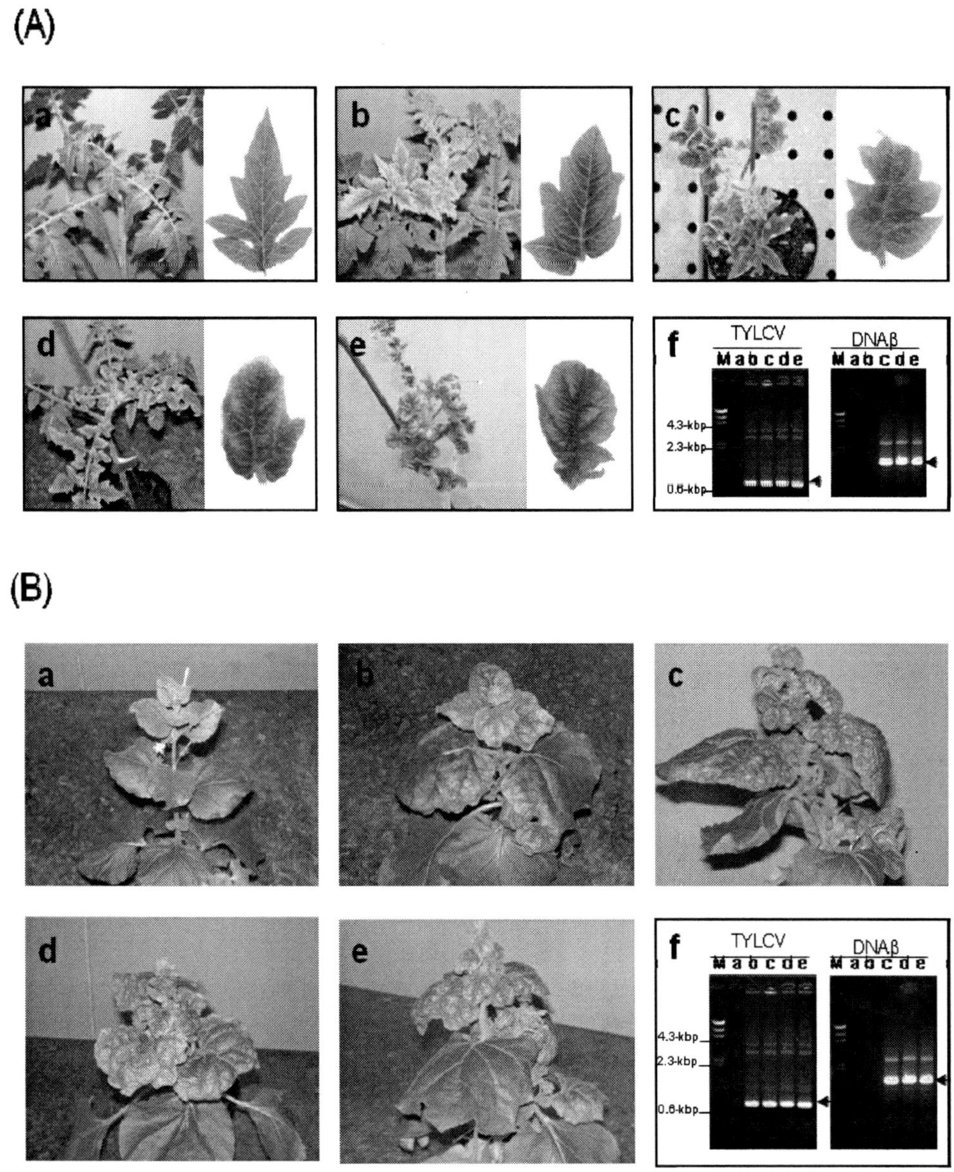

Figure 4. Symptoms of TYLCV-IL [JR:Toc:07] with different betasatellites in tomato and *N. benthamiana*. (A) The leaves of tomato plants inoculated with (a) mock-agroinoculation; (b) TYLCV-IL[JR:Toc:07] (35 dpi); (c) TYLCV-IL[JR:Toc:07] and TbLCJB-[JR:Iba:06] (35 dpi); (d) TYLCV-IL[JR:Toc:07] and HYVMB-[JR:Nar:06] (35 dpi); (e) TYLCV-IL[JR:Toc:07] and ToLCPHB-[PH:Lag1:08] (35 dpi). (f) PCR analysis of the respective inoculated tomato plants (a-e). M, λ digested by *Hin*dIII. Arrows indicate TYLCV-IL[JR:Toc:07] and DNAβ bands. (B) The leaves of *N. benthamiana* plants inoculated with (a) mock-agroinoculation; (b) TYLCV-IL[JR:Toc:07] (35 dpi); (c) TYLCV-IL[JR:Toc:07] and TbLCJB-[JR:Iba:06] (35 dpi); (d) TYLCV-IL[JR:Toc:07] and HYVMB-[JR:Nar:06] (35 dpi); (e) TYLCV-IL[JR:Toc:07] and ToLCPHB-[PH:Lag1:08] (35 dpi). (f) PCR analysis of the respective inoculated *N. benthamiana* plants (a-e). M, λ digested by *Hin*dIII. Arrows indicate TYLCV-IL[JR:Toc:07] and DNAβ bands (Ito et al., 2009).

On the other hand, inoculation of *N. benthamiana* plants with TYLCV-IL[JR:Toc:07] alone produced symptoms of downward leaf curling, leaf crinkling, yellowing and stunting (Figure 4Bb). The combined inoculation with HYVMB-[JR:Nar:06] and ToLCPHB-[PH:Lag1:08] induced more severe symptoms than that with TbLCJB-[JR:Iba:06]; that is, TYLCV-IL[JR:Toc:07] with TbLCJB-[JR:Iba:06] produced symptoms of downward leaf curling, severe leaf crinkling and crumpling, blight yellowing and severe stunting (Figure 4Bc), while TYLCV-IL[JR:Toc:07] with either HYVMB-[JR:Nar:06] or ToLCPHB-[PH:Lag1:08] produced symptoms of severe downward leaf curling, severe leaf crinkling and crumpling, yellowing, severe stunting and vein swelling (Figure 4Bd and 4Be). The presence of viral genomic DNA and DNAβ in infected tomato and *N. benthamiana* plants was confirmed by PCR analysis and Southern blotting. Taken together, TYLCV-IL[JR:Toc:07] appears to functionally interact with and trans-replicate diverse betasatellites, thereby enhancing its pathogenicity in tomato and *N. benthamiana.*

To date, TYLCV has spread across 32 prefectures in Japan and has caused significant agricultural losses in the production of tomato fruits and *Eustoma grandiflorum* flowers. However, betasatellites have not yet been detected in TYLCV-infected tomato crops. Another begomoviruses, TbLCV and HYVMV are associated with betasatellites, TbLCJB-[JR:Iba:06] and HYVMB-[JR:Nar:06], respectively, and induce yellow dwarf symptoms in tomato crops in Japan. Co-agroinoculation of tomato with TYLCV-IL[JR:Toc:07] and either HYVMB-[JR:Nar:06] or TbLCJB-[JR:Iba:06] caused more severe symptoms than agroinoculation with TYLCV-IL[JR:Toc:07] alone. Thus, co-infection of TYLCV-IL[JR:Toc:07] and HYVMB-[JR:Nar:06] and/or TbLCJB-[JR:Iba:06] into a single plant in a field leads to simultaneous replication in the same cell, most likely enhancing the pathogenisity of the virus.

It has was showed that monopartite begomoviruses not associated with betasatellites, like TYLCV-IL[JR:Toc:07], can functionally interact with heterogeneous betasatellites. The repeat sequences (iteron) in bipartite begomoviruses comprise the sequence-specific binding site for the replication-associated protein (Rep) (Arguello-Astorga et al., 1994; Chatterji et al., 2000; Hanley-Bowdoin et al., 2000). The predicted iteron present in TYLCV-IL [JR:Toc:07] (GGTGTC) was not found in HYVMB-[JR:Nar:06] (GCTACGC), TbLCJB-[JR:Iba:06] (GCTAC) or ToLCPHB-[PH:Lag1:08] (GGAGAC), suggesting a more relaxed specificity for Rep binding of this begomovirus during replication. It has been reported that two betasatellites were replicated by two distinct DNAβ-associated monopartite begomoviruses (ToLCJAV and AYVV-[Java]). However, the iterons in ToLCJAV and AYVV-[Java] were not found in these two betasatellites (Kon et al., 2006). Ageratum yellow vein China virus, another DNAβ-associated monopartite begomovirus, trans-replicated the replication of betasatellites associated with four other Chinese begomoviruses (Xiong et al., 2007). Taken together, DNAβ has the capability to use Rep from a diverse range of begomoviruses for replication. Recently, Saunders et al. (2008) have also reported that DNAβ can functionally interact with members of distinct geminivirus species. The region between the *Bgl*II^{1047} and *Bgl*II^{1146} sites of ageratum yellow vein virus (AYVV) DNAβ was important for its replication. The AYVV DNAβ sequence in this region contained an inverted repeat and multiple copies of the sequence GTCTCC and its complement GGAGAC. Although this sequence differed from the predicted AYVV iteron GGTACTCA, these DNAβ components appear to play a role in replication.

CONCLUSION

HYVMV-Nara1 and HYVMV-Nara2, a satellite DNA (honeysuckle yellow vein mosaic betasatellite [Japan:Nara:2006]; HYVMB-[JR:Nar:06]) and defective DNAs were isolated from honeysuckle (*Lonicera japonica*), a perennial ornamental and weed showing characteristic yellow vein mosaic symptoms in Nara Prefecture, Japan. Tobacco leaf curl Japan virus-Honeysuckle[Ibaraki] (TbLCJV-Hs[Iba]) and a satellite DNA (tobacco leaf curl Japan betasatellite [Japan:Ibaraki:2006]; TbLCJB-[JR:Iba:06]) were isolated and cloned from honeysuckle plants exhibited typical yellowing of veins and small elliptical shaped enations along veins on the under side of the leaves in Ibaraki Prefecture, Japan. The genome organizationes of the three viruses are the same as those of other Old World monopartite begomoviruses. HYVMV-Nara1 and HYVMV-Nara2 have hybrid genomes, which are likely to have formed recombination between HYVMV and TbLCJV. TbLCJV-Hs[Iba] or HYVMV-Nara2 could infect and cause yellowing, leaf crinkling and stunting symptoms when partial tandem dimeric constructs were agro-inoculated on tomato plants. Moreover, in the presence of DNAβ, both TbLCJV-Hs[Iba] or HYVMV-Nara2 produced more severe stunting symptoms in tomato plants. Therefore, these viruses along with their satellites are causal agents of tomato yellow dwarf disease in Japan, and honeysuckle acts as a potential reservoir host. Tomato yellow leaf curl diseases are a very acute problem for tomato cultivation throughout the world. TYLCV is the causative agent of tomato yellow leaf curl disease. TYLCV-IL[JR:Toc:07] is not associated with betasatellites. Therefore co-agroinoculation with TYLCV-IL[JR:Toc:07] and two betasatellites associated with tomato yellow dwarf disease in Japan (TbLCJB-[JR:Iba:06] and HYVMB-[JR:Nar:06]), as well as with Tomato leaf curl Philippines betasatellite [Philippines:Laguna1:2008] (ToLCPHB-[PH:Lag1:08]) has been carried out. TYLCV-IL [JR:Toc:07] trans-replicated tobacco leaf curl Japan betasatellite-[Japan:Ibaraki:2006], honeysuckle yellow vein mosaic betasatellite-[Japan:Nara:2006] and tomato leaf curl Philippines betasatellite [Philippines:Laguna1:2008], inducing more severe tomato yellow leaf curl disease-related symptoms than TYLCV-IL[JR:Toc:07] alone in the tomato and *Nicotiana benthamiana* plants.

ACKNOWLEDGMENTS

We are thankful for support received from the Grant-in-Aid for Academic Frontier Promotion program and Grant-in-Aid for Scientific Research of the Ministry of Education, Culture, Sports, Science and Technology of Japan and the Japan Society for the Promotion of Science Young Scientist Scheme. We sincerely acknowledge Dr. N, Bajet for providing the infected samples from the Philippines.

REFERENCES

Alberter, B., Rezaian, A.M. & Jeske, H. (2005). Replicative intermediates of ToLCV and its satellite DNAs. *Virology 331,* 441-448.

Arguello-Astorga, G. R., Guevara-Gonzalez, R., Herrera-Estrella, L. R. & Rivera-Bustamante, R. F. (1994). Geminivirus replication origins have a group-specific organization of iterative elements: A model for replication. *Virology 203*, 90-100.

Briddon, R. W., Bull, S. E., Amin, I., Idris, A. M., Mansoor, S., Bedford, I. D., Dhawan, P., Rishi, N., Siwatch, S. S., Abedel-Salam, A. M., Brown, J. K., Zafar, Y. & Markham, P. G. (2003). Diversity of DNAβ, a satellite molecule associated with some monopartite begomoviruses. *Virology 312*, 106-121.

Briddon, R. W., Mansoor, S., Bedford, I. D., Pinner, M. S., Saunders, K., Stanley, J., Zafar, Y., Malik, K. A. & Markham, P. G. (2001). Identification of DNA components required for induction of cotton leaf curl disease. *Virology 285*, 234-243.

Chatchachawankanphanich, O. & Maxwell, D. P. (2002). *Tomato leaf curl Karnataka virus* from Bangalore, India, appears to be a recombinant begomovirus. *Phytopathology 92*, 637-645.

Chatterji, A., Chatterji, U., Beachy, R. N., & Fauquet, C. M. (2000). Sequence parameters that determine specificity of binding of the replication-associated protein to its cognate site in two strains of *Tomato leaf curl virus*-New Delhi. *Virology 273,* 341-350.

Czosnek, H. & Laterrot, H. (1997). A worldwide survey of tomato yellow leaf curl viruses. *Arch. Virol. 142,* 1391-1406.

Fauquet, C. M. & Stanley, J. (2005). Revising the way we conceive and name viruses below the species level: a review of geminivirus taxonomy calls for new standardized isolate descriptors. *Arch. Virol. 150,* 2151-2179.

Galvao, R. M., Mariano, A. C., Lus, D. F., Alfenas, P. F., Andrade, E. C., Zerbini, F. M., Almeida, M. R. & Fontes, E. P. B. (2003). A naturally occurring recombinant DNA-A of a typical bipartite begomovirus does not require the cognate DNA-B to infect *Nicotiana benthamiana* systemically. *J. Gen. Virol. 84,* 715-726.

Hanley-Bowdoin, L., Settlage, S. B., Orozco, B. M., Nagar, S. & Robertson, D. (2000). Geminivirus: Models for plant DNA replication, transcription, and cell cycle regulation. *Crit. Rev. Biochem. Mol. Biol. 35,* 105-140.

Ikegami, M., Osaki T. & Inouye, T. (1987). Single-stranded DNA in tobacco leaf curl virus. *Ann. Phytopathol. Soc. Japan 53,* 745-751.

Ito, T., Kimbara, J., Sharma, P. & Ikegami, M. (2009). Interaction of tomato yellow leaf curl virus with diverse betasatellites enhances symptom severity. *Arch. Virol. 154,* 1233-1239.

Jose, J. & Usha, R. (2003). Bhendi yellow vein mosaic disease in India is caused by association of a DNAβ satellite with a begomovirus. *Virology 305*, 310-317.

Kirthi, N., Maiya, S. P., Murthy, M. R. N. & Savithri, H. S. (2002). Evidence for recombination among tomato leaf curl virus strains/species from Bangalore, India. *Arch. Virol. 147,* 255-272.

Kitamura, K., Murayama, A. & Ikegami, M. (2004). Evidence for recombination among isolates of *Tobacco leaf curl Japan virus* and *Honeysuckle yellow vein mosaic virus*. *Arch. Virol. 149*, 1221-1229.

Kitamura, K., Ogawa, T., Sharma, P. & Ikegami, M. (2008). First report of *Honeysuckle yellow vein mosaic virus* on tomato affected by yellow dwarf disease in Japan. *Plant Pathol. 57*, 391.

Khan, A. J., Idris, A. M., Al-Saady, N. A., Al-Mahruki, M. S., Al-Subhi, A. M. & Brown, J. K. (2008). A divergent isolate of *Tomoto yellow leaf curl virus* from Oman with an associated DNAβ satellite: an evolutionary link between Asia and the Middle Eastern virus-satellite complexes. *Virus Genes 36,* 169-176.

Kon, T., Hidayat, S. H., Hase, S., Takahashi, H. & Ikegami, M. (2006). The natural occurrence of two distinct begomoviruses associated with DNAβand a recombinant DNA in a tomato plant from Indonesia. *Phytopathology 96*, 517-525.

Kon, T., Kuwabara, K., Hidayat, S. H. & Ikegami, M. (2007). A begomovirus associated with ageratum yellow vein disease in Indonesia: evidence for natural recombination between tomato leaf curl Java virus and Ageratum yellow vein virus-[Java]. *Arch. Virol. 152*, 1147-1157.

Liu, Y., Robinson, D. J. & Harrison, B. D. (1998). Defective forms of cotton leaf curl virus DNA-A that have different combinations of sequence deletion, duplication, inversion and rearrangement. *J. Gen. Virol. 79,* 1501-1508.

Matsuda, N., Sharma, P., Bajet, N. B. & Ikegami, M. (2008). Molecular characterization of a new strain of tomato leaf curl Philippines virus and its associated satellite DNAβ molecule: further evidence for natural recombination amongst begomoviruses. *Arch. Virol. 153*, 961-967

Mansoor, S., Briddon, R. W., Zafar, Y. & Stanley, J. (2003). Geminivirus disease complexes: an emerging threat. *Trends Plant Sci. 8*, 128-134.

Martin, D. P., Williamson, C. & Posada, D. (2005). RDP2: Recombination detection and analysis from sequence alignments. *Bioinfomatics 21*, 260-262.

Morilla, G., Krenz, B., Jeske, H., Bejarano, E. R. & Wege, C. (2004). Tete a tete of Tomato yellow leaf curl virus and Tomato yellow leaf curl Sardinia virus in single nuclei. *J. Virol. 78,* 10715-10723.

Navot, N., Pichersky, E., Zeidan, M., Zamir, D & Czosnek, H. (1991). Tomato yellow leaf curl virus: A whitefly-transmitted geminivirus with a single genomic component. *Virology 185*, 151-161.

Ogawa, T., Sharma, P., & Ikegami, M. (2008). The begomoviruses Honeysuckle yellow vein mosaic virus and Tobacco leaf curl virus with DNAβ satellites cause yellow dwarf disease of tomato. *Virus Research 137*, 235-244.

Osaki, T. & Inouye, T. (1978). Resemblance in morphology and intranuclear appearance of viruses isolated from yellow dwarf diseased tomato and leaf curl diseased tobacco. *Ann. Phytopathol. Soc. Japan 44,* 167-178.

Osaki, T., Kobatake, H. & Inouye, T. (1979). Yellow vein mosaic of honeysuckle (*Lonicera japonica* Thunb.), a disease caused by tobacco leaf curl virus in Japan. *Ann. Phytopathol. Soc. Japan 45*, 62-69.

Padidam, M., Sawyer, S. & Fauquet, C. M. (1999). Possible emergence of new geminiviruses by frequent recombination. *Virology 265*, 218-225.

Saeed, M., Behjatnia, S. A., Mansoor, S., Zafar, Y., Hasnain, S. & Rezaian, M. A. (2005). A single complementary-sense transcript of geminiviral DNAβsatellite is determinant of pathogenicity. *Mol. Plant-Microbe Interact. 18*, 7-14.

Saunders, K., Bedford, I. D., Briddon, R. W., Markham, P. G., Wong, S. M. & Stanley, J. (2000). A unique virus complex causes Ageratum yellow vein disease. *Proc. Natl. Acad. Sci. USA 97,* 6890-6895.

Saunders, K., Bedford, I. D., Yahara, T. & Stanley, J. (2003). Aetiology: The earliest recorded plant virus disease. *Nature 422*, 831.

Saunders, K., Norman, A., Gucciardo, S. & Stanley, J. (2004). The DNAβsatellite component associated with ageratum yellow vein disease encodes an essential pathogenicity protein (βC1). *Virology 324,* 37-47.

Saunders, K., Briddon, R. W. & Stanley, J. (2008). Replication promiscuity of DNA-β satellites associated with monopartite begomoviruses; detection mutagenesis of the *Ageratum* yellow vein virus DNA-β satellite localizes sequences involved in replication. *J. Gen. Virol. 89*, 3156-3172.

Sharma, A., Murayama, A., Osaki, T., Ooi, K., Yahara, T. & Ikegami, M. (1997). Genotypic variability in AC1 ORF region of tobacco leaf curl geminivirus from naturally infected wild plants. *Ann. Phytopathol. Soc. Japan 63*, 298-303.

Sharma, A., Murayama, A., Osaki, T. & Ikegami, M. (1998a). Phylogenic and structural analysis of the intergenic region of tobacco leaf curl geminivirus. *Ann. Phytopath.ol Soc. Japan 64,* 187-190.

Sharma, A., Murayama, A., Osaki, T., & Ikegami, M. (1998b). Characterization of virus-specific DNA forms from tomato tissue infected by tobacco leaf curl virus: Evidence for a single genomic component producing defective DNA molecules. *Plant Pathol. 47*, 787-793.

Shimizu, S. & Ikegami, M. (1999). Complete nucleotide sequence and the genome organization of tobacco leaf curl geminivirus from Japan. *Microbiol. Immunol. 43*, 989-992.

Xiong, Q., Fan, S., Wu, J., & Zhou X (2007). Ageratum yellow vein China virus is a distinct begomovirus species associated with a DNAβ molecule. *Phytopathology 97,* 405-411

Zhou, X., Xie, Y., Tao, X., Zhang, Z., Li, Z. & Fauquet, C. M. (2003). Characterization of DNA beta associated with begomoviruses in China and evidence for co-evolution with their cognate viral DNA-A. *J. Gen. Virol. 84,* 237-247.

In: Emerging Geminiviral Diseases and their Management ISBN 978-1-61668-620-8
Editors: P. Sharma, R. K. Gaur et al. pp. 147-165

Chapter 7

THE EMERGENCE OF YELLOW LEAF CURL DISEASE IN CHILLI PEPPER IN INDONESIA

***Sri Hendrastuti Hidayat*[1], *Sri Sulandari*[2] *and Noor Aidawati*[3]**
[1]Bogor Agricultural University (IPB), Bogor 16680, West Java, Indonesia;
[2]Gadjah Mada University, Yogyakarta 55281, Indonesia;
[3]University of Lambung Mangkurat, Banjarbaru 70714, South Kalimantan, Indonesia.

ABSTRACT

Whitefly transmitted geminiviruses (*Geminiviridae: Begomovirus*) are economically important pathogens causing serious losses in food crops globally. Chilli pepper-infecting Begomoviruses cause serious problem throughout Indonesia. The disease was first reported in 1999 in West Java and has occurred ever since in all chilli pepper growing area, causing severe to complete loss of chilli pepper crops. Losses have been so extensive in some areas, such as southern part of Sumatra, that chilli pepper was not grown at that time period. Identification of the causal agent was initiated based on biological and molecular characterization. The virus was tentatively called *pepper yellow leaf curl Indonesia virus* and sweetpotato whitefly, *Bemisia tabaci* (Hemiptera: Aleyrodidae), was demonstrated as the only agent responsible for its transmission and dispersal. The epidemics of the disease was associated with the occurrence of a long and severe dry season which effcet the development of whitefly population. Since resistant chilli pepper cultivars are not available, the major control measures have primarily involved numerous applications of insecticides and modifications of cultural practices.

INTRODUCTION

Chilli pepper is an important and essential component of the daily Indonesian diet. It is also an important commercial crop grown year-round mainly by small farmers both in high and lowlands under rain-fed as well as irrigated conditions. In 2007, it was cultivated on a total area of more than 200 thousand ha producing about 1.13 million of fresh weight with an average yield of 10 ton per ha [Directorate General of Horticulture, Ministry of Agriculture Indonesia, 2008].

The most commonly-grown chilli pepper species in Indonesia are *Capsicum annuum* and *Capsicum frutescens* with fruit types belong to red pepper ("cabai besar" or "cabai keriting") and bird's eye pepper ("cabai rawit"). Farmers grow hybrid varieties or local and open-pollinated (improved) varieties, their preference varies from one location to another. In Indonesia, the majority of chilli pepper parcels are intercropped with mostly one crop.The hybrid chilli pepper was intercropped either with shallot, tomato, or cabbage, whereas local type chilli pepper was mainly intercropped with shallot [Mustafa *et al.*, 2006].

The production of chilli pepper is limited by a wide range of diseases including fungal, viral, bacterial and nematode diseases.The most important diseases reported in the last five years are anthracnose, bacterial wilt, pytophthora blight, and various viral diseases [Duriat, 1996; Taufik *et al.*, 2005]. In the late 1990s a new disease inducing symptoms involving foliar chlorosis and curling, reduced leaf size, inhibited fruit set and abnormal fruit development was observed in Central Java. Since then the symptoms have become more widespread and census data for the period from 2001 – 2003 shows that the disease has undergone a 4.6 fold increase between 2001 and 2002 and a 2.5 fold increase between 2002 and 2003 [Indonesian Ministry of Agriculture 2003, unpublished data]. Their striking appearance, bright yellow leaves is now widespread over Java, Sumatra, Kalimantan and Bali (Figure 1). Disease transmission by *Bemisia tabaci* was demonstrated by Rusli *et al.* [1999] and Sulandari *et al.* [2006] and the causal agent is known belong to Begomovirus group of Geminiviridae which tentatively named *pepper yellow leaf curl Indonesia virus* [Hidayat *et al.*, 2006; Tsai *et al.*, 2006].

(a) (b)

Figure 1. Characteristic appearance of severe infection of pepper yellow leaf curl Indonesia virus (PYLCIV) in chilli pepper fields in (a) Central Java and (b) Bali.

WHITEFLY TRANSMITTED GEMINIVIRUSES REPORTED FROM INDONESIA

Geminivirus is a family of plant viruses that contains a circular single-stranded DNA (ssDNA) genome encapsidated in geminate particles. They are classified into four genera, i.e. Mastrevirus, Curtovirus, Begomovirus, and Topocuvirus, based on their vector relationship, host range and genome organization [van Regenmortel *et al.*, 2000]. Members of the genus Begomovirus are transmitted by the whitefly *Bemisia tabaci* Genn.(Hemiptera: Aleyrodidae) and infect dicotyledonous plants. Diseases caused by whitefly-transmitted geminiviruses (WTGs) have become a major problem to crops in tropical and subtropical areas throughout the world [Idris & Brown, 1998; Samretwanich *et al.*, 2000].

Diseases caused by WTGs in Indonesia has been reported to cause a serious constraints especially those infecting tobacco [Trisusilowati *et al.*, 1990], tomato [Sukamto *et al.*, 2005; Kon *et al.*, 2006; Santoso *et al.*, 2008], chilli pepper [Sulandari *et al.*, 2006; Hidayat *et al.*, 2006; Trisno *et al.*, 2009], and also a weed species *Ageratum conyzoides* [Haerani & Hidayat, 2003; Kon *et al.*, 2007].

Tobacco Leaf Curl Disease

The first geminivirus originity from Indonesia was reported by Thung [1932 cited in originating from Trisusilowati *et al.*, 1990]. The virus caused leaf curling on tobacco plants and called *tobacco leaf curl virus* (TLCV) or known as "virus krupuk". In 1984, TLCV caused serious damage in Bojonegoro, East Java, with up to 30% disease incidence [Poerbokoesoemo 1984 cited in Trisusilowati *et al.*, 1990]. The disease may affect the yield quality, especially when the leaves were targeted for cigar wrapping. The symptoms of TLCV infection include leaf curling, vein banding, uneven leaf surface, and rigid leaves (Figure 2). Twin icosahedral particles, a unique character of geminivirus group, was associated with symptomatic leaf tissues [Trisusilowati *et al.*, 1990].

Figure 2. Symptoms of Begomovirus infection in (a) chilli pepper; (b) *Ageratum conyzoides*; (c) tomato; (d) tobacco.

The virus is easily transmitted by its insect vector, *B. tabaci*. Transmission experiment with adults *B. tabaci* emerging from viruliferous nymphs showed that insect molting (nymph

– pupa – adult) did not eliminate the virus from the insect body, therefore the transmission of TLCV was persistent and transtadial [Aidawati *et al.*, 2002].

Recent molecular analysis showed that Begomovirus associated with leaf curl disease in tobacco from East Java has the closest relationship with *ageratum yellow vein virus-Zimbabwe*, a strain of TLCV from Southern Africa [Hidayat *et al.*, 2008a].

Tomato Leaf Curl Disease

A high incidence of leaf curl disease in tomato plants in Indonesia has been observed in the last 5 years and it has become a major problem in tomato growing areas especially in West Java and Central Java [Aidawati *et al.*, 2005]. It has been known that several Begomoviruses were associated with tomato leaf curl disease in Java, Indonesia. Affected tomato plants are stunted especially when infection occurred during the early stage of plant growth. The leaves are small, curled, and sometimes chlorotic; and most of the flowers drop prematurely (Figure 2). Affected plants produce either no fruit or a few small fruits. Young plants lose vigor after infection, and rarely produce any marketable fruits. Symptoms may vary with different tomato varieties and virus isolates [Aidawati *et al.*, 2005]. Sudiono *et al.* [2001] conducted host range studies under greenhouse conditions using artificial inoculation by viruliferous *B. tabaci*. The following plant species were found to be susceptible *to tomato leaf curl virus* (ToLCV): *L. pimpinellifolium*, *Nicotiana benthamiana*, *C. annuum*, *C. frutescens*, *A. conyzoides*, *Gomphrena globosa*; whereas *Solanum nigrum*, *Cucumis sativus*, and *Phaseolus vulgaris* were found to be resistant.

Sequence analysis showed an interesting facts in which all Begomoviruses associated with tomato leaf curl disease in Java formed separate groups from those of other tomato infecting Begomoviruses. Sukamto *et al.* [2005] reported the presence of DNA B component in an isolate of tomato leaf curl virus collected from Bandung, West Java which demonstrate the property of its bipartite genome. Based on sequence comparisons and phylogenetic analysis tomato Begomoviruses from Java had a closest relationship with *ageratum yellow vein virus* (AYVV), *squash leaf curl virus* (SCLV), and also *tomato leaf curl virus* from Malaysia [Sukamto *et al.*, 2005; Kon *et al.*, 2006; Santoso *et al.*, 2008].

Ageratum Yellow Mosaic Disease

A. conyzoides is a weed species widely distributed throughout South East Asia, and frequently exhibits striking yellow vein symptoms. It has been known that the symptom is associated with *ageratum yellow vein virus* (AYVV) infection [Saunders *et al.*, 2000]. In Indonesia *A. conyzoides* is often grows near cultivated crops such as chilli pepper and tomato, and yellow vein symptom is easily observed associated with the weed (Figure 2). Specific geminivirus DNA was successfully amplified from symptomatic leaves using specific primers for geminivirus [Rojas *et al.*, 1993], and whitefly *B. tabaci* was effectively transmit the virus to tomato, chilli pepper, and tobacco [Haerani & Hidayat, 2003]. Furthermore, a Begomovirus and a satellite DNAβ component have been isolated from *A. conyzoides* L. plants with yellow vein symptoms growing in Java, Indonesia [Kon *et al.*, 2007]. The Begomovirus is most closely related to *Tomato leaf curl Java virus* (ToLCJV) (91 and 98% in the total nucleotide and coat protein amino acid sequences, respectively), although the products of ORFs C1 and C4 are more closely related to those of *Ageratum yellow vein virus*-[Java] (91 and 95% identity, respectively). For this reason, the Begomovirus is considered to be a strain of ToLCJV and is referred to as ToLCJV-Ageratum [Kon *et al.*, 2006; Sukamto *et al.*, 2005]. Therefore, it is believed that *A. conyzoides* acts as a reservoir for important Begomovirus diseases that infect tomato plants, and probably chilli pepper in Indonesia.

Pepper Yellow Leaf Curl Disease

A virus infection causing yellow mosaic, upward leaf curling, and stunting was observed infecting chilli pepper throughout production area of Bogor, West Java since 1998 and of Central Java and Jogyakarta since 2000 (Figure 2). Begomovirus association with this disease was confirmed based on transmission study using *B. tabaci* and amplification of DNA fragment using universal primers for geminivirus. Host range study showed that the virus has an intermediate host range including plants belong to the family of Solanaceae (pepper, tomato, eggplant, tobacco, and ground cherry), Leguminosae (soybean, yard long bean, mungbean, and *Crotalaria* sp.), and Compositae (sun flower, *A. conyzoides,* and *Hyptis* sp.). The species belonging to the families of Cucurbitaceae (cucumber, and watermelon), Malvaceae (cotton, and *Sida* sp), and Amaranthaceae (*Amaranthus* sp, and *Gomphrena globosa*) were resistant to the virus [Rusli *et al.*, 1999; Sulandari *et al.*, 2006].

Several field isolates associated with yellow leaf curl diseases in chilli pepper have now been identified based on their partial nucleotide sequences. Although the sequence homology among those field isolates is considered high variability in their sequence characters was evidenced [Hidayat *et al.*, 2006; Trisno *et al.*, 2009].

In this book chapter we will discuss into detail the current status of the disease in Indonesia, especially those occurred in Java, and strategy applied to reduce the disease incidence.

Present Status of Pepper Yellow Leaf Curl Indonesia Virus (PYLCIV)

Emergence of pepper yellow leaf curl disease in Indonesia was first reported in 1999 in West Java [Rusli *et al.*,1999], and in 2003 the disease had been widely spread in Java with the highest incidence and severity occurred particularly in Central Java [Sulandari *et al.*, 2006]. The emergence of the disease was thought to be associated with upsurge of whitefly population on the crops. The whitefly vector spreaded the virus very quickly throughout the chilli pepper growing area. In 2004 the disease was reported in southern part of Sumatra [Sudiono *et al.*, 2005] and an epidemic occurred in West Sumatra beginning 2005, imposing a great problem for chili pepper cultivation [Trisno *et al.*, 2009]. The disease, based on visual symptoms, is now widely spread over Java, Sumatra, and Bali, but less common in Kalimantan and Sulawesi, and has not been observed in Nusa Tenggara, Maluku and Irian Jaya (Figure 3).

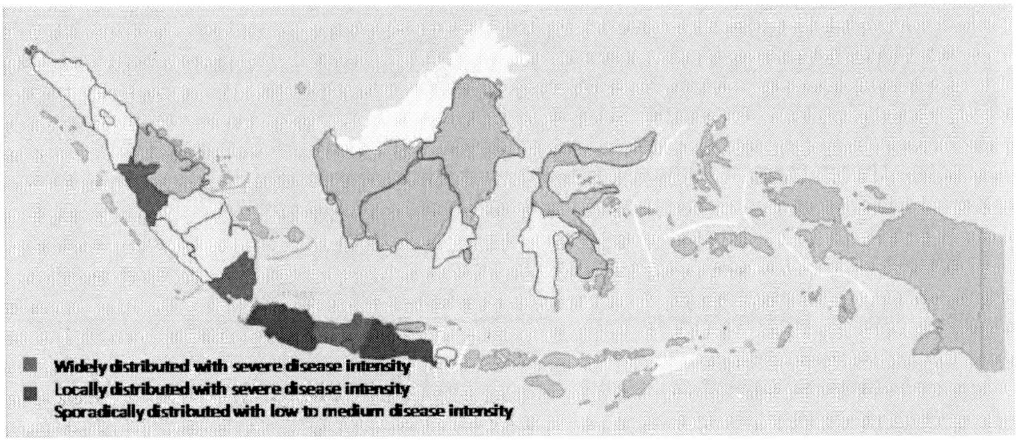

Figure 3. Distribution of pepper yellow leaf curl disease in Indonesia (Data compiled in August 2009).

Infection of WTGs was also reported on other vegetable crops such as tomato and cucurbitaceae. However, the incidence of the disease on chilli pepper was considered as the most devastating problems.This is particularly true due to economic value of chilli pepper commodities in Indonesia. Chilli pepper are an important source of cash income for small crop vegetable farmers in Indonesia. Crop surveys [Mustafa *et al.*, 2006] suggest that chilli production occupies at least 155,000 hectares and invloves in excess of 500,000 farmers throughout the nation particularly in Java and Sumatra which account for approximately 80% of the total area under production. Loses in commercial production in Central Java ranging from 20 to 100% while yield trials undertaken in 2003 recorded average crop losses in excess of 85% [De Baro *et al.*, 2008; Sumardiyono *et al.*, 2003]. Trisno *et al.* [2009] reported that yield loses in West Sumatra reached 90% during the first year of the epidemic.

Table 1. Disease incidence (%) caused by the most important viruses in chilli pepper in Indonesia based on field survey in Central Java and Yogyakarta in 2006 and 2007

Virus*	2006	2007
Cucumber mosaic cucumovirus (CMV)	19.4	27.2
Chilli veinal mottle potyvirus (ChiVMV)	27.6	12.8
Tobaco mosaic tobamovirus (TMV)	5.2	0
Pepper mild mottle potyvirus (PMMV)	16.2	0.6
Potato Y potyvirus (PVY)	33.7	0.6
Pepper yellow leaf curl begomovirus (PYLCV)	48.4	96.4

*Virus detection from field samples was conducted using ELISA or PCR.

Field survey conducted in a consecutive years in 2006 and 2007 in chilli pepper growing area in Central Java and Yogyakarta [Hidayat, unpublished data] suggests co-infection of PYLCV with other viruses (Table 1) although it was obvious that PYLCV infection was dominant over CMV, ChiVMV, TMV, PMMV, and PVY. Virus interactions during such mixed infections could play an important role leading to the development of complex diseases.

Mixed infection with two or more geminiviruses are common in tropical and subtropical areas [Torres-Pacheco *et al.*, 1996; Harrison *et al.*,1997; Sanz *et al.*, 2000; Pita *et al.*, 2001]. The type of interactions between co-infecting viruses can range from synergism where symptoms are stronger than those observed for the separate viruses [Pruss *et al.*, 1997], to interference, where one virus affects the infection cycle of the second virus [Mendez-Lozano *et al.*, 2003]. In an experiment conducted in the greenhouse, Hidayat *et al.* [2005] reported that when two isolates of PYLCV, one isolate was collected from Segunung, West Java and the other from Lembang, West Java, were inoculated together to test plants the symptoms on infected plants is slightly different but less severe than those on plants inoculated with each isolate separately. Possible interactions include also events such as complementation [Berrie *et al.*, 2001] or recombination [Saunders *et al.*, 2001; Mendez-Lozano *et al.*, 2003]. There has not been evidenced for pepper infecting Begomovirus in Indonesia for such interactions. However, Kon *et al.* [2006] successfully identified and characterized two distinct Begomoviruses, two satelite DNAβs, and a recombinant DNA from a single tomato plants with leaf curl symptoms collected from Bandung, West Java. Another example of natural recombination of two Begomoviruses has been observed with the *Tomato yellow leaf curl Sardinia virus* (TYLCSV) and TYLCV in Spain, the resulting recombinant appearing better fit than the two original isolates [Monci *et al.*, 2002]. Recombination is considered to have a crucial role in virus evolution, especially in geminivirus populations where it strongly contributes to its genetic diversification [Fondong *et al.*, 2000; Chatchawankanphanich & Maxwell, 2002; Monci *et al.*, 2002].

Polymerase chain reaction (PCR)-based procedures has been used for detection of Begomovirus from leaf samples collected from different chilli pepper growing area in Indonesia [Rojas *et al.*, 1993, Hidayat *et al.*, 2006; Trisno *et al.*, 2009]. Attempts to characterize molecular property of PYLCIV has been initiated by sequencing part of the virus genome. Partial sequences of the virus genome has been submitted to GenBank i.e. PYLCIV isolated from *L. esculentum* collected from Bandung, Lembang, West Java (PYLCIV-LBI; GenBank No. AB189845), PYLCIV isolated from *C. frutescens* collected from Segunung, Bogor, West Java (PYLCIV-Bgr; GenBank No. AB246170), and PYLCIV isolated from *C. annuum* collected from Bandung, Lembang, West Java (PYLCIV-LBI.1; GenBank No. 189850). Homology of this three strains of PYLCIV is very high: PYLCIV-Bgr has 93% and 98% homology to PYLCIV-LBI.1 and PYLCIV-LBI, respectively, whereas PYLCIV-LBI has 95% homology to PYLCIV-LBI.1 [Hidayat *et al.*, 2006]. The last two virus strains were collected from the same geographic location in West Java but from different host plants. A few sequence differences between them showed the high genetic diversity nature of the geminivirus. Genome type of the Begomoviruses causing yellow leaf curl disease in Indonesia may involve both monopartite and bipartite types. Using specifically designed primers to amplify DNA B of Asian tomato geminiviruses, DNABLC1/DNABLV2 and DNABLC2/DNABLV2, Tsai *et al.* [2006] was not able to detect the DNA B from infected chilli collected from Bogor, West Java. On the other hand, Ikegami [unpublished data] was able to amplify DNA B fragment from isolate of *pepper yellow leaf curl Indonesia virus* (PepYLCIDV) using similar primer pair (DNABLC1/DNABLV2) [Green *et al.*, 2001]. Therefore, we may expect that several distinct Begomoviruses may involve in yellow leaf curl disease in chilli pepper in Indonesia. Similar situation has been identified for tomato yellow leaf curl (ToYLCV) and tomato leaf curl (ToLC) diseases, in which the viruses were correspondingly called tomato yellow leaf curl virus (ToYLCV) or tomato leaf curl virus (ToLCV) [Nakhla & Maxwell, 1998].

Preliminary analysis on the common region of the genome of several strains of PYLCIV revealed a conservation structure of the intergenic region showing TATA sequences (TAATATTAC), repeat sequences, mostly GGAGACA in 3 positions, and stem-loop structure. However, variability in the common region is observed for the number of bases which resulted in different stem-loop structures. For instance, three strains of PYLCIV from West Sumatra have stem-loop structures consists of 31 bases which are similar with PYLCIV-Bgr, whereas the other 9 strains of PYLCIV from West Sumatra have different structures consists of 27, 29, 30, or 33 bases. Again, the variability in stem-loop structures indicates the possible genetic diversity among PYLCIV [Hidayat *et al.*, 2006, Trisno *et al.*, 2009]. Further sequence and genome analysis involving different strains of PYLCIV is being carried out to characterize the molecular nature of the virus causing devastating problems in chilli pepper in Indonesia.

TRANSMISSION AND DISPERSAL OF THE DISEASE

The vector for PYLCIV is the tobacco whitefly *Bemisia tabaci* and no means of transmission of PYLCIV other than *B. tabaci* has been found [Rusli *et al.*,1999]. Therefore, the spread of PYLCIV in the fields appears to be due entirely to the whitefly vector. The damage to chilli pepper by PYLCIV has always been directly correlated with fluctuations in the whitefly population [Yuliani, unpublished]. The interaction between PYLCIV and its vectors, *B. tabaci* has been studied extensively [Sulandari *et al.*, 2007] and summarized below.

The Role of Insect Vector in Disease Transmission

Transmission of PYLCIV was observed with one adult *B. tabaci*, but the efficiency of transmission increased almost twofold as the number of adults was increased to three per plant. Transmission of geminivirus by single whitefly has been reported earlier for TLCV on tobacco [Aidawati *et al.*, 2002], *Squash leaf curl virus* (SLCV) on squash [Cohen *et al.*, 1983], *Texas pepper virus* (TPV) on chilli pepper (Lotrakul *et al.*, 2000), and *Sinaloa tomato leaf curl virus* (STLCV) on tomato [Idris & Brown, 1998]. Maximum transmission efficiency of PYLCIV (100%) was reached with at least 10 adults per plant in the transmission experiment. Acquisition and inoculation feeding period of the whitefly to transmit the virus was identified to be optimum in the range of 3 to 6 hours. The virus needs at least 9 hours in the vector to complete latent period, with the retention period is until the insect die. The transmission efficiency will increase with longer acquisition and inoculation feeding period of the insect and the higher number of insect per plant.

PYLCIV is not transmitted through the eggs. Adults *B. tabaci* emerging from eggs of viruliferous insects were not able to cause infection. The same results were reported by Butter & Rattaul [1977], Mehta *et al.* [1994], Idris & Brown [1998], and Aidawati *et al.* [2002] with *tomato leaf curl virus*, ToYLCV, STLCV, and ToLCV, respectively.

Transmission of PYLCIV by female whiteflies was relatively higher than those by male whiteflies. When female whiteflies were introduced all test plants showed symptom of PYLCIV infection with relatively shorter incubation period, i.e. 6 – 8 d. These findings were in line with those of Cohen & Nitzany [1966], and Rathi & Nene [1974]. Transmission efficiency of female whiteflies is assumed to be correlated to their feeding behaviour. Female whiteflies tend to feed more actively to support their reproduction role. Symptom development of PYLCIV on tomato plants inoculated serially via viruliferous *B. tabaci* showed that PYLCIV persisted in the insect body through the end of the insect's life cycle.

It is now recognized that significant biological differences exist among populations of *B. tabaci* from different locations. More than 20 biotypes of *B. tabaci* has been determined based on various characters that was identified through various methods [Costa & Brown, 1991; Brown *et al.*, 1995; Bedford *et al.*, 1994]. Preliminary studies has been conducted to characterize the biotypes diversity of *B. tabaci* in Indonesia.

Biotype Determination of *B. tabaci* Populations in Indonesia

Crop damage by *B. tabaci* was first recorded in soybean in West Java, Central Java, and Southern part of Sumatra in early 1980s [Saranga, unpublished]. In the early 2000s high populations of *B. tabaci* was reported in chilli pepper, tomato, and soybean fields in West Java [Yuliani, unpublished]. Whiteflies population collected from different crops, i.e. tomato, brocolli, chillipepper, eggplant, cucumber, soybean and edamame was evaluated by Hidayat *et al.* [2008b] using silverleaf-induction test, and PCR-RAPD analysis. It was evidenced that *B. tabaci* from brocolli was able to induce silverleaf, whereas the other *B. tabaci* population was not. Genetic analysis based on PCR-RAPD characters showed there were two genetic types of *B. tabaci* i.e. *B. tabaci* biotype B (*B. tabaci* from brocolli) and *B. tabaci* biotype non B (*B. tabaci* from chillipepper, eggplant, cucumber, soybean and edamame).

Biotypes determination was also characterized recently by analyzing differences of mtCOI gene of various *B. tabaci* populations [Aidawati, unpublished]. *B. tabaci* biotype B from Indonesia (originally from broccoli) showed genetic differences compared to other *B. tabaci* biotype B from other geographic location. Four population of *B. tabaci* biotype non B (*B. tabaci* from chillipepper, cucumber, eggplant and soybean) share closed genetic similarity with those from China, Pakistan, Turkey, Malaysia, Singapore, and Thailand; whereas *B. tabaci* biotype non B from edamame showed genetic differences with other *B. tabaci* biotype non B. Most recently, De Baro *et al.* [2008] uses DNA macrosatelittes to identify geographic pattern of *B. tabaci* populations involving a *B. tabaci* from south east Asia. It was suggested that there was an invasion between 1994 and 1999 of *B. tabaci* from central Thailand into the Indonesia islands of Sumatra, then Java and Bali. The invasion is supposedly associated with the invasion of PYLCV. The capability to transmit PYLCIV could be varies among different whiteflies populations. Therefore, the distribution of different biotypes of *B. tabaci* across the country should give better understanding for the epidemic incidence of PYLC disease.

Current Management Practices for Controlling the Disease

Various methods, involving chemical application and several cultural practices, have been tested and evaluated for control of PYLC disease in research station as well as in farmers field. A satisfactory level of disease management can apparently be obtained through an integrated approach which combine several methods.

Production and use of Healthy Transplants

Chilli production in Indonesia is characterized mainly by small producers (farmers) which are planted year-round. Transplants are usually produced next to production fields although in the last few years farmers in Central Java started to buy the transplants from local nurseries belong to local growers. Ideally, transplants production should take place in an area isolated from PYLCIV infection. Unfortunately, most of the transplant producers build their screen house nearby chilli pepper production area.

A range of protective cultural strategies was performed for the production of healthy chilli pepper seedlings. The most common practices include: 1) seedbeds under low tunnels of cheesecloth or plastic polyethylene sheets; 2) open seedbeds surrounded with plastic polyethylene sheets; 3) open seedbeds treated with insecticides. Growers experienced reduction of the number of infected plants in the first week after transplanting when seedlings produced under protective practices. Avoiding the exposure of young plants to whitefly populations in the open-fields crops is believed to delay virus infection [Nakhla & Maxwell, 1998]. Early seedling stage is more susceptible than latest stages. Experiment conducted in the green house showed that the number of infection is higher in 2-leaves seedling stage than those in 6-leaves seedling stage [Ganefianti *et al.*, unpublished].

Cultural Practices

Covering the soil with plastic mulches and raised beds is common practices in chilli pepper fields. Mulch was most effective in reducing the growth of weeds which may play an important role as an alternative hosts for the virus or whiteflies. It was also reported that plastic mulches might reduce *B. tabaci* populations [Nakhla & Maxwell, 1998] and, consequently, the incidence of PYLC disease. In some chilli growing area such as Brebes, Central Java or Lembang, West Java, farmers grow chilli pepper in intercropping system with shallot or leafy vegetables, respectively. Intercropping system involving plant species, which is a good host for the vector but a non-host for the virus, might attract the vector and hence reduce the incidence of the disease [Al-Musa, 1982]. Cultural practices such as crop-free periods, altering dates, and crop rotation, have been suggested; however growers were reluctant to adopt such cultural practices that require significant changes in their conventional practices and might be time consuming.

Cultural practices intentionally employed for controlling PYLC disease is a system using border plants. A system using a border of maize and other crop plants has demonstrated some value in reducing virus incidence (Figure 4). It functions both as a physical barrier and as a reservoir for beneficial insects. Farmers grow maize, yard-long bean, *Crotalaria* sp, and other crops in the hedging, with unknown results or effectiveness. Beans have an advantage over maize, in that they create a denser wall, longer crop duration, and generate significant revenue in their own right (at about one third the unit value of chillies). The use of border plants as physical barriers does not appear to be sufficiently effective unless used in combination with insecticides or resistant cultivars. However, this practice is very popular among growers due to its capability to successfully delayed the onset of PYLC disease especially for the first 30 days after transplanting [farmers, personal communication].

(a) (b) (c)

Figure 4. Physical barrier using border plants for management of pepper yellow leaf curl disease. The most common crops grown arround the chilli pepper are: (a) maize; (b) yard long beam; (c) *Crotalaria* sp. (Pictures are courtesy of Directorate of Plant Protection, Ministry of Agriculture, Republic of Indonesia).

Breeding for Resistance

It was generally believed that the most effective and environmentally sound management practices for viral diseases would be planting of resistant cultivars. At present, no chilli pepper cultivars with resistance to PYLCIV are commercially available. Therefore, most of the chilli pepper produced in Indonesia are completely vulnerable to the indigenous Begomovirus complexes. Several commercial and university breeding programs are attempting to develop resistance to PYLCV. Some chilli pepper genotypes have been evaluated under green house condition and showed to become promising germplasms for development of cultivars with resistance or tolerance to PYLCV [P. Gniffke, AVRDC, personal communication; Ganefianti *et al.*, unpublished data].

Eleven commercial cultivars of chilli pepper were evaluated in 2005 by artificial inoculation using *B. tabaci* in screenhouse, and all of them were infected by PYLCV with disease incidence reach 100% [Sulandari *et al.*, 2006]. Incubation period of the virus in the host tissue was around 2 weeks in average, but interestingly there was a delay on symptom development in chilli pepper cultivar Bara. Disease severity in this cultivar was also mild in comparison to the other cultivars. This phenomena of delay symptom development might be an indicator for its resistance. Fourteen breeding lines were evaluated during late planting season in 2008 in the PYLC disease endemic area around Yogyakarta. Three breeding lines seem to perform better under high infection pressure compared to the others, i.e. having disease incidence less than 50% [Hidayat *et al.*, unpublished].

One of the important challenges when working with different chilli pepper genotypes was variation of symptom expression. Difficulties in determination of visual scoring for resistance, in addition to different severity of virus infection may cause conflicting results regarding the response of the chilli pepper genotypes to PYLCIV. Symptoms diversity was commonly found among different chilli pepper genotypes (Figure 5). The putative existence of different strains of PYLCIV may also contribute to variation response of chilli pepper genotypes. Three categories of response to PYLCIV infection were defined and employed for grouping genotypes response (Table 2). It is obvious that detection method is becoming a very important component in breeding program. Dot immunobinding assay and dot blot hybridization could provide a sensitive and accurate detection method [Aidawati *et al.*, 2008; Hidayat *et al.*, unpublished]. Using this molecular detection approach, infection of begomoviruses was detected in the genotypes showing no symptoms.

(a) (b) (c)

Figure 5. Symptom diversities of pepper yellow leaf curl Indonesia virus in chilli pepper: (a) yellow leaf curl; (b) bright yellow; (c) yellow mosaic.

Learning from the case of TYLCV in several countries where commercial tomato cultivars are available, it is believed that vector management is an essential complement to the use of resistant cultivars. Polston & Anderson [1997] reported that commercially available resistance to TYLCV can be overcome when moderate to high populations of viruliferous whiteflies are present and when plants are inoculated within the first few weeks in the field.

Table 2 Classification of plant response for screening resistance to *pepper yellow leaf curl Indonesia virus*

Plant Response	Symptom	Disease incidence (%)	Virus accumulation
Susceptible	Severe	>50 - 100	Easily detected
Moderate Resistant	Mild	>20 - 50	Moderately detected
Resistant	Symptomless	0 - 20	Low or Not detected

Management of Vector Populations

Two strategies most frequently implemented to protect crops from insect-transmitted viruses, including whitefly-transmitted geminiviruses, are insecticides and resistant varieties [Polston & Anderson, 1997]. Insecticides are the most often employed and most expensive approach to geminivirus management. However, control of *B. tabaci* to a level that results in a significant reduction in virus infection has been difficult. Some of the factors causing the difficulties in managing whiteflies are: (1) a wide host range which exceeds 500 species [Byrne & Bellows,1991]; (2) the presence of immature stages and adults on the underside of leaves; (3) the active mobility of adults; and (4) the ability to rapidly develop resistance to most classes of existing insecticides [Makkouk & Laterrot, 1983; Byrne & Bellows, 1991]. Several insecticides, oils, and detergents are used to reduce whitefly populations and incidence of infected plants both in transplant-production sites and in the fields. Relatively selective insecticide, imidacloprid, is the most commonly chosen synthetic insecticides by farmers. However, the incidence of PYLC disease and populations of whiteflies is still high.

Farmers in some places in Java and Sumatra occasionally make their own formulation, i.e. extracts of indigenous plant species. Although the effectiveness of these compounds in reducing whitefly populations and incidence of infected plants was unrecorded, the farmers still include their application as part of PYLC disease management practices. Experience indicates that chemical measures can at best provide only partial control of PYLC disease; . Under some circumstances, e.g. when *B. tabaci* occurred in high populations, the use of insecticides have been less successful. Therefore, in potential epidemic cases, wide scale chemical control of *B. tabaci* might be recommended for the early stages of plant growth.

Use of Plant Growth Promoting Rhizobacteria

As discussed above, management strategies for plant viruses typically include the use of resistant varieties, the integration of selected cultural practices, the application of insecticied to control insect vector, and combinations thereof [Hull, 2002]. Two other approaches to managing viruses are cross protection and development of genetically engineered plants that express a viral structural or nonstructural protein [Fitchen & Beachy,1993]. Management of viral disease can also be accomplished through induction of a plant's natural defense mechanism. An alternative method to induce plant defenses is through the use of nonpathogenic microorganism, an approach has been referred to as induced systemic resistance [van Loon *et al.*, 1998]. This approach has been applied widely for fungal and bacterial diseases, but recently it was also evaluated for viral diseases such as *tobacco mosaic virus* (TMV), *tobacco necrosis virus* (TNV), *cucumber mosaic virus* (CMV), and *tomato mottle begomoviruses* (ToMoV) [Maurhofer *et al.*, 1994; Murphy *et al.*, 2000; Raupach *et al.*, 1996]. Raupach *et al.* [1996] were the first to show that treatment of cucumber or tomato plants with plant growth-promoting rhizobacteria (PGPR) resulted in induced systemic resistance against systemic infection by CMV. Murphy *et al.* [2000] also showed that treatment of tomato plants with selected PGPR strains in some cases reduced ToMoV disease severity and incidence of ToMoV. We conducted greenhouse trial in 2005 involving 4 strains of PGPR for evaluating their potential usage in controlling pepper yellow leaf curl disease. In general, treatment of chilli pepper plants with tested PGPR strains during seedling period delayed PYLCIV symptom development (Figure 6). Incubation period of PYLCIV on chilli pepper plants treated with *B. subtilis* SB3, *P. fluorescens* PG01, and *Pf* ES32 was significantly lengthened than those control plants without any PGPR treatments. Delay in symptom development affected the disease severity in early infection period (Table 3). Treatment of PGPR caused significant reduction in disease severity in 3 to 4 weeks after inoculation. However, its effect was declining beginning in 5 weeks after inoculation. This situation suggests that PGPR-mediated resistance against a virus that systemically infects its host was temporary. Periodic application of PGPR throughout plant growth phase might be necessary for inducing resistance of the plants.

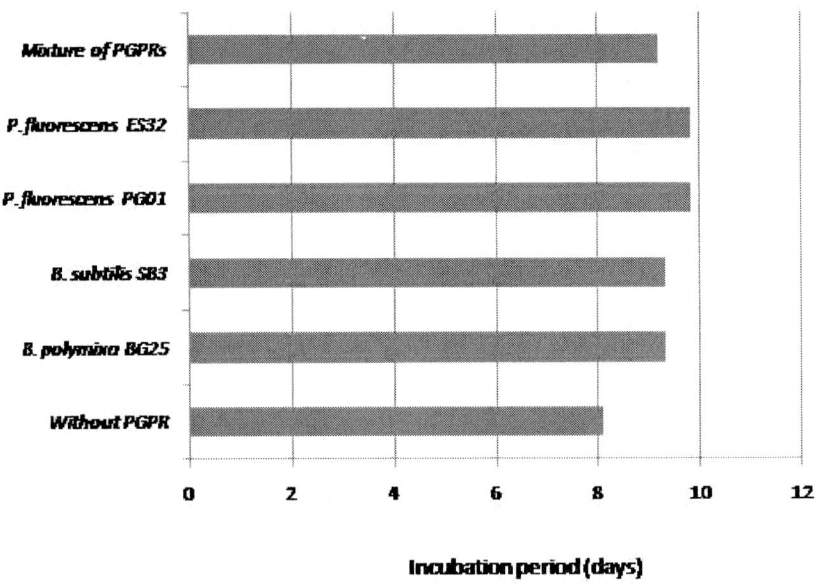

Figure 6. The effect of plant growth promoting rhizobacteria application on incubation period of *Pepper yellow leaf curl Indonesia virus*.

Conclusion

The impact of infection of *pepper yellow leaf curl Indonesia virus* in chilli pepper growing area in Indonesia in the last 5 years has become apparent. Yield loss in chilli pepper production reflect its significant impact especially for small village producers. There is no doubt for the occurrence of wider distribution and dispersal of the disease throughout the country due to climate shift in the last few years in which more area experienced long dry season. This weather condition prefers the development of whitefly population which in turn increase the possible dissemination of the virus. Application of insecticides and various cultural practices have been tested for control of PYLC disease. Although they seem to be promising, satisfactory level of disease management can apparently be obtained through an integrated approach, i.e. combining several control methods.

Table 3 Disease severity (%) based on symptom observation on 3, 4, 5, 6, 7, and 8 days post inoculation (dpi) of PYLCIV by insect vector, *B. tabaci*, following plant growth promoting rhizobacteria (PGPR) treatments [a)]

Treatment of PGPR	Disease severity (%) [b)]						
	3 dpi	4 dpi	5 dpi	6 dpi	7 dpi	8 dpi	9 dpi
Without PGPR	34,03 a	52,78 a	64,58 a	69.44 a	72,92 a	70,83 a	67,36 a
Bacillus polymixa BG25	29,17 a	41,67 b	54,86 a	61.81 a	70,14 ab	67,36 a	64,58 a
B. subtilis SB3	31,94 a	42,36 b	56,25 a	63.89 a	68,05 b	66,67 a	64,58 a
Pseudomonas fluorescens PG01	29,17 a	38,89 b	57,64 a	63.89 a	68,75 ab	66,67 a	65,28 a
P. fluorescens ES32	29,17 a	37,50 b	56,94 a	61.81 a	69,44 ab	66,67 a	65,97 a
Mixture of PGPRs	32,64 a	41,67 b	60,42 a	67.36 a	68,75 ab	65,97 a	65,28 a

[a)] PGPR was applied during seedling stage, i.e. by seed and transplants treatment;
[b)] Values followed by the same letter in similar column are not significantly different at level 0.05 according to Tukey test.

REFERENCES

Aidawati, N., Hidayat, S.H., Suseno, R., & Sosromarsono, S. (2002). Transmission of an Indonesian isolate of Tobacco leaf curl virus by *Bemisia tabaci* Genn. (Hemiptera: Aleyrodidae). *Plant Pathology Journal, 18*, 231-236.

Aidawati, N., Hidayat, S.H., Suseno, R., Hidayat, P., & Sujiprihati, S. (2005). Identification of geminivirus infecting tomato based on polymerase chain reaction-restriction fragment length polymorphism. *Jurnal Mikrobiologi Indonesia, 10*, 29-32.

Aidawati, N., Hidayat, S.H., Hidayat, P., & Sujiprihati, S. (2008). Response of Tomato Genotypes to Begomovirus Infection and Improved Diagnostic Technique for Begomovirus. *Hayati, 14*, 93-97.

Al-Musa, A. (1982). Incidence, economic importance and control of tomato yellow leaf curl in Jordan. *Plant Disease, 66*, 561-563.

Bedford, I.D., Briddon, R.W., Brown, J.K., Rosell, R.C., & Markham, P.G. (1994). Geminivirus transmission and biological characterisation of *Bemisia tabaci* (Gennadius) biotypes from different geographic regions. *Annal Applied of Biology, 125*, 311-325.

Berrie, L.C., Rybicki, E.P., & Rey, M.E.C. (2001). Complete nucleotide sequence and host range of South African cassava mosaic virus: further evidence for recombinations amongst begomoviruses. *Journal of General Virology, 82*, 53-58.

Brown, J.K., Coasts, S., Bedford, I.D., Markham, P.G., Bird, J., & Frohlich, D.R. (1995). Characterization and distribution of esterase electromorphs in the whitefly, *Bemisia tabaci* (Genn.) (Homoptera: Aleyrodidae). *Biochemical Genetics, 33*, 205-214.

Butter, N.S., & Rataul, H.S. (1977). The virus-vector relationship of the tomato leaf curl virus (TLCV) and its vector, *Bemisia tabaci* Gennadius (Homoptera ; Aleyrodidae). *Phytoparasitica, 5*, 173-186.

Byrne, D.N., & Bellows, T.S. (1991). Whitefly biology. *Annual Review of Entomology, 36*, 431-457.

Chatchawankanphanich, O., & Maxwell, D.P. (2002). Tomato leaf curl Karnataka virus from Bangalore, India, appears to be a recombinant begomovirus. *Phytopathology,* 92, 637-645.

Cohen, S., & Nitzany, F.E. (1966). Transmission and host range of the tomato yellow leaf curl virus. *Phytopathology*, *56*, 1127-1131.

Cohen, S., Duffus, J.E., Larsen, R.C., Liu, H.Y., & Flock, R.A. (1983). Purification, serology, and vector relationships of Squash leaf curl virus a whitefly transmitted geminivirus. *Phytopathology*, *3*, 1669-1673.

Costa, H.S., & Brown, J.K. (1991).Variation in biological characteristics and esterase patterns among population with silverleaf symptom induction. *Entomologica Experimenta et. Applicata*, *61*, 211-219.

De Barro, P.J., Hidayat, S.H., Frohlich, D., Subandiyah, S., & Ueda, S. (2008). A virus and its vector, pepper yellow leaf curl virus and *Bemisia tabaci*, two new invaders of Indonesia. *Biological Invasions*, *10*, 411-433.

Duriat, A.S. (1996). Cabai merah: Komoditas Prospektif dan Andalan pp 1-3. In: *Teknologi produksi cabai merah* (eds: A.S. Duriat, W.H. Widjaja, T.A. Soetiarso, & L. Prabaningrum). Balai Penelitian Tanaman Sayuran Pusat Penelitian dan Pengembangan Hortikultura, Badan Penelitian dan Pengembangan Pertanian.

Fitchen, J.H., & Beachy, R.N. (1993). Genetically engineered protection against viruses in transgenic plants. *Annual Review Microbiology*, *4*, 739-763.

Fondong, V.N., Pita, J.S., Rey, M. E. C., deKochko, A., Beachy, R.N., & Fauquet, C. M. (2000). Evidence of synergism between African cassava mosaic virus and a new double-recombinant geminivirus infecting cassava in Cameroon. *Journal of General Virology*, *81*, 287-297.

Green, S.K., Tsai, W.S., Shih, S.I., & Black, L.L. (2001). Molecular characterization of begomoviruses associated with leaf curl disease of tomato in Bangladesh, Laos, Malaysia, and Vietnam. *Plant Disease*, *85*, 1286.

Haerani, & Hidayat, S.H. (2003). Geminivirus causing yellow disease on *Ageratum conyzoides*: Molecular detection and transmission study on some solanaceae plants. *Jurnal Fitopatologi Indonesia*, *72*, 65-70.

Harrison, B.D., Zhou, X., Otim-Nape, G.W., Liu, Y., & Robinson, D.J. (1997). Role of a novel type of double infection in the geminivirus-induced epidemic of severe cassava mosaic in Uganda. *Annals of Applied Biology*, *131*, 437-448.

Hidayat, S.H., Chatchawankanpanich, O., & Aidawati N. (2008a). Molecular identification and sequence analysis of tobacco leaf curl begomovirus from Jember, East Java, Indonesia. Hayati, *15*, 13-17.

Hidayat, P., Aidawati, N., Hidayat, S.H., & Sartiami, D. (2008b). Indicator plant and PCR-RAPD for biotype determination of *Bemisia tabaci* Gennadius (*Hemiptera: Aleyrodidae*). *Jurnal HPT Tropika*, *8*, 1-7.

Hidayat, S.H., Rachmawati, R., Lestari, P., Aidawati, N. (2005). Interaksi geminivirus pada kejadian inokulasi ganda. *Jurnal Perlindungan Tanaman Indonesia*, *11*, 21-29.

Hidayat, S.H., Chatchawankanpanich, O., Rusli, E., & Aidawati, N. (2006). Begomovirus associated with pepper yellow leaf curl disease in West Java, Indonesia. *Jurnal Mikrobiologi Indonesia*, *11*, 87-90.

Hull, R. (2002). Matthew's Plant Virology 4th edition. Orlando: *Academic Press*.

Idris, A.M., & Brown, J.K. (1998). *Sinaloa tomato leaf curl geminivirus*: Biological and molecular evidence for a new subgroup III virus. *Phytopathology, 88*, 648 – 657.

Ikegami, M., Morinaga, T., & Miura, K. (198)8. Potential gene products of bean golden mosaic virus have higher sequence homologies to those of tomato golden mosaic virus than those of cassava latent virus. *Virus Genes, 1*, 191-203.

Kon, T., Hidayat, S.H., Hase, H., Takahashi, H., & Ikegami, M. (2006). The natural occurrence of two distinct begomoviruses associated with DNAβ and a recombinant DNA in a tomato plant from Indonesia. *Phytopathology, 96*, 517-525.

Kon, T., Kuwabara, K., Hidayat, S.H., & Ikegami, M. (2007). A begomovirus associated with ageratum yellow vein disease in Indonesia: evidence for natural recombination between tomato leaf curl Java virus and Ageratum yellow vein virus-[Java]. *Archives of Virology, 152*, 1147-1157.

Lotrakul, P., Valverde, R.A., de La Torre, R., Sim, J., & Gomez, A. (2000). Occurrence of a strain of Texas pepper virus in Tabasco and Habanero pepper in Costa Rica. *Plant Disease, 84*, 168 – 172.

Makkouk, K.M., & Latterot, H. (1983). Epidemiology and control of tomato yellow leaf curl virus. pp 315-321. In: Plant Virus Epidemiology (eds: R.T. Plumb and J.M. Thresh). Blackwell Scientific Publications, Oxford, UK.

Mehta, P. J., Wyman, J.A., Nakhla, M.K., & Maxwell, D.P. (1994). Transmission of tomato yellow leaf curl geminivirus by *Bemisia tabaci* (*Homoptera: Aleyrodidae*). *Journal of Economic Entomology, 87*, 1291-1297.

Mendez- Lozano, J., Torres-Pacheco, I., Fauquet, C.M., & Rivera-Bustamante, R.F. (2003). Interactions between geminiviruses in a naturally occurring mixture: Pepper huasteco virus and Pepper golden mosaic virus. *Phytopathology, 93*, 270-277.

Monci, F., Sanchez-Campos, S., Navas-Castillo, J., & Moriones, E. (2002). A natural recombinant between the geminiviruses Tomato yellow leaf curl Sardinia virus and Tomato yellow leaf curl virus exhibits a novel pathogenic phenotype and is becoming prevalent in Spanish populations. *Phytopathology, 303*, 317-326.

Maurhofer, M., Hase, C., Meuwly, P., Métraux, J.-P., & Défago, G. (1994). Induction of systemic resistance of tobacco to tobacco necrosis virus by the root-colonizing *Pseudomonas fluorescens* strain CHA0: Influence of the *gacA* gene and of pyoverdine production.*Phytopathology, 84*, 139-146.

Murphy, J.F., Zehnder, G.W., Schuster, D.J., Sikora, E. J., Polston, J. E., & Kloepper, J.W. (2000). Plant growth-promoting rhizobacterial mediated protection in tomato against *Tomato mottle virus*. *Plant Disease, 84*, 779-784.

Mustafa, U., Ali, M., & Kuswanti, H. (2006). Indonesia. pp 145-196. In: Chilli (*Capsicum* spp.) food chain analysis: Setting research priorities in Asia (ed: M. Ali). Technical Bulletin no.38. AVRDC-The World Vegetable Center, Shanhua, Tainan, Taiwan, ROC.

Nakhla, M.K., & Maxwell, D.P. (1998). Epidemiology and management of tomato yellow leaf curl disease. pp 565-583. In: *Plant Virus Disease Control* (eds: A. Hadidi, R.K. Kheterpal, and H. Koganezawa). APS Press, St.Paul, Minnesota.

Pita, J.S., Fondong, V.N., Sangare, A., Otim-Nape, G.W., Ogwal, S., & Fauquet, C. M. (2001). Recombination, pseudorecombination and synergism of geminiviruses are determinant keys to the epidemic of severe cassava mosaic disease in Uganda. *Journal of General Virology, 82*, 655-665.

Polston, J.E., & Anderson, P.K. (1997). The emergence of whitefly-transmitted geminiviruses in tomato in the western hemisphere. *Plant Disease*, *81*, 1358-1369.

Pruss, G., Ge, X., Shi, X. M., Carrington, J. C., & Vance, V.B. (1997). Plant viral synergism: The potyviral genome encodes a broad-range pathogenicity enhancer that transactivates replication of heterologous viruses. *Plant Cell*, *9*, 859-868.

Rathi, Y.P.S., & Nene, Y.L. (1974). Sex of *Bemisia tabaci* Genn. In relation to the transmission of mung bean yellow mosaic virus. *Acta Botanica Indica*, *2*, 74-76.

Raupach, G.S., Liu, L., Murphy, J.F., Tuzun, S., & Kloepper, J.W. (1996). Induced systemic resistance of cucumber and tomato against cucumber mosaic cucumovirus using plant growth-promoting rhizobacteria (PGPR). *Plant Disease*, *80*, 891-894.

Rusli, E.S., Hidayat, S.H., Suseno, R., & Tjahjono, B. (1999). Geminivirus on pepper: Symptom variation and transmission study. *Buletin Hama dan Penyakit Tumbuhan*, *11*, 126=-131.

Rojas, M.R., Gilbertson, R.L, Russel, D.R, & Maxwell, D.P. (1993). Use of degenerate primers in the polymerase chain reaction to detect whitefly-transmitted geminiviruses. *Plant Disease*, *77*, 340-347.

Saunders, K., Bedford, I.D, Briddon R.W., Markham, P.G., Wong, S.M., & Stanley, J. (2000). A unique virus complex causes Ageratum yellow vein disease. *Proceedings of the National Academic of Science*, *97*, 6890-6895.

Saunders, K., Bedford, I. D., & Stanley, J. (2001). Pathogenicity of a natural recombinant associated with Ageratum yellow vein disease: Implications for geminivirus evolution and disease aetiology. *Virology*, *282*, 38- 47.

Samretwanich, K., Chiemsombat, P., Kittipakorn, K., & Ikegami, M. (2000). *Tomato leaf curl geminivirus* associated with cucumber yellow leaf disease in Thailand. *Journal of Phytopathology*, *148*, , 615-617.

Sanchez-Campos, S., Navas-Castillo, J., Camero, R., Soria, C., Diaz, J.A., & Meriones, E. (1999). Displacement of *Tomato yellow leaf curl virus* (TYLCV)-Sr by TYLCV-Is in tomato epidemics in Spain. *Phytopathology*, *89*, 1038-1043.

Santoso, T.J., Hidayat, S.H., Duriat, A.S., Herman, M., Sudarsono. (2008). Identiity and sequence diversity of begomovirus associated with yellow leaf curl disease of tomato in Indonesia. *Microbiology Indonesia*, *2*, 1-7

Sanz, A.I., Fraile, A., Garcia-Arenal, F., Zhou, X. P., Robinson, D. J., Khalid, S., Butt, T. & Harrison, B. D. (2000). Multiple infection, recombination and genome relationships among begomovirus isolates found in cotton and other plants in Pakistan. *Journal of General Virology*, *81*, 1839-1849.

Sudiono, Hidayat, S.H., Suseno, R., & Sosromarsono, S. 2001. Molecular detection and host range study of geminivirus infecting tomato. pp 208-217 In: Proceeding of Indonesian Phytopathological Society Seminar, Bogor, Indonesia 22-24 August 2001.

Sudiono, Yasin, N., Hidayat, S.H., Hidayat, P. (2005). Penyebaran dan deteksi molekuler virus gemini penyebab penyakit kuning pada tanaman cabai di Sumatra. *Jurnal Hama dan Penyakit Tropika*, *5*, 113 – 121.

Sukamto, Kon, T., Hidayat, S.H., Ito, K., Hase, S., Takahashi, H., & Ikegami, M. (2005). Begomovirus associated with leaf curl disease of tomato in Java, Indonesia. *Journal of Phytopathology*, *153*, 562-566.

Sulandari, S., Suseno, R., Hidayat, S.H., Harjosudarmo, J., & Sosromarsono, S. (2006). Detection and host range study of virus associated with pepper yellow leaf curl disease. *Hayati*, *13*, 1-6.

Sulandari, S., Hidayat, S.H., Suseno, R., Harjosudarmo, J., & Sosromarsono, S. (2007). Transmission of pepper yellow leaf curl virus by the insect vector *Bemisia tabaci* Genn. (Hemiptera: Aleyrodidae). *Journal of ISSAAS*, *13*, 10-17.

Sumardiyono, Y.B., Sulandari, S., & Hartono, S. (2003) Epidemi Penyakit Daun Keriting Kuning Cabai. *Jurnal Perlindungan Tanaman Indonesia*, *9*, 1-3

Stenger, D. C., Duffus, J.E., & Vilalon, B. (1990). Biological and genomic properties of geminivirus isolated from pepper. *Phytopathology*, *80*, 704 – 709.

Tsai, W.S., Shih, S.L., Green, S.K., Rauf, A., Hidayat, S.H., & Jan, F.J. (2006). Molecular characterization of *pepper yellow leaf curl Indonesia virus* in leaf curl and yellowing diseased tomato and pepper in Indonesia. *Plant Disease*, 90, 247.

Taufik, M., Astuti, A.P., & Hidayat, S.H. (2005). Survei infeksi *Cucumber mosaic virus* dan *Chilli veinal mottle virus* pada tanaman cabai dan seleksi ketahanan beberapa kultivar cabai. *Agrikultura*, *16*, 146 -152.

Torres-Pacheco, I.T., Tiznado, J.A.G., Brown J.K., Becerra-Flora, A., & Bustamante, R.F.R. (1996). Detection and distribution of geminiviruses in Mexico and the Southern United States. *Phytopathology*, *86*, 1186-1192.

Trisno, J., Hidayat, S.H., Habazar, T., Manti, I., & Jamsari. (2009). Detection and sequence diversity of Begomovirus associated with yellow leaf curl disease of pepper (Capsicum annuum in West Sumatra, Indonesia. *Microbiology Indonesia*, *3*, 61-66.

Trisusilowati, E.B., Suseno, R., Sosromarsono, S., Barizi, Soedarmadi, & Nur, M.A. (1990). Transmission, serological aspects and morphology of the tobacco krupuk virus. *Indonesian Journal of Agriculture*, *2*, 75-79.

van Loon, L.C., Bakker, P.A.H.M., & Pieterse, M. J. (1998). Systemic induced resistance by rhizosphere bacteria. *Annual Review of Phytopathology*, *36*, 453-483.

van Regenmortel, M.H.V., Fauquet, C.M., Bishop, D.H.L., Carsten, E.B., Estes, M.K., Lemon, S.M., Maniloff, J., Mayo, M.A., Mc Geoch, D.J., Pringle, C.R., & Wickner, R.B. (2000). Virus Taxonomy. Classification and Nomenclature of Viruses. 7th The International Committee on Taxonomy of Viruses. San Diego: Academic Press.

In: Emerging Geminiviral Diseases and their Management ISBN 978-1-61668-620-8
Editors: P. Sharma, R. K. Gaur et al. pp. 167-181

Chapter 8

EMERGENCE OF BEGOMOVIRUS DISEASES IN CUCURBITS IN INDIA

Bikash Mandal*

Indian Agricultural Research Institute, New Delhi-110012, India.

ABSTRACT

Begomoviruses in cucurbits have emerged as a serious constraint in India over last three decades. Several begomoviruses and their variants have been reported to affect cucurbits from various countries belonging to Africa, America and Asia of which 13 are well characterised and recognized as species. Although, a great diversity of cucurbit infecting begomoviruses have been recorded in Asia, only two begomovirus species, *Squash leaf curl China virus* (synonyms, Pumpkin yellow vein mosaic virus) and *Tomato leaf curl New Delhi virus* (ToLCNDV) are known to affect cucurbits in India. Yellow vein mosaic of pumpkin, the first cucurbit begomovirus diseases was recorded in India during 1950s. Begomovirus disease problems in cucurbits emerged in India during 1980s onward. ToLCNDV has emerged as the most important begomovirus causing leaf curl, yellow mosaic/mottle and stunting diseases in numerous cucurbits in northern India.

INTRODUCTION

Cucurbits, the important group of vegetable crops under family, Cucurbitaceae are widely grown throughout India. Fruits and tender leaves/shoots of cucurbits are generally consumed as cooked vegetables and some are consumed fresh as salads and fruits. Dry seeds are used in confectionary. Cucurbits are good sources of vitamins and minerals and some are source of medicine. The family Cucurbitaceae contains 118 genera and 825 species. Cucurbits are the

* Correspondence concerning this article should be addressed to: Bikash Mandal, Plant Virology Unit, Division of Plant Pathology, Indian Agricultural Research Institute, New Delhi-110012, India. E-mail: leafcurl@rediffmail.com.

largest group of summer season vegetables grown in India. The important cucurbits that are grown in India are ash gourd (*Benincasa hispida*), bitter gourd (*Momordica charantia*), bottle gourd (*Lagenaria siceraria*), cucumber (*Cucumis sativus*), chayote (*Sechium edule*), Indian squash (*Praecitrullus fistulosus*), ivy gourd (*Coccinia cordifolia*), long melon (*Cucumis melo* var. *utilissimus*), muskmelon (*Cucumis melo*), pointed gourd (*Trichosanthes dioica*), pumpkin (*Cucurbita moschata*), sponge gourd (*Luffa cylindrica*), summer squash (*Cucurbita pepo*), snake gourd (*Trichosanthes cucumerina*), snapmelon (*Cucumis melo* var. *momordica*), ridge gourd (*Luffa acutangula*), winter squash (*Cucurbita maxima*) and watermelon (*Citrullus lanatus*) (Nayar and More, 1998). About 42.9 million hectares are under cucurbit cultivation in India with productivity of 10.52 t/h.

A large number of viruses belonging to 14 genera are known to affect cucurbits worldwide. *Bgomovirus, Crinivirus, Cucumovirus, Potyvirus* and *Tospovirus* are the important genera that affect cucubits in various parts of the world. In India, begomoviruses are major threat to numerous crops (Varma and Malathi, 2003). Begomoviruses (Family *Geminiviridae*) are characterised by tween-quasiisometric virion of 20x30 nm and are exclusively transmitted by whitefly, *Bemisia tabaci* in persistent manner. Although, a very few begomovirus species are known to affect cucurbits in India, there is a wide spread emergence of begomovirus diseases in numerous cucurbit crops.

Chronological Records of Virus Diseases in Cucurbits in India

Virus disease of cucurbits was recorded in India during late 1930s. *Cucumber green mottle mosaic virus*- a member of *Tobamovirus* causing mild green mottle disease in bottle gourd was the first cucurbit-infecting virus reported from India (Capoor and Varma, 1948). During 1950s, yellow vein mosaic disease of pumpkin was observed in Pune (Maharashtra). The virus was transmitted by whitefly, *B. tabaci* and named as Pumpkin yellow vein mosaic virus (PYVMV) (Varma, 1955). During this period, geminivius or begomovirus was not known. Subsequently, PYVMV was characterized as the first bonafide begomovirus infecting cucurbit in India (Muniyappa et al., 2003). During 1960 to 1977, three more RNA viruses, *Cucumber mosaic virus* (CMV), *Papaya ring spot virus* (PRSV) and *Watermelon mosaic virus* (WMV) causing mosaic disease in numerous cucurbits, ash gourd, bottle gourd, chayote, cucumber, long melon, muskmelon, pointed gourd, pumpkin, snake gourd, sponge gourd, summer squash and watermelon were identified (Bhargava and Joshi, 1960; Reddy and Nariani, 1963; Mitra and Nariani, 1965; Ram et al., 1971, Raychaudhuri, 1973; Dubey et al., 1974; Shankar and Nariani, 1974; Mayee et al., 1976; Ahlawat and Kulashreshtha, 1977; Bhargava and Bhargava, 1977). During 1990s, bud necrosis; a new destructive viral disease emerged in watermelon, which was identified to be a tospovirus- Watermelon bud necrosis virus (Singh and Krishnareddy, 1998). In 2004, a potyvirus- *Zucchini yellow mosaic virus* was identified in bottle gourd (Verma et al., 2004).

Till 1970s, there was no record of begomovirus associated disease except yellow vein mosaic of pumpkin in India. Begomovirus disease problems were known in many different cucurbits only 1980 onward (Figure 1). During 1980s and 1990s, begomovirus diseases, yellow mosaic, curling and leaf distortion were recorded in bitter gourd, cucumber, sponge gourd and muskmelon (Giri et al., 1981; Raj and Singh, 1996; Varma 1990; Varma and Giri, 1998; Shivkumar, 1999). The pumpkin yellow vein mosaic disease was mainly confined to central-western India until 1980. Now, the disease has spread in both northern and southern India (Capoor and Ahmad, 1975; Muniyappa et al., 2003; Sohrab, 2006; Singh et al., 2009). In 1990, severe epidemic of leaf curl in muskmelon appeared in northern India (Varma, 1990). Epidemic occurrence of yellow mosaic of sponge gourd (*Luffa cylindrical*) is common in northern India (Sohrab et al., 2003; Sohrab 2006).

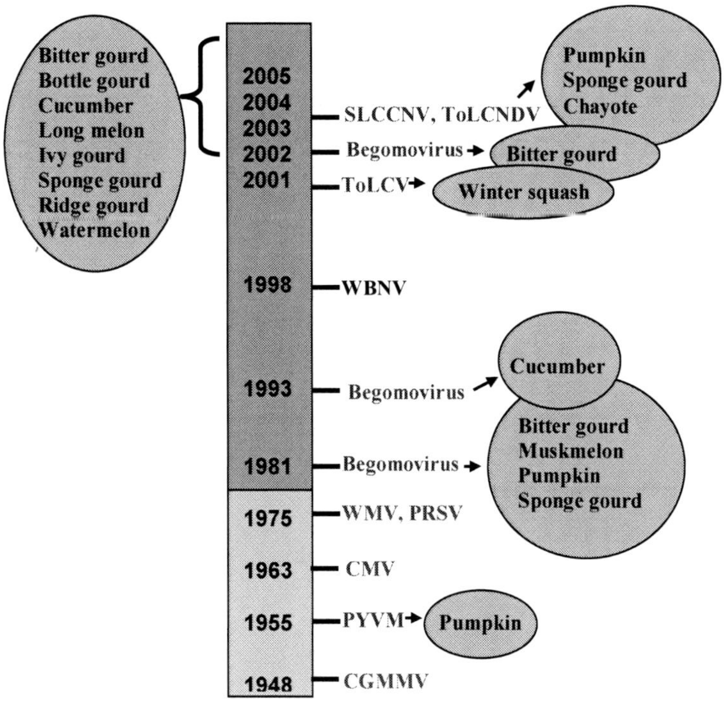

Figure 1. Chronological record of viruses infecting cucurbits in India.ToLCNDV: *Tomato leaf curl New Delhi virus*; ToLCV: Tomato leaf curl virus; SLCCNV: *Squash leaf curl China virus*; PYVMV: Pumpkin yellow vein mosaic virus; CMV: *Cucumber mosaic virus*, PRSV: *Papaya ring spot virus;* WMV: *Watermelon mosaic virus* (WMV); CGMMV: *Cucumber green mottle mosaic virus*; WBNV: Watermelon bud necrosis virus.

Availability of molecular tools led to identification of specific begomovirus species in cucurbits eg., *Tomato leaf curl New Delhi virus* (ToLCNDV) in yellow mosaic of chayote (Mandal et al., 2004) and sponge gourd (Sohrab et al., 2003) and *Squash leaf curl China virus* (SLCCNV) [synonymous to PYVMV] in yellow vein mosaic of pumpkin (Muniyappa et al., 2003). During 2003- 2005, ToLCNDV was detected (Figure 3) in ten different cucurbits-bitter gourd, bottle gourd, chayote, cucumber, ivy gourd, long melon, pumpkin, ridge gourd, sponge gourd and watermelon in northern India (Sohrab et al., 2006; Mandal et al., 2004; Phaneendra et al., 2007). So far, only two begomovirus species ToLCNDV and SLCCNV are known to affect cucurbits in India. The chronological records (Figure 1) of viruses in cucurbits show emergence of begomovirus diseases after 1980 and ToLCNDV emerged as most significant begomovirus species affecting several cucurbits in India.

Figure 2. Symptoms of cucurbit exhibiting begomovirus like diseases in northern India during 2004-2006. Chlorotic stunt of bottle gourd (*Lagenaria siceraria*), (B) Yellow mottle of bitter gourd (*Momordica charantia*), (C) yellow leaf curl of cucumber (*Cucumis sativus*), (D) yellow leaf curl of long melon (*Cucumis melo* var. *utilissimus*), (E) mosaic distortion of ivy gourd (*Coccinia cordifolia*), (F) yellow vein mosaic of pumpkin (*Cucurbita moschata*), (G) blister yellow mosaic of pumpkin, (H) leaf curl of pumpkin, (I) yellow mosaic of sponge gourd (*Luffa cylindrica*), (J) yellow mosaic of chayote (*Sechium edule*).

Global Distribution of Begomoviruses in Cucurbits

Begomoviruses are constraints in cucurbits worldwide. At least, 21 different begomoviruses are known to affect 13 different cucurbit crops in 16 countries belonging to America, Africa, and Asia (Table 1). Not all these reported cucurbit infecting begomoviruses are recognized species; only 13 of them are recognized as species by the International Committee on Taxonomy of Viruses (Fauquet et al., 2006). Except TYLCV, all these cucurbit-infecting begomoviruses contain bipartite genome. Phylogenetic affinities of these virus species correlate to the continental origin (Figure 4) and they are grouped into three major clades- Asian, African and American.

Table 1. Begomoviruses infecting cucurbits

Crop	Disease	Virus	Country	References
Ash gourd	Leaf curl	Tomato leaf curl virus	Thailand	Samretwanich et al., 2000b
Bitter gourd	Curly mottle	Tomato leaf curl virus	India	Khan et al., 2002
Bottle gourd	Yellow leaf	Tomato leaf curl New Delhi virus	Thailand	Ito et al., 2008
Buffalo gourd	Curly mottle	Watermelon curly mottle virus	USA	Rosemeyer et al., 1986
Chayote	Yellow mosaic	Chayote yellow mosaic virus	Nigeria	Thottapillay et al.,1997
	Yellow mosaic	Tomato leaf curl New Delhi virus	India	Mandal et al., 2004
Cucumber	Curly mosaic	Tomato leaf curl New Delhi virus	Thailand	Samretwanich et al.,2000a
	Leaf curl	Melon leaf curl virus	USA	Duffus et al., 1985
	Yellow mosaic	Cucumber yellow mosaic virus	India	Raj and Singh, 1996
Cantaloupe	Leaf curl	Tomato leaf curl virus	Thailand	Samretwanich et al., 2000b
	Leaf curl	Squash leaf curl virus	USA	Dodds et al., 1984
	Curly mottle	Watermelon curly mottle virus	USA	Brown et al., 1986
	Leaf curl	Melon leaf curl virus	USA	Duffus et al., 1985
Muskmelon	Leaf curl	Melon chlorotic leaf curl virus	Guatemala	Brown et al., 2001
	Leaf curl	Cucurbit leaf curl virus	USA	Brown et al., 2000
	Leaf curl	Cucurbit leaf crumple virus	USA	Guzman et al., 2000
	Curly mottle	Watermelon curly mottle virus	USA	Brown et al., 1986
	Yellow leaf	Tomato leaf curl New Delhi virus	Thailand	Ito et al., 2008
Pumpkin	Yellow vein	Squash leaf curl China virus	India	Muniyappa et al., 2003
	Leaf crumple	Cucurbit leaf crumple virus	USA	Brown et al., 2000
	Leaf curl	Squash leaf curl virus	USA	Ingham and Lazarowitz.,1993
	Yellow leaf	Tomato leaf curl New Delhi virus	Thailand	Ito et al., 2008
Ridge gourd	Yellow mosaic	Luffa yellow mosaic virus	Vietnam	Revill et al., 2003
Sponge gourd	Yellow mosaic	Tomato leaf curl New Delhi virus	India	Sohrab et al., 2003
Squash	Leaf curl	Squash leaf curl China virus	China	Hong et al.,1995
	Leaf curl	Squash leaf curl China virus-B	Vietnam	Revill et al., 2002
	Leaf curl	Squash leaf curl China virus-K	Vietnam	Revillet al., 2003
	Leaf curl	Squash leaf curl Philippines virus	Philippines	Kon et al., 2003
	Leaf curl	Squash leaf curl virus	Egypt	Abdel-Salem et al., 2006
	Leaf curl	Squash leaf curl virus	USA	Flock and Mayhew, 1981
	Mild mottle	Squash yellow mild mottle virus	Nicaragua	Ala-Poikala et al., 2005
	Leaf curl	Squash yellow mild mottle virus	Costa Rica	Karkashian et al., 2001
	Leaf curl	Squash leaf curl virus	Egypt	Idris et al., 2005
	Leaf curl	Squash leaf curl Israel virus	Israel	Antignus et al., 2002
	Leaf curl	Squash leaf curl Yunnan virus	China	Xie and Zhou. 2003
	Mild leaf curl	Squash mild leaf curl virus	USA	Brown et al ., 2002
	Leaf curl	Tomato yellow leaf curl virus	Cuba	Zubaiur et al., 2004
Watermelon	Leaf curl	Melon chlorotic leaf curl virus	Costa Rica	Ramirez et al., 2004
	Leaf curl	Cucurbit leaf crumple virus	USA	Brown et al., 2000
	Chlorotic stunt	Watermelon chlorotic stunt virus	Sudan/Iran	Kheyr-pour et al ., 2000
	Leaf curl	Squash leaf curl virus	USA	Isakeit et al., 1994
	Leaf curl	Tomato leaf curl India virus	Pakistan	Mansoor et al., 1999
	Leaf curl	Melon leaf curl virus	USA	Duffus et al., 1985
	Curly mottle	Watermelon curly mottle virus	USA	Brown et al., 1986

The greatest diversity of cucurbit begomoviruses are in Asia. A dozen of begomoviruses and their variants affecting cucurbits were recorded from ten Asian countries; however, only seven of them were recognized as species. A majority of these begomovirus species are distributed in southeast Asian countries: *Luffa yellow mosaic virus* in Vietnam (Revill et al., 2003), SLCCNV (Hong et al., 1995) and *Squash leaf curl Yunnan virus* (Xie and Zhou, 2003) in China, *Squash leaf curl Philippines virus* in Philippines (Kon et al., 2003) and Taiwan (Tsai et al., 2007) and ToLCNDV in Thailand (Samertwanich et al., 2000a).

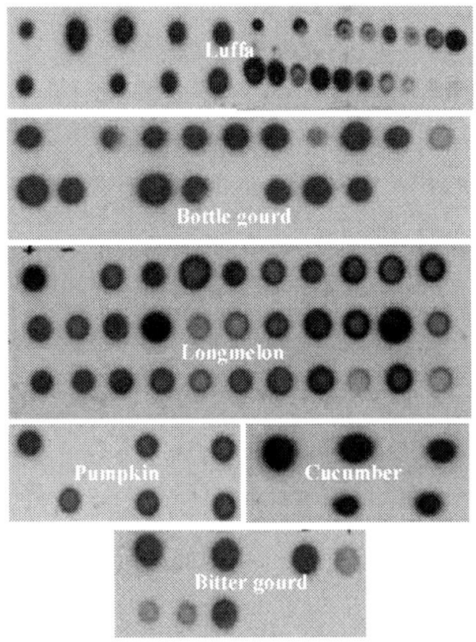

Figure 3. Detection of begomovirus in field samples of different cucurbits from northern India during 2004-06 by dot blot hybridization using DNA probe to coat protein gene of *Tomato leaf curl New Delhi virus*.

In West Asia and Africa, three begomovirus species are distributed, *Chayote yellow mosaic virus* (ChaYMV) (Thottapilly et al., 1997), *Watermelon chlorotic stunt virus* (WmCSV) (Kheyr-Pour et al., 2000) and *Squash leaf curl virus* (SLCV) (Antignus et al., 1000; Idris et al., 2006). In Sudan and Iran, severe outbreak of chlorotic stunt disease in melon and watermelon was occurred in cucurbit growing regions during 1993-1996 (Lecoq et al., 1994; Dafalla *et al.*, 1998). The causal virus was identified as WmCSV, which was detected in most cucurbits grown in these regions (Kheyr-Pour *et al.,* 2000).). In Nigera, ChaYMV a distinct begomovirus species was shown to be associated with yellow mosaic disease of chayote (Thottapillay *et al.*, 1997). SLCV, a new world begomovirus originally described from North America has now spread in Middle East (Al-Shahwan et al., 2002; Antignus et al., 2003; Idris et al., 2006; Al-Musa et al., 2008).

North America is the another centre for diverse cucurbit begomoviruses, where at least, four distinct begomovirus species are distributed, *Cucurbit leaf curl virus* (CuLCuV) (Brown *et al.*, 2000), SLCV (Lazarowitz and Lazdins, 1991), *Squash mild leaf curl virus* (SMLCV) (Brown *et al.*, 2002) and *Squash yellow mild mottle virus* (SYMMoV) (Karkashian et al., 2001). SLCV is the first cucurbit begomovirus recorded in USA. Leaf curl disease was first observed in 1977-78 in California in banana squash (*C. maxima*) (Cohen et al., 1983). Subsequently, SLCV was recorded in Arizona, California and Texas (Brown and Nelson, 1986; Isakeit et al., 1994; Dodds et al., 1995; Duffus et al., 1985, Brown et al, 1986). A new begomovirus tentatively designated as Watermelon curly mottle virus deferent from SLCV was recorded in Arizona causing vein banding and mottling in watermelon and muskmelon (Brown et al., 1986). A new bipartite begomovirus species, CuLCuV (Synonyms, Cucurbit leaf crumple virus) causing leaf curl and crumpling of muskmelon, watermelon in Imperial valley of California was reported (Brown *et al.,* 2002).

In Central America, three begomoviruses are known to affect cucurbits. In Guatemala and Costa Rica, *Melon chlorotic leaf curl virus* (MCLCuV) causing chlorotic leaf curl symptom in melon was recorded in 2000 (Brown et al., 2001). In Nicaragua, yellow mottle disease in cushaw squash (*Cucurbita argyrosperma*) was recorded and the virus was identified as Squash yellow mottle virus based on partial coat protein gene sequence (Ala-Poikala et al., 2005). In Havana Province, Cuba, association of *Tomato yellow leaf curl virus* was recorded with the mild yellow leaf curl of squash (*Cucurbita pepo*) during 2003 (Zubaiur et al., 2004). CuLCuV and SLCV were recorded from Mexico (Ramirez-Arredondo et al., 1998; Brown and Bird, 1992).

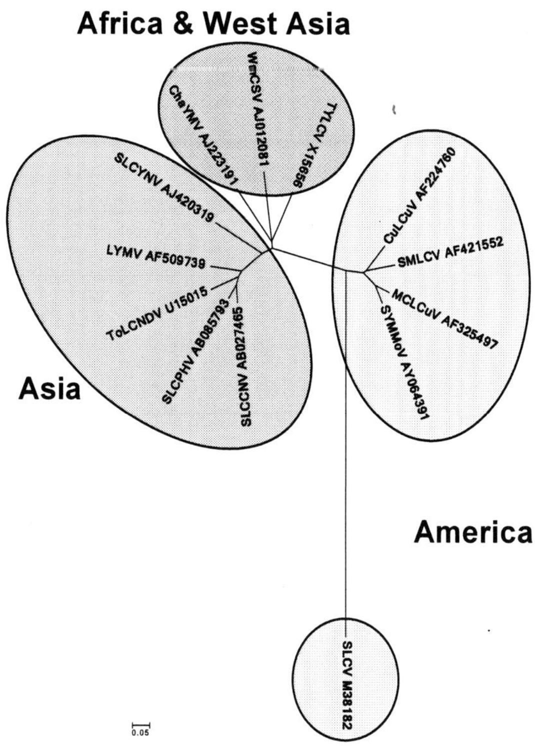

Figure 4. Phylogenetic relationships among the cucurbit infecting begomovirus species. The tree is constructed by using MEGA4 programme. The evolutionary history was inferred using the Neighbor-Joining method. The bootstrap consensus tree inferred from 500 replicates is taken to represent the evolutionary history of the taxa analyzed. Branches corresponding to partitions reproduced in less than 50% bootstrap replicates are collapsed. The tree is drawn to scale, with branch lengths in the same units as those of the evolutionary distances used to infer the phylogenetic tree. *Chayote yellow mosaic virus* (ChaYMV), *Cucurbit leaf curl virus* (CuLCrV), *Luffa yellow mosaic virus* (LYMV), *Melon chlorotic leaf curl virus* (MCLCuV), *Squash leaf curl virus* (SLCV), *Squash leaf curl China virus* (SLCCNV), *Squash leaf curl Philippines virus* (SLCPHV), *Squash leaf curl Yunnan virus* (SLCYNV), *Squash mild leaf curl virus* (SMLCV), *Squash yellow mild mottle virus* (SYMMoV), *Watermelon chlorotic stunt virus* (WmCSV), *Tomato leaf curl New Delhi virus* (ToLCNDV), *Tomato yellow leaf curl virus* (TYLCV).

BEGOMOVIRUS DISEASES IN CUCURBITS IN INDIA

Bitter Gourd

Bitter gourd (*Momordica charantia*) is cultivated throughout India. During 2001, 100% incidence of leaf curl/distortion of bitter gourd was observed in Lucknow (northern India) (Khan et al., 2002). The disease was transmitted by *B. tabaci* and the virus was identified as a begomovirus based on amplification of part of DNA-A and Southern hybridization with the probe to DNA-A of Tomato leaf curl virus and Cotton leaf curl virus. During 2004-2006, high incidence of severe leaf curl as well as yellow mottling (Figure 2 B) was observed in bitter gourd in the cucurbit growing areas in Delhi, Haryana and Rajasthan. Samples collected from these places were positive in PCR and dot-blot hybridization using the primer and probes to the coat protein gene of ToLCNDV. Not many viral diseases are known in bitter gourd in India, only a mosaic disease (Potyvirus?) was reported during 1970s (Bhargava and Bhargava, 1977). The current works indicate emergence of begomovirus diseases in bitter gourd, however, no specific begomovirus species has been identified so far in bitter gourd in India. In Pakistan, ToLCNDV has been identified to affect bitter gourd (Tahir and Haider, 2005).

Bottle Gourd

Bottle gourd (*Lagenaria siceraria*), a popular vegetable in India was known to be commonly affected by several RNA viruses eg., CGMMV and WMV, CMV, PRSV-W and ZYMV (Capoor and Varma 1948; Bhargava and Bhargava 1977; Dubey *et al.*, 1974; Verma et al., 2004); but not by any *Begomovirus*. During 2003-2006, bottle gourd was observed to be affected by a chlorotic stunt disease in Delhi and Haryana. The infected plants developed chlorosis, leaf curling, reduction of leaf lamina and stunting of plants (Figure 2A.). Infection at seedling stage resulted suppression of further growth of the shoot and plant eventually died. The association of *Begomovirus* in the infected sample was confirmed by electron microscopy, transmission, nucleic acid hybridization and sequence similarities; and the virus was identified as a minor variant ToLCNDV (Unpublished results).

Cucumber

Cucumber (Cucumis sativus) is cultivated throughout India. CMV is commonly known to affect cucumber in India (Bhargava and Bhargava 1977; Munshi et al., 2008). Occurrence of begomovirus with a yellow green mosaic disease of cucumber was first observed in 1993-94 in India. Association of begomovirus with the disease was established based on whitefly transmission and hybridization with the probe to DNA-B of Indian tomato leaf curl virus (Raj and Singh, 1996). High incidence of yellow leaf curl disease (Figure 2G) in cucumber was observed at experimental fields of Indian Agricultural Research Institute, New Delhi since 2003. Severely infected plants were stunted containing small pale yellow curly leaves and were unproductive. The begomovirus was detected in the field samples of cucumber using the

DNA-probe to coat protein gene of ToLCNDV (Figure 3). The begomovirus species infecting cucumber in India has not been identified so far. However, in Thailand, ToLCNDV has been identified to be associated with the yellow leaf curl disease of cucumber, which emerged in similar time as in India (Ito et al., 2008).

Long Melon

Long melon (*Cucumis melo* var. utilissimus) is cultivated in summer season and used as salad. Long melon is reported to be naturally infected by CGMMV and an unidentified nepovirus (Raychaudhuri, 1973; Rani et al., 1971). There is no record of begomovirus disease in long melon in India. During 2004, yellow leaf curl disease (Figure 2D) of long melon was observed in Yamuna river-bank area (Delhi). Nearly, 100% of plants were affected by the disease at the end of the growing season. The infected plants developed mild yellow mosaic, curling and smalling of leaves. The association of *Begomovirus* was confirmed by electron microscopy and dot blot hybridization using probe to ToLCNDV (Figure 3).

Luffa

Two species of *Luffa* are cultivated, sponge gourd (*Luffa acutangula)* and ridge gourd (*L. cylindrica)*. India is considered as the center of origin of *Luffa*. Sponge gourd is widely cultivated during summer in northern India. A yellow mosaic disease (SgYMD) was known to affect sponge gourd since 1980s (Varma and Giri, 1998). The affected plants developed yellow spots on the newly emerged leaves followed by mosaic, mild curling, leaf distortion, small leaves and distorted fruits (Figure 2I). The virus was transmitted by whiteflies to other cucurbits, bottle gourd, cucumber and muskmelon (Sivakumar, 1999). The virus was efficiently transmitted by whitefly; even a single whitefly was capable of transmitting the disease to *Luffa*. The virus could also be transmitted by sap inoculation, (Sohrab, 2006). The two other begomoviruses i.e. MCLCuV and SLCV occurring in Arizona were reported to be sap transmissible but the rate of transmission was also very low (Duffus *et al.*, 1985; Brown and Nelson, 1984). The virus from sponge gourd, however, could be efficiently sap transmitted up to 100% of inoculated ridge gourd, sponge gourd and *N. benthamiana* plants. The virus however is not sap transmissible to and from tomato, whereas, whitefly transmission from sponge gourd resulted in development of severe leaf curl disease in tomato. The comparisons of the complete nucleotide sequence of DNA A showed that the virus associated with the SgYMD was a severe strain of ToLCNDV (Sohrab, 2006).

Pumpkin

Pumpkin (*Cucurbita pepo; C. moschata*) is an economically important vegetable crop, which is widely cultivated in India. The cultivation season vary from region to regain, in the hills of northern India, it is cultivated during April-May, in southern and central India, it is grown throughout the year. Yellow vein mosaic (PYVMD) is an important disease of pumpkin in India. The disease appears in epidemic proportion and causes significant damage to pumpkin production. The disease is characterized by development of yellow veins on young leaves followed by bright yellow mosaic. The affected plants are stunted and flowers drop prematurely resulting in yield loss up to 100%. Virus-vector relationships, host-range and serology of the virus associated with *C. moschata* have been studied from southern India (Muniyappa et al., 2003). The virus was efficiently transmitted by whitefly in a persistent manner. The minimum acquisition access, inoculation access and latent periods are 30 min, 10 min and 6h, respectively. The virus persisted in whitefly at least for 8 days. The vector-transmission varies based on the biotypes of whitefly. The B-biotype, which was introduced to southern India in 1999, was shown to be more efficient vector compared to the indigenous biotype in transmitting the virus (Maruthi et al., 2007). The virus has a narrow host range infecting only four cucurbits, *C. moschata, C. maxima, C. pepo* and *L. siceraria* under experimental conditions. The complete nucleotide sequence data of DNA-A of the isolates from northern and southern India showed that the virus associated with the PYMD is very similar to SLCCNV reported from China (Sohrab, 2006; Maruthi et al., 2007; Singh et al., 2008; Singh et al 2009). So far, infectivity of the cloned DNAs has not been demonstrated and therefore, the disease etiology has not yet been conclusively established.

A yellow mosaic disease of pumpkin (*C. maxima*) was recorded in Lucknow (northern India) during 1999-2000, which was transmitted by *B. tabaci*. Based on partial sequence data, the virus was provisionally identified as a monopartite begomovirus, *Tomato leaf curl Karnatak virus* (Singh et al., 2001). Recently, pumpkin in northern India has been shown to be affected by ToLCNDV causing leaf curl disease (Figure 2H) (Phaneendra et al., 2007; Maruthi et al., 2007). Yellow vein mosaic and leaf curl, both the disease symptoms were observed in single plants under field conditions in northern India indicating mixed infection of ToLCNDV and PYVMV. Another begomovirus associated disease of pumpkin characterised by blister yellow mosaic (Figure 2G) was observed in northern India during 2005. The associated virus species with leaf curl and blister yellow mosaic of pumpkin have not been characterised.

CONCLUSION

Begomoviruses are major threat in numerous crops in India due to favorable environmental conditions, intensive cultivation, susceptibility of the genotype and abundance of vector, *B. tabaci* (Varma and Malathi, 2003). During last three decades, however cucurbit-infecting begomoviruses (Singh *et al.*, 2001; Khan *et al.*, 2002; Sohrab *et al.*, 2003; Muniyappa *et al.*, 2003; Mandal *et al.*, 2004) have emerged as a serious threat to cucurbit cultivation, particularly in northern India. Although, two begomovirus species, SLCCNV and ToLCNDV are known to affect cucurbits in India, only ToLCNDV is most significant being affecting a large number of cucurbits. Leaf curl of tomato caused by ToLCNDV is known in India for nearly 60 years (Vasudeva and Samraj 1948), but its occurrence in cucurbits was noticed only after 1980s. One of the major reasons of ToLCNDV to be prolific in cucurbits in the recent years could be due to the abundance of whitefly throughout the year, whereas during pre 1990s whiteflies were mainly found in the monsoon season (Varma and Malathi, 2003). Cucurbit such as bottle gourd, cucumber and pumpkin are preferred hosts for breeding of whitefly. During dry summer, these cucurbits provide cool and humid microenvironments due to coverage of foliage especially in the irrigated field and as a result, support a very high population of whitefly. Cucurbits are grown in northern India from February to November in overlapping with tomato and potato. Therefore, cultivation practice makes the crop-host available throughout and that helps perpetuation of ToLCNDV. Abundance of whitefly further provides opportunity for association of virus and their variants with these crop-hosts. Mutation, recombination and genome reassortment result into genetic diversity of viruses. Emergence of new cucurbit infecting begomovirus and expansion of host range has been shown through genome reassortment (Brown et al., 2002; Idris et al., 2008). Therefore, another reason of emergence of ToLCNDV in cucurbits may be that the virus has evolved to adopt new hosts. In India, apical leaf curl- a hitherto unknown disease of potato emerged in northern plain during 1990s (Garg et al., 2001) and ToLCNDV was identified as the causal virus (Usharani et al., 2004). This is another example of emergence of ToLCNDV in new host in northern India. Interestingly, PYVMV known to occur since last more than 60 years has narrow host range and so far not known to emerge in any cucurbits in India. The two begomovirus species infecting cucurbits have been identified in India based on genome sequence. Infectivity of the cloned DNAs has not been demonstrated and therefore the etiology of these diseases remains inconclusive.

REFERENCES

Abdel-Salam, A.M., Abdallah, N.A., Soliman, D.Z.R. & Rezk, A.A.S. (2006). The incidence of squash leaf curl begomovirus (SqLCV) in Egypt. *Arab Journal of Biotechnology, 9*, 375-388.

Ahlawat, Y.S. & Kulshreshtha, U.K. (1977). A new graft transmissible disease of *Sechium edule*. *Indian Phytopathology, 30*, 268-269.

Ala-poikela, M., Svensson, E., Rojas, A., Horka, T., Paulin, T.,Valkonen, J.P.T. & Kvarnheden, A. (2005). Genetic diversity and mixed infections of begomoviruses infecting tomato, pepper and cucurbit crops in Nicaragua. *Plant Pathology, 54*, 448-459.

Al-Musa, A., Anfoka, G., Misbeh, S., Abhary, M & Ahmad, F.H. 2008. Detection and Molecular Characterization of Squash leaf curl virus (SLCV) in Jordan. *Journal of Phytopathology, 156*, 311-316.

Al-Shahwan, I.M., Abdalla, O.A. & Al-Saleh, M.A. (2002). Squash leaf curl virus (SqLCV) and other begomoviruses in Saudi Arabia. *Dirasat Agricultural Sciences, 29*, 28-36.

Antignus, Y., Lachman, O., Pearlsman, M., Omer, S., Yunis, H., Messika, Y., Uko, O. & Koren, A. (2003). Squash leaf curl geminivirus- a new illegal immigrant from the Western Hemisphere and a threat to cucurbit crops in Israel. *Phytoparasitica, 31*, 415.

Antignus,Y., Lachman, O. & Pearlsman, M. (2002). Squash leaf curl Israel virus coat protein gene, complete cds. GenBank Accession AY206998. www.ncbi.nlm.nih.gov

Bhargava, B. & Bhargava, K.S. (1977). Cucurbit mosaic virus in Gorakhpur. *Indian Journal of Agricultural Sciences, 47*, 1-5.

Bhargava, K.S. & Joshi, R.D. (1960). Detection of Watermelon mosaic virus in U.P. *Current Science, 29*, 443-444.

Brown, J. K., & Bird, J. (1992). Whitefly-transmitted geminiviruses in the Americas and the Caribbean Basin: Past and present. *Plant Disease, 76*, 220-225.

Brown, J.K. & Nelson, M.R. (1986). Whitefly-borne viruses of melons and lettuce in Arizona. *Phytopathology, 76*, 236-239.

Brown, J.K., Idris, A.M. and Rogan, D. (2001). Melon Chlorotic leaf curl virus, a new begomovirus associated with *Bemisia tabaci* infestations in Guatemala. *Plant Dis. 85*: 1027.

Brown, J.K., Idris, A.M., Olsen, M.W., Miller, M.E., Isakeit, T. & Anciso, J. (2000). Cucurbit leaf curl virus, a new whitefly transmitted geminivirus in Arizona, Texas and Mexico. *Plant Disease, 84*, 809.

Brown, J.K., Idris, M., Alteri, C. & Stanger, D.C. (2002). Emergence of new cucurbit infecting begomovirus species capable of forming viable reasortments with related viruses in the Squash leaf curl virus cluster. *Phytopathology, 92*, 734-742.

Capoor S.P. & Varma P.M. (1948). A mosaic disease of *Lagenaria vulgaris* Ser in the Bombay Province. *Current Science, 17*, 274-275.

Capoor, S.P. & Ahmad, R.J. (1975). Yellow vein mosaic disease of field pumpkin and its relationship with the vector, *Bemisia tabaci. Indian Phytopathology, 28*, 241-246.

Cohen, S., Duffus, J.E., Larsen, R.C., Liu, H.Y. & Flock, R.A. (1983). Purification, serology and vector relationships of Squash leaf curl virus, a whitefly-transmitted geminivirus. *Phytopathology, 73*, 1669-1673.

Dafalla, G. A., Gronenborn, B., Kheyr-Pour, A., & Lecoq, H. (1998). Watermelon chlorotic stunt virus: A new emerging epidemic in export melons in Sudan. In: 2nd *Int. Workshop Bemisia Geminiviruses*, Page 37, San Juan, Puerto Rico.

Dafalla, G.A., Lecoq, H., Kheyr- Pour, A. & Gronenborn, B. (1994). Disease and pest outbreaks in Sudan, a whitefly-transmitted geminivirus associated with yellowing disease of watermelon in Sudan, Arab, and Near East. *Plant protection Newsletter, 19*, 39.

Dodds, J.A. Lee, J.G., Nameth, S.T. & Faemmlen, F.F. (1984). Aphid- and whitefly-transmitted cucurbit viruses in Imperial County, California. Phytopathology, 74, 221-225.

Dubey, G.S., Nariani., T.K. & Prakash, N. (1974). Identification of Snake gourd mosaic virus. *Indian Phytopathology, 27*, 470-474.

Duffus, J.E., Hsing-Yes-Liu. & Mathew, R. (1985). Melon leaf curl- a new geminivirus with host and serological variation from squash leaf curl virus. *Phytopathology*, 1312.

Fauquet, C.M., Mayo, M.A., Maniloff, J., Desselberger, U. & Ball L.A. (2005). *Virus Taxonomy: Eighth Report of the International Committee on Taxonomy of Viruses.* Academic Press, 1162 pp.

Flock, R.A. & Mayhew, D. (1981). Squash leaf curl, a new disease of cucurbits in California. *Plant Disease, 65*, 75-76.

Garg, I.D., Khurana, S.M.P, Kumar, S. & Lakra, B.S. (2001). Association of a geminivirus with potato apical leaf curl in India and its immuno-electron microscopic detection. *Journal of Indian Potato Association, 28*, 227-232.

Giri, V.K., Chenulu, V.V. & Mishra, M. D. (1981). A new graft transmissible disease of *Momordica charantia.* Discussion on current problems viruses and virus like plant diseases, Kalingpong Oct. 19-21, IARI, Regional Station Kalingpong.

Guzman, P., Sudarshana, M.R., Seo, Y.S., Rojas, M.R., Natwick, E., Turini, T., Mayberry, K. & Gilbertson, RL. (2000). A new bipartite geminivirus (begomovirus) causing leaf curl and crumpling in cucurbits in the Imperial Valley of California. *Plant Disease, 84*, 488.

Hong, Y., Wang, X., Tian, B. & Cai, J. (1995). Chinese squash leaf curl virus, a new whitefly transmitted geminivirus. *Science-in-China-Series-B, 38*, 179-186.

Idris, A.M., Abdel-Salam, A. & Brown, J.K. (2006). Introduction of the New World Squash leaf curl virus to Squash (*Cucurbita pepo)* in Egypt: a potential threat to important food crops. *Plant Disease, 90*, 1262.

Idris, M., Mills-Lujan, K., Martin, K. & Brown, J.K. (2008). Melon chlorotic leaf curl virus: characterization and differential reassortment with closest relatives reveal adaptive virulence in the Squash leaf curl virus clade and host shifting by the host-restricted Bean calico mosaic virus. *Journal of Virology, 82*, 1959-1967.

Ingham, D.J. & Lazarowitz, S.G. (1993). A single missense mutation in the BR1 movement protein alters the host range of Squash leaf curl virus. *Virology, 196*, 694-702.

Isakeit, T., Robertson, N.L., Brown, J.K., Gilbertson, R.L. (1994). First report of Squash leaf curl virus on watermelon in Texas. *Plant Disease, 78*, 1010.

Ito, T., Sharma, P., Kittipakorn, K. & Ikegami, M. (2008). Complete nucleotide sequence of a new isolate of tomato leaf curl New Delhi virus infecting cucumber, bottle gourd and muskmelon in Thailand. *Archives of Virology, 153*, 611-613.

Karkashian, J.P., Maxwell, D.P. & Ramirez, P. (2001). Squash yellow mild mottle virus-[CR] isolate 98-631 segment DNA-A, complete sequence. GenBank Accession AY064391.

Khan, J.A., Siddiqui, M.R. & Singh, B.P. (2002). Association of begomovirus with bitter melon in India. *Plant Disease, 86*, 328.

Kheyr-pour, A., Bananej, K., Dafalla, G.A., Caciagli, P., Noris. E., Ahoonmanesh, A., Lecoq, H. & Gronenborn, B. (2000). Watermelon chlorotic stunt virus from Sudan and Iran: sequence comparisons and identification of a whitefly-transmission determinant. *Phytopathology, 90*, 629-635.

Kon, T., Dolores, M., Bajet, N.B., Hase, S., Takahashi, H. & Ikegami, M. (2003). Molecular characterization of a strain of Squash leaf curl China virus from the Philippines. *Journal of Phytopathology, 15*, 535-539.

Lazarowitz, S.G. & Lazdins, I.B. (1991). Infectivity and complete nucleotide sequence of the clone genomic components of a bipartite Squash leaf curl geminivirus with a broad host rage phenotype. *Virology, 180*: 58-69.

Lecoq, H., Dafalla, G.A., Mohamed, Y.F., Pitrat, M., Omar, S.E. & Ali, H.A. (1994). A survey on cucubit viruses in Kordofan, east and central region of Sudan. *University Journal of Agriculture Science, 1*, 47-71.

Mandal, B., Mandal, S., Sohrab, S.S., Pun K.B. & Varma A. (2004). A new yellow mosaic disease of chayote in India. *Plant Pathology, 53*, 797.

Mansoor, S., Khan, S.H., Hussain, M., Mushtaq, N., Zafar, Y., Malik, K.A. (2000). Evidence that watermelon leaf curl disease in Pakistan is associated with tomato leaf curl virus-India. *Plant Disease, 84*, 102.

Martinez-Zubiaur, Y., Fonseca, D., Quiñones, M., & Palenzuela, I. (2004). Presence of tomato yellow leaf curl virus infecting squash (*Cucurbita pepo*) in Cuba. *Plant Disease, 88*, 572.

Maruthi, M.N., Rekha, A.R. & Muniyappa, V. (2007). Pumpkin yellow vein mosaic disease is caused by two distinct begomoviruses: complete viral sequences and comparative transmission by an indigenous Bemisia tabaci and the introduced B-biotype. *Bulletin-OEPP/EPPO-Bulletin, 37*, 412-419.

Mayee, C.D, Nandpuri, K.S. & Lal, T. (1976). A Mosaic disease of muskmelon in Punjab. *Vegetable Science, 3*, 87-90.

Mitra, D.K. & Nariani, T.K. (1965). A mosaic disease of Tori (*Luffa acutangula*). *Indian Phytopathology, 18*, 233-236.

Muniyappa, V., Maruthi, M.N., Babitha, C.R., Colvin, J., Briddon, R.W. & Rangaswamy, K.T. (2003). Characterization of Pumpkin yellow vein mosaic virus from India. Annals of Applied Biology, 142, 323-331.

Munshi A.D., Panda, B., Mandal, B., Bisht, I.S., Rao, E. S. & Kumar, R. (2008). Genetics of resistance to *Cucumber mosaic virus* in Cucumis sativus var. hardwickii R. *Alef. Euphytica, 164*, 501–507.

Nayar, N.M. & More T.A. (1998). *Cucurbits*. Oxford and IBH publishing House Pvt. Ltd., New Delhi.

Phaneendra, Mandal, B. & Jain, R.K. (2007). Association of *Tomato leaf curl New Delhi virus* with pumpkin leaf curl in northern India. In the International conference on "Emerging and re-emerging viral diseases of the Tropics and Sub-Tropics", held in Dec 11-14, 2007, New Delhi-12.

Raj, S.K. & Singh, B.P. (1996). Association of geminiviral infection with yellow mosaic disease of *Cucumis Sativus*. Diagnosis by nucleic acid probes. *Indian Journal of Experimental Biology, 34*, 603-605.

Ram, S., Sharma, G.S. & Varma, H.N. (1971). Epidemiology of virus diseases of cucurbits. *Proceedings of Indian National Science Academy, 37*, 345-351.

Ramirez-Arredondo, J.A, Cardenas, I.A., Sanchez, F.D., Garzon-Tiznado, J.A. (1998). Virus transmitted by whitefly (*Bemisia tabaci* Gennadius) on pepper and zucchini squash in the Mayo Valley Sonora, Mexico. *Agricultura-Tecnica-en-Mexico. 24*, 37-43.

Rani, S., Verma, G.S. & Verma, H.N. (1971). Epidemiology of virus disease of cucurbits. *Proceedings of Indian National Science Academy. 37*, 345-351.

Raychaudhuri, M. (1973). *Studies on cucurbit viruses. Ph.D. Thesis*. P.G. School, IARI, New Delhi.

Reddy, H.R. & Nariani, T.K. (1963). Studies on mosaic disease of vegetable marrow (*Cucurbita pepo*). *Indian Phytopathology, 16*, 260-267.

Revill, P.A., Ha, C.V., Porchun, S.C., Vu, M.T. & Dale, J.L. (2003). The complete nucleotide sequence of two district geminiviruses infecting cucurbits in Vietnam. *Archives of Virology, 148*, 1523-141.

Rosemeyer, M.E. Brown, J. K. & Nelson, M.R. (1986). Five viruses isolated from field grown buffalo gourd. (*Cucurbita foetidissima*), a potential crop for semi-arid lands. *Plant Disease, 70*, 405-409.

Samretwanich, K., Chiemsombat, P., Kittipakorn, K. & Ikegami, M. (2000b). Yellow leaf disease of cantaloupe and wax gourd from Thailand caused by Tomato leaf curl virus. *Plant Disease, 84*, 200.

Samretwanich, K., Chiemsombat, P., Kittipakorn, K., Ikegami, M. (2000a). Tomato leaf curl geminivirus associated with cucumber yellow leaf disease in Thailand. *Journal of Phytopathology, 148*, 615-617.

Shankar, G, Nariani, T.K. & Prakash, N. (1972). Studies on Pumpkin mosaic virus (PMV) with particular reference to purification, electron microscopy and serology. *Indian Journal of Microbiology, 12*, 154-165.

Shankar, G. & Nariani, T.K. (1974). A mosaic disease of watermelon (*Citrullus Lanatus*). *Currrent Science, 43*, 281-282.

Singh, A.K., Mishra, K.K., Chattopadhyay, B. & Chakraborty, S. (2009). Biological and molecular characterization of a begomovirus associated with yellow vein mosaic disease of pumpkin from northern India. *Virus Genes, 39*, 359-370.

Singh, R., Raj, S.K., & Chanadra, G. (2001). Association of a monopartite begomovirus with yellow mosaic disease of pumpkin (*Cucurbita maxima*) in India. *Plant Disease, 85*, 1029.

Singh, R., Raj, S.K., Prasad, V. (2008). Molecular Characterization of a Strain of *Squash leaf curl China Virus* from North India. *Journal of Phytopathology, 156*, 222-228.

Sivakumar, A.S. (1999). Characterization of Luffa leaf distortion mosaic virus. M.Sc. Thesis, PG. School, IARI, New Delhi.

Sohrab, S.S. (2006). *Variability in the geminiviruses infecting cucurbits. Ph.D.Thesis*, Jamia Milla Islamia. New Delhi.

Sohrab, S.S., Mandal, B., Pant, R.P. & Varma, A. (2003). First report of association of Tomato leaf curl New Delhi virus with yellow mosaic disease of *Luffa cylindrica* in India. *Plant Disease, 87*, 1148.

Tahir, M. & Haider, M.S. (2005). First report of *Tomato leaf curl New Delhi virus* infecting bitter gourd in Pakistan. *Plant Patology, 54*, 807.

Thottappilly, G. (1997). Chayote yellow mosaic virus DNA-A, complete sequence. GenBank Accession AJ223191. www.ncbi.nlm.nih.gov.

Tsai, W.S., Shih, S.L., Green S.K. & Jan, F.J. (2007). Occurrence and Molecular Characterization of *Squash leaf curl Phillipines virus* in Taiwan. *Plant Disease, 91*, 907.

Usharani, K.S., Surendranath, B., Paul-Khurana, S.M., Garg, I.D., & Malathi, V.G. (2004). Potato leaf curl- a new disease of potato in northern India caused by a strain of *Tomato leaf curl New Delhi virus*. *Plant Pathology, 53*, 235.

Varma, A. & Giri, B.K. (1998). Virus disease of cucurbits in India. In: *Cucurbits*, Nayar, N.M. and More T.A. (Eds). Oxford and IBH publishing house Pvt. Ltd., New Delhi.

Varma, A. & Malathi, V.G. (2003). Emerging geminivirus problem: A serious threat to sustainable crop production. *Annals of Applied Biology, 142*, 145-164.

Varma, A. (1990). Changing pattern of plant diseases caused by whitefly-transmitted geminiviruses. In: *Abstract of the 8th International congress of Virology*, Berlin, P 31.

Vasudeva, R.S. & Lal, T.B. (1943). A mosaic disease of bottle gourd. *Indian Journal of Agricultural Sciences, 3*, 182-191.

Vasudeva, R.S. & Samraj, J. (1948). A leaf curl disease of tomato. *Phytopathology, 38*, 364-369.

Verma, P.M. (1955). Ability of whitefly to carry more than one virus simultaneously, *Currrent Science, 24*, 317-318.

Verma. R., Ahlawat Y.S., Tomar, S.P.S., Prakash, S. & Pant, R.P. (2004). First report of Zucchini yellow mosaic virus in bottle gourd (*Lageneria siceraria*) in India. *Plant Disease, 88*, 426.

Xie, Y, & Zhou, X.P. (2003). Molecular characterization of Squash leaf curl Yunnan virus, a new begomovirus and evidence for recombination. *Archives of Virology, 148*, 2047-2054.

In: Emerging Geminiviral Diseases and their Management ISBN 978-1-61668-620-8
Editors: P. Sharma, R. K. Gaur et al. pp. 183-193

Chapter 9

AN OVERVIEW OF *TOMATO YELLOW LEAF CURL VIRUS* AND ITS VECTOR INSECT *BEMISIA TABACI* IN JAPAN

Shigenori Ueda[1,*], Bong Choon Lee[1,2], Keisuke Kijima[3], Hirosato Kodama[4], Junichi Shirakawa[1], Shinichiro Okazaki[5], Masahiro Taniguchi[4], Kenkichi Kanmiya[6] and Masatoshi Onuki[1]

[1]National Agricultural Research Center for Kyushu Okinawa Region, Kumamoto, Japan;
[2]National Institute of Crop Science, RDA, Milyang, Korea;
[3]Okinawa Prefectural Agricultural Research Center, Itoman, Okinawa, Japan;
[4]Okinawa Prefectural Plant Protection Center, Naha, Okinawa, Japan;
[5]Oita Prefectural Agricultural, Forestry and Research Center, Usa, Oita, Japan;
[6]Kurume University, Kurume, Fukuoka, Japan.

ABSTRACT

Tomato yellow leaf curl disease (TYLCD) caused by *Tomato yellow leaf curl virus* (TYLCV) is one of the most devastating viral diseases worldwide. TYLCD was first reported in Japan in 1996. Both TYLCV-Israel and TYLCV-Mild strains have become widely distributed and have caused serious reductions in tomato and *Eustoma grandiflorum* crops throughout Japan. *Bemisia tabaci* is both the insect vector of TYLCV and a serious insect pest. An invasive population of the Q biotype of *B. tabaci* was identified in 2004, and has largely displaced the B biotype. This chapter provides an overview of TYLCV and *B. tabaci*, including indigenous begomoviruses and distinct populations of *B. tabaci* in Japan.

* Correspondence concerning this article should be addressed to: Shigenori Ueda, National Agricultural Research Center for Kyushu Okinawa Region, 2421 Suya, Koshi, Kumamoto 861-1192, Japan. Tel.: +81-96-242-7730; Fax: +81-96-249-1002; E-mail: sued@affrc.go.jp.

INTRODUCTION

Tomato yellow leaf curl disease (TYLCD) caused by *Tomato yellow leaf curl virus* (TYLCV; genus *Begomovirus*, family *Geminiviridae*) is one of the most devastating viral diseases affecting tomato plants worldwide (Salati et al., 2002, Mansoor et al., 2003). TYLCV is a monopartite begomovirus species. Based on the recent taxonomy and nomenclature of geminiviruses, the TYLCV species consists of five strains, and is separate from other TYLCV-like viruses such as *Tomato yellow leaf curl Sardinia virus* (TYLCSV) (Fauquet et al., 2008, Abhary et al., 2007). In Japan, TYLCV first caused significant yield losses in tomato production in 1996 (Kato et al., 1998, Haga & Doi, 2002, Onuki et al., 2004). Both the Israel (TYLCV-IL) and the Mild (TYLCV-Mld) strains of TYLCV have been identified in Japan (Kato et al., 1998, Ueda et al., 2004, 2005).

The vector insect of TYLCV, *Bemisia tabaci* (Gennadius) (Hemiptera: Aleyrodidae), is the most economically important and widely distributed insect pest throughout the tropics and subtropics (Brown et al., 1995, De Barro et al., 2000). Taxonomic studies on the *B. tabaci* complex have shown that it is a cryptic species, for which genetic variation can be demonstrated in the absence of morphological variation. Such genetic variants can be differentiated using molecular markers (Brown, 2000, Costa & Brown, 1991, Perring, 2001). In the southwestern United States, the B biotype displaced the indigenous A biotype in the field within several years of its introduction (Brown & Bird, 1995), possibly because of differences in susceptibility to insecticides that were used to control whitefly at the time. Some populations of *B. tabaci* biotypes, such as the B and Q biotypes, have developed high levels of resistance to certain insecticides. Currently, both populations have proliferated and have become vector insects for TYLCV throughout Japan.

DISTINCT TYLCV ISOLATES WERE INTRODUCED INTO JAPAN

TYLCV was first recorded in 1996 in Shizuoka and Aichi Prefectures in the Tokai district in central Japan, and in Nagasaki Prefecture in Kyushu, southwest Japan (Kato et al., 1998). Analyses of complete nucleotide sequences from various TYLCV isolates in Japan revealed that isolates from Shizuoka (TYLCV-Mld[Shizuoka]) and Aichi (TYLCV-Mld[Aichi]) are closely related to the TYLCV-Mild isolate (TYLCV-Mld), while those from Nagasaki (TYLCV-IL[Nagasaki]) are closely related to TYLCV-IL (Kato et al., 1998, Onuki et al., 2004, Ueda et al., 2004) (Figure 1). Since then, TYLCV quickly spread to Kyushu. In 2003–2004, a genetically distinct isolate of TYLCV-IL, TYLCV-IL[Tosa], was identified in Kochi Prefecture in Shikoku. It was thought that TYLCV-IL[Tosa] was the third isolate introduced into Japan, because its genome showed a chimeric structure as a recombinant between TYLCV-IL and TYLCV-Mild (Ueda et al., 2005). In 2009, TYLCV had spread across 32 prefectures including Okinawa Prefecture, which is located in a subtropical region of Japan (Figure 1). TYLCV has also spread into the major tomato growing area of Japan. Both TYLCV-IL and TYLCV-Mld have been detected in some districts, such as Kanto and Tokai.

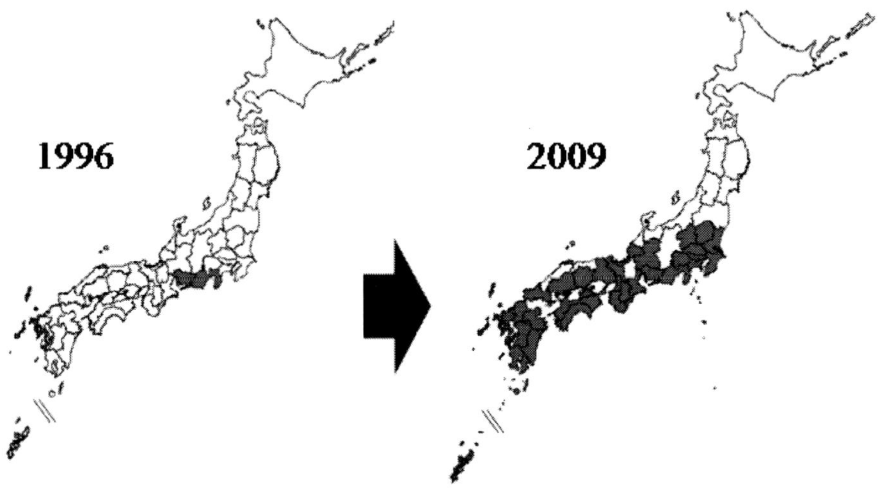

Figure 1. Map showing prefectures in which TYLCV was detected in 1996 and 2009 in Japan.

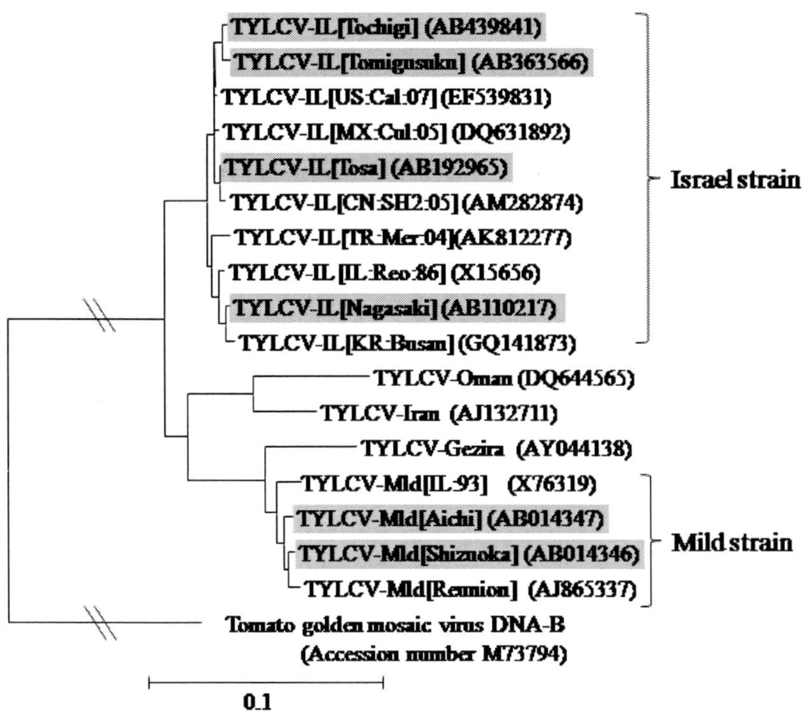

Figure 2. Neighbor-joining phylogenetic tree showing relationships among complete DNA-A sequences of 17 TYLCV isolates. DNA-B sequence of a TGMV isolate (Acc. no. M73794) was used as the outgroup. Isolates reported from Japan are shaded.

TYLCV has a high, RNA-like, rate of nucleotide substitution, which is independent of its frequent recombination. The mean genomic substitution rate of TYLCV was estimated to be 2.88×10^{-4} nucleotide substitutions per site, per year (subs/site/year) (Duffty & Holmes, 2008). Recently, sister isolates of TYLCV-IL[Tosa] were recorded from some foreign countries, and from Okinawa and Tochigi Prefectures in Japan (Figure 2). TYLCVs isolated from eastern North America and the Caribbean were thought to have been introduced from Asia (Duffy & Holmes, 2007). TYLCV was first introduced into Okinawa Prefecture in 2007 (Ueda et al., 2008a) (Figure 3A). Since then, we have collected 23 isolates from 12 localities, and analyzed their phylogenetic relationships based on partial nucleotide sequences of the TYLCV genome. All isolates were grouped into the TYLCV-IL[Tosa] cluster (data not shown). In Spain, a natural recombinant between TYLCV and TYLCSV exhibiting a novel pathogenic phenotype has become prevalent (Monci et al., 2002). TYLCV-like viruses such as *Tomato yellow leaf curl Axarquia virus* (TYLCAxV) and *Tomato yellow leaf curl Malaga virus* (TYLCMalV) also represent novel recombinant begomoviruses (Abhary et al., 2007). It is unclear how this genetic group (TYLCV-IL[Tosa] and sister isolates) have been distributed worldwide. Their wide distribution suggests that they may have obtained some advantageous characteristics for adapting to host plants.

Figure 3. Tomato yellow leaf curl disease. (A) Outbreak of TYLCD in greenhouse in Okinawa Prefecture in 2008. (B) Typical symptoms of TYLCV-IL[Nagasaki]-infected tomato plants.

DISEASE SYMPTOMS

Leaves of TYLCV-infected tomato plants show chlorotic symptoms and curled-up margins; the plants are stunted and the flowers abscise (Figure 3). In Japan, popular commercial tomato cultivars such as Momotaro are highly susceptible to both TYLCV-IL and TYLCV-Mld. Some resistant commercial cultivars are highly tolerant to TYLCV-IL and do not show obvious disease symptoms. However, depending on their resistance locus, some cultivars infected with TYLCV-Mld show visible disease symptoms in the field. In 1999, leaf curl disease of *Eustoma grandiflorum* caused by TYLCV was first recorded in Nagasaki Prefecture (Uchikawa et al., 2002), and by 2009 it had been reported in 11 prefectures. We compared disease symptoms of *E. grandiflorum* (var. Halleymistypink) infected with TYLCV-IL and TYLCV-Mld. There were few visible symptoms on TYLCV-Mld-infected leaves of *E. grandiflorum*. However, plants infected with TYLCV-IL were stunted, and leaves and flower petals showed severe symptoms of leaf curling and vein thickening (Figure 4). It is possible that *E. grandiflorum* is more resistant to the TYLCV-Mld strain.

Figure 4. Symptoms of *Eustoma grandiflorum* plants infected with TYLCV-IL (A) and TYLCV-Mild (B).

TYLCV is phloem-limited in systemically infected tomato leaves, and is restricted to sieve tubes and nuclei of companion and phloem parenchyma cells (Morilla et al., 2004). To observe leaf vein tissue formation of TYLCV-infected tomato plants, paraffin embedded sections were prepared and stained with Safranin and Fast Green. Abnormal cell formations developed in parts of both internal and external phloem parenchyma cells of infected leaf vein tissue. The spongy cells surrounding the phloem in infected leaf vein tissue were smaller than those in healthy leaf vein tissue (Figure 5). One of our research aims is to investigate relationships between pathogenicity and tissue formation, comparing resistant and susceptible cultivars.

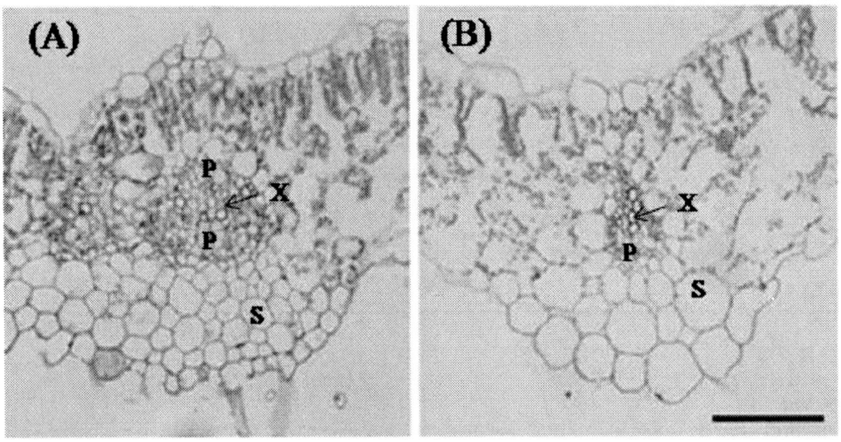

Figure 5. Effects of virus infection on tomato leaf vein tissue. Transverse sections of leaves from TYLCV-infected plant (A) and healthy plant (B). Abnormal tissue formations were observed in parts of phloem parenchyma cells and spongy cells of infected leaf vein tissue. P: phloem, S: spongy cell, X: xylem. Scale bar = 100 μm.

Populations of the Vector *B. tabaci* in Japan

B. tabaci comprises a species complex containing a number of biotypes and genetic variants (haplotypes). (Brown, 2000, Costa & Brown, 1991, Perring, 2001). The B biotype of *B. tabaci* is the most widespread worldwide, due to its characteristics of extreme polyphagy, high fecundity, and insecticide resistance (Costa et al., 1993, Brown et al., 1995). The B biotype transmits begomoviruses to a large number of crop ornamental, and weed species (Brown & Bird, 1995).

In Japan, the B biotype of *B. tabaci* was first detected in 1989, and was associated with imported poinsettia plants (Ohto, 1990). Since then, it has dispersed rapidly throughout Japan (Matsui 1992, 1995). The Q biotype was detected in Japan in 2004 (Ueda & Brown, 2006, Matsuura, 2006). The use of neonicotinoids or pyriproxyfen insecticides positively selects for the Q biotype, which exhibits greater resistance to these insecticides than does the B biotype (Nauen et al., 2002, Horowitz et al., 2005). By 2009, the Q biotype had become widely distributed throughout Honshu, Shikoku, Kyushu, and Okinawa islands (Kijima et al., unpublished) in Japan (over 40 prefectures). In many regions, the displacement of the B biotype by the Q biotype has become particularly noticeable (Higuchi et al., 2007). This might be because chemical applications to cultivated crops select for the Q biotype. At present, the Q biotype has become the major vector population for TYLCV in Japan.

Japanese honeysuckle *Lonicera japonica* (honeysuckle, family: *Caprifoliaceae*), an evergreen perennial weed vine native to eastern Asia, has been identified as the natural wintering host of *Tobacco leaf curl Japan virus* (TbLCJV), *Honeysuckle yellow vein mosaic virus* (HYVMV), and related Far East Asian begomovirus species (Kitamura et al., 2004, Ueda et al., 2008b). The indigenous Japanese honeysuckle population of *B. tabaci* (designated as the *B. tabaci* JpL biotype), which was earlier described as *B. lonicerae* Takahashi, feeds on Japanese honeysuckle growing in wooded areas (Miyatake, 1980). The JpL-biotype of the *B.*

tabaci was considered to be the vector insect for indigenous begomoviruses in Japan (Osaki et al 1976, 1979).

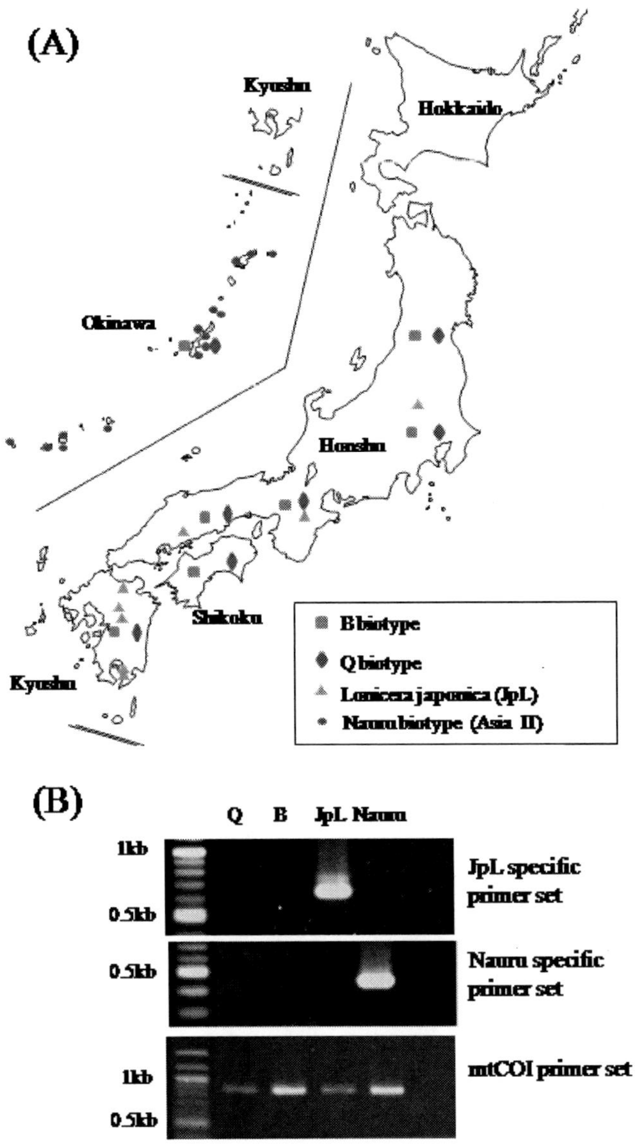

Figure 6. Four major populations (biotypes or genetic groups) of *B. tabaci* in Japan. (A) Map showing distribution of populations (Adapted from Ueda et al., 2009). (B) PCR-based detection of JpL and Nauru biotypes using the partial mtCOI sequences. To identify JpL biotype, primers BTJPL212 (5'-AGTTTTTGGTAGACTAGGA-3') and BTJPL213 (5'- GAAGTAATAAAGATTCAAG–3') were used for amplification of specific PCR fragments. To identify Nauru biotype, primers BTNauru224 (5'-GCAGGTTTAGTCCGCTTAGA -3') and BTNauru225 (5'- CTATACTATGACCAAGAGCC –3') were used for amplification of specific PCR fragments. Partial mtCOI sequences of *B. tabaci* were obtained by PCR amplification using primers MTD-10 (5'-TTGATTTTTGGTCATCCAGAAGT-3') and MTD-12 (5'- TCCAATGCACTAATCTGCCATATTA –3') (Frohlich et al., 1999). PCR was performed as described by Brown (2000). 100-bp ladder (Takara) was used as a marker (M).

Analyses of global phylogenetic relationships separate *B. tabaci* into 12 major genetic groups, designated as follows: Mediterranean/Asia Minor/Africa (includes the B biotype), Mediterranean (includes the Q biotype), Indian Ocean, sub-Saharan Africa silverleafing, Asia I, Australia, China, Asia II, Italy, New World, sub-Saharan Africa non-silverleafing, and Uganda sweet potato (Boykin et al., 2007). In Japan, little was known about the relationship between *B. tabaci* distribution and its molecular genetic background until invasion and dispersal of TYLCV. Using the Bayesian method, we analyzed sequences of mitochondrial cytochrome oxidase I (mtCOI) of various *B. tabaci* populations collected from most inhabited regions of Japan. The phylogenetic tree of *B. tabaci* separated into 12 major clades, together with an independent clade consisting only of JpL-biotype populations. In addition, we identified populations belonging to the Asia I, Asia II, and China genetic groups, together with invasive populations such as the B and the Q biotypes (Ueda et al., 2009). Those populations were distributed in the subtropical region of Japan (Figure 6A). Several distinct populations of *B. tabaci* have been reported from eastern Asia (De Barro et al., 2005, Hsieh et al., 2006, Li, 2006). The An, Nauru, and some other biotypes have been identified in Taiwan and China's mainland (Hsieh et al., 2006). Previously, we developed a PCR-RFLP method to identify the Q biotype in a number of field samples (Ueda, 2006). Here, we describe PCR primers designed from obtained unique mtCOI sequences that identify two major native populations (Figure 6B). It would be useful to survey distribution, prevalence, and variations of *B. tabaci* populations in Japan for early detection of new invasions.

Conclusion

Our research goal is to control TYLCV and *B. tabaci* without reducing the yields of affected crops and without imposing environmental burdens. Several management strategies have been improved for control of TYLCV and *B. tabaci* in Japan. However, due to global warming, the winter climatic conditions of the southwest region of Japan might become suitable for propagation of the B and the Q biotypes of *B. tabaci*. Infected wild tomato plants could overwinter in fields, as is the case in subtropical regions. Wild populations of the Nauru biotype (the putative Asian native strain) grow on many kinds of plants (De Barro et al., 1998, Hsieh et al., 2006, Ueda et al., 2009). To prevent whiteflies from penetrating into greenhouses, improved equipment such as fine-mesh screens (0.4–0.6 mm mesh size) and UV-absorbing plastic sheets have become popular, together with application of insecticides. Implementation of a mandatory host-free period in summer is a good strategy for farmers to reduce viral resource pressure. Recently, seven seed companies have released commercial TYLCV-resistant cultivars in Japan. The combination of such approaches and the fundamental knowledge we have obtained will help to establish effective and sustainable management practices to control TYLCV. As well, we need to control TYLCVs and insecticide-resistant populations of *B. tabaci* to avoid further spread of these diseases in Japan and around the world.

Acknowledgments

B. C. Lee was supported by a fellowship from the Korean government for long-term overseas studies.

References

Abhary, M., Patil, B. L. & Fauquet, C. M. (2007). Molecular biodiversity, taxonomy, and nomenclature of tomato yellow leaf curl-like viruses. In: Czosnek H (ed.), *Tomato yellow leaf curl virus disease*. Dordrecht: Springer; 2007; 85-118.

Boykin, L. M., Shatters, Jr, R. G., Rosell, R. C., McKenzie, C. L., Bagnall, R.A., De Barro, P. & Frohlich, D. R. (2007). Global relationships of *Bemisia tabaci* (Hemiptera: Aleyrodidae) revealed using Bayesian analysis of mitochondrial COI DNA sequences. *Molecular Phylogenetics and Evolution. 44*, 1306–1319.

Brown, J. K. (2000). Molecular markers for the identification and global tracking of whitefly vector-begomovirus complexes. *Virus Research. 71*. 233–260.

Brown, J. K. & Bird, J. (1995). Introduction of an exotic whitefly (Bemisia) vector facilitates secondary spread of Jatropha mosaic virus, a Geminivirus previously vectored exclusively by the "Jatropha" biotype. In: D. Gerling & R. T. Mayer (eds), Bemisia 1995: *Taxonomy, Biology, Damage, Control and Management*. Intercept, Andover, U.K., pp. 351–353.

Brown, J. K., Frohlich, D. R. & Rosell, R.C. (1995). The sweetpotato or silverleaf whiteflies: biotypes of *Bemisia tabaci* or a species complex? *Annual Review of Entomology. 40*, 511–534.

Costa, H.S. & Brown, J. K. (1991). Variation in biological characteristics and esterase patterns among populations of *Bemisia tabaci*, and the association of one population with silverleaf symptom induction. *Entomologia Experimentalis et Applicata. 61*, 211–219.

Costa, H.S., Brown, J. K., Sivasupramanium, S. & Bird, J, (1993). Regional distribution, insecticide resistance and reciprocal crosses between the A and B biotypes of *Bemisia tabaci*. *Insect Science and its Application* 14, 255–266.

De Barro, P. J., Liebregts, W. & Carver, M. (1998). The distribution and identity of biotypes of Bemisia tabaci (Gennadius) (Hemiptera: Aleyrodidae) in member countries of the Secretariat of the Pacific Community. *Australian Journal of Entomology 37*, 214–218.

De Barro, P. J., Driver, F., Trueman, J. W. H., & Curran, J, (2000). Phylogenetic relationships of world populations of *Bemisia tabaci* (Gennadius) using ribosomal ITS1. *Molecular Phylogenetics and Evolution. 16*. 29–36.

De Barro, P. J., Trueman, J. W. H. & Frohlich, D. R. (2005). *Bemisia argentifolii* is a race of *Bemisia tabaci* (Hemiptera: Aleyrodidae): the molecular genetic differentiation of *Bemisia tabaci* populations around the world. *Bulletin of Entomological Research 95*, 193–203.

Duffy, S. & Holmes, E. C. (2007). Multiple introduction of the Old World begomovirus *Tomato yellow leaf curl virus* into the New World. *Applied and Environmental Microbiology. 73*, 7114-7117.

Duffy, S. & Holmes, E. C. (2008). Phylogenetic evidence for rapid rates of molecular evolution in the single-stranded DNA begomovirus *Tomato yellow leaf curl virus*. *Journal of Virology. 82*. 957-965.

Fauquet, C. M, Bridoon R.M., Brown, J. K., Moriones, E., Stanley, J., Zerbini, M. & Zhou, X. (2008). Geminivirus strain demarcation and nomenclature. *Archives of Virology. 153*. 783-821.

Frohlich, D. R., Torres-Jerez, I, Bedford, I.D., Markham, P.G. & Brown, J.K. (1999). A phylogeographical analysis of the *Bemisia tabaci* species complex based on mitochondrial DNA markers. *Molecular Ecology. 8*. 1683–1691.

Haga, H. & Doi, M. (2002). The frequent occurrences and control of Tomato yellow leaf curl in Shizuoka Prefecture. *Shokubutuboueki 56*.153-156. (in Japanese)

Higuchi, S., Maeda, M., Gyoutoku, Y., Ueda, S. & Komaki, K. (2007). Distribution and seasonal fluctuations in the population of different biotypes of *Bemisia tabaci*(Gennadius)in Kumamoto Prefecture. *Kyushu Plant Protection Research. 53*. 59-65. (abstract in English)

Horowitz, A. R., Kontsedalov, S., Khasdan, V. & Ishaaya, I. (2005). Biotype B and Q of *Bemisia tabaci* and their relevance to neonicotinoid and pyriproxyfen resistance. *Archives of Insect Biochemistry and Physiology. 58*, 216–225.

Hsieh, C. H., Wang, C. H. & Ko, C. C. (2006). Analysis of *Bemisia tabaci* (Hemiptera: Aleyrodidae) species complex and distribution in Eastern Asia based on mitochondrial DNA markers. *Annals of the Entomological Society of America 99*. 768–775.

Kato, K, Onuki, M, Fuji, S. & Hanada, K. (1998). The first occurrence of *Tomato yellow leaf curl virus* in tomato (*Lycopersicon esculentum* Mill.) in Japan. *Annual Phyopathological Society of Japan. 64*. 552–559.

Kitamura, K., Murayama, A. & Ikegami, M. (2004). Evidence for recombination among isolates of *Tobacco leaf curl Japan virus* and *Honeysuckle yellow vein mosaic virus*. *Archives of Virology 149*. 1221–1229.

Li, Z. X. (2006). Molecular phylogenetic analysis reveals at least five genetic races of *Bemisia tabaci* in China. *Phytoparasitica. 34*. 431-440.

Mansoor, S., Briddon, R.W., Zafar, Y. & Stanley, J. (2003) Geminivirus disease complexes: an emerging threat. *Trends in Plant Science. 8*. 128–134.

Matsui, M. (1992). Irregular ripening of tomato fruit caused by the sweetpotato whitefly, *Bemisia tabaci* (Gennadius) in Japan. *Japanese Journal of Applied Entomology and Zoology. 36*. 47–49. (Abstract in English)

Matsui, M. (1995). The occurrence of a new whitefly, *Bemisia argentifolii* and its control measures in Japan. *Shokubutuboueki 49*. 111–114. (in Japanese)

Matsuura, A. (2006). The occurrence and control of the Q biotype of *Bemisia tabaci* in Miyazaki prefecture. *Kongetsunonougyou. 50*. 57-61. (in Japanese)

Miyatake, Y. (1980). A list of the whiteflies of Japan with their host plant and distribution data (Homoptera: Aleyrodidae) *Rostria 32*. 291–327. (in Japanese)

Monci, F., Sanchez-Campos, S., Navas-Castillo, J. & Moriones, E. (2002). A natural recombinant between the geminiviruses *Tomato yellow leaf curl Sardinia virus* and *Tomato yellow leaf curl virus* exhibits a novel pathogenic phenotype and is becoming prevalent in Spanish populations. *Virology 303*. 317-326.

Morilla, G., Krenz, B., Jeske, H., Bejarano, E. R. & Wege, C. (2004). Tête à tête of *Tomato yellow leaf curl virus* and *Tomato yellow leaf curl Sardinia virus* in single nuclei. *Journal of Virology 78*. 10715–10723.

Nauen, R., Stumpf, N. & Elbert, A, (2002). Toxicological and mechanistic studies on neonicotinoid cross resistance in Q-type *Bemisia tabaci* (Hemiptera: Aleyrodidae). *Pest Management Science. 58*. 868–875.

Ohto, K., (1990). Occurrence of the sweetpotato whitefly, Bemisia tabaci (Gennadius), on the poinsettia. *Shokubutubouеki 44*. 264–266. (in Japanese)

Onuki, M, Ogawa, T., Uchikawa, K., Kato, K. & Hanada, K. (2004). Molecular characterization and strain-specific detection of the *Tomato yellow leaf curl virus* occurring in Kyushu, Japan. *Bulletin of the National Agricultural Research Center for Kyushu Okinawa Region. 44*, 55–77 (abstract in English)

Osaki, T., Kobatake, H. & Inoue, T. (1976). Tomato yellow dwarf disease transmitted by *Bemisia tabaci*. *Shokubutu Boueki 30*. 458-462. (abstract in English)

Osaki, T., Kobatake, H. & Inoue., T. (1979). Yellow vein mosaic of honeysuckle (*Lonicera japonica* Thunb.), a disease caused by tobacco leaf curl virus in Japan. *Annals of the Phytopathological Society of Japan 45*: 62-69.

Perring, T. M. (2001). The *Bemisia tabaci* species complex. Crop Protection. 20, 725–737.

Salati, R., Nahkla, M. K., Rojas, M. R., Guzman, P., Jaquez, J., Maxwell, D. P., & Gilbertson, R. L. (2002). Tomato yellow leaf curl virus in the Dominican Republic: characterization of an infectious clone, virus monitoring in whiteflies, and identification of reservoir hosts. *Phytopathology 92*. 487–496.

Uchikawa, K., Ueda, S. Onuki, M. & Hanada, K. (2002). Occurrence of leaf curl of *Eustoma russellianum* caused by tomato yellow leaf curl virus. *Japanese Journal of Phytopathology. 68*. 50. (in Japanese)

Ueda, S. (2006). Simple and rapid detection to distinguish Q biotype of *Bemisia tabaci* by mtCOI PCR-RFLP. Kyushu Plant Protection Research. 52. 44-48. (abstract in English)

Ueda, S., Kimura, T., Onuki, M., Hanada, K. & Iwanami, T. (2004). Three distinct groups of isolates of *Tomato yellow leaf curl virus* in Japan and construction of infectious clone. *Journal of General Plant Pathology. 70*, 232–238.

Ueda, S., Takeuchi, S., Okabayashi, M., Hanada, K., Tomimura, K. & Iwanami, T. (2005). Evidence of a new *Tomato yellow leaf curl virus* in Japan and its detection using PCR. *Journal of General Plant Pathology. 71*, 319–325.

Ueda, S & Brown, J. K. (2006). First report of the Q biotype of *Bemisia tabaci* in Japan by mitochondrial cytochrome oxidase I sequence analysis. *Phytoparasitica. 34*, 405–411.

Ueda, S., Onuki, M., Kijima, K., Futagami, K., Kinjo, K., Murayama, Y., Taniguchi, M. & Kawano, S. (2008a). Introduction and molecular characterization of Tomato yellow leaf curl virus in Okinawa, Japan. *Japan Agricultural Research Quarterly. 43*. 19-24.

Ueda, S., Onuki, M., Hanada, K., & Takanami, Y. (2008b). Unique grouping of the Far East Asian begomovirus complex based on sequence analyses of the DNA-A genome and associated DNAβ satellite molecules isolated from tomato, honeysuckle and Eupatorium plants in Japan. *Archives of Virology. 153*. 417-426.

Ueda, S., Kitamura, T., Kijima, K., Honda, K., & Kanmiya, K. (2009). Distribution and molecular characterization of distinct Asian populations of *Bemisia tabaci* (Hemiptera: Aleyrodidae) in Japan. *Journal of Applied Entomology. 133*. 355-366.

In: Emerging Geminiviral Diseases and their Management ISBN 978-1-61668-620-8
Editors: P. Sharma, R. K. Gaur et al. pp. 195-208

Chapter 10

GEMINI VIRUS-A THREAT TO GREEN LIFE

V. Bajpai[1], P.K. Jain[1], P. Sharma[2], Pradeep Sharma[3] and R. K. Gaur[1]
[1]Department of Science, MITS, Lakshmangarh, (Sikar) Rajasthan-332311, India;
[2]Department of Microbiology, Bundelkhand University Jhansi (U.P) 284128, India;
[3]Directorate Research of Wheat Karnal, Haryana, India.*

ABSTRACT

Whitefly-transmitted Gemini viruses of the genus *Begomovirus* are important pathogens of a wide range of crop eco-systems. Geminiviruses are a family of plant viruses that contain a circular single-stranded DNA (ssDNA) and replicate through a rolling-circle mechanism, en-capsidated in geminate particles. The Geminiviridae family includes a large number of viruses. The particles of gemini viruses are quasi-isometric. They are called gemini (twin) viruses, because they are usually found in pairs. Each particle has a diameter of just 15 – 20 nm. This group of viruses consists of approximately 38 species of extremely narrow host ranges. The group is peculiar and noted for its nucleic acid and is grouped as whitefly transmitted (WFT) viruses. Gemini viruses belong to the smallest virus particles able to multiply without a helper virus. They have a circular DNA with a molecular weight of 0.7– 0.8x106 (2,500 base pairs). The genome of Gemini viruses consists of two molecules of DNA of almost equal size, but different sequences. The isolated circular DNA alone is not infectious. In infected host cells, the nucleus holds the chief amount of viral DNA. Symptoms vary from mosaic, yellow leaf, curling, rolling of leaves, thickening of veins, reduction of fruit set, yellow mottling, crinkling, severe stunting, reduced yields and at times death of plants. On infection typical mosaic-like leaf patterns of light and dark green occur. The infection spreads often over the whole leaf beginning at the veins. Leaves that had been infected during their development are usually deformed or volute.

INTRODUCTION

Gemini virus group is the large plant virus group. The Geminiviridae family includes a large number of viruses that infect plants and produce in many cases very significant

* Corresponding author: R.K.Gaur E-mail: gaurrajarhi@hotmail.com

reductions in economically important crops of both monocotyledonous and dicotyledonous plants. It contains viruses with DNA as the genome; however unlike caulimo viruses, the particles of Gemini viruses are quasi-isometric. They are called Gemini (twin) viruses, because they are usually found in pairs. Each particle has a diameter of just 15 – 20 nm (Harrison et. al.1999) Gemini viruses belong to the smallest virus particles able to multiply without a helper virus. Gemini viruses contain ss DNA as the genome and recent evidence suggests that the Gemini viruses are multipartite viruses. They are, basically, characterized by two distinctive features: (i) The morphology of the virion particle which is geminate, 18–30 nm in size, and has the appearance of two quasi-icosahedral moieties with a total of 22 pentameric capsomers, and (ii) The nature of their genetic material, consisting of one or two single stranded DNA (ssDNA) molecules, 2.5–3.0 kb in length. Gemini viruses have small spherical particles many of which occur in pairs both in vivo and in purified preparations. The individual particles of each pair of chloris striate mosaic virus (CSMV) look pentagonal while the paired particles consist of two incomplete icosahedral and possess T =1 symmetry with 22 pentameric capsomeres (Hatta, 1976).

About 15 years ago, when plant scientists first seriously studied geminiviruses. The viruses, so called because they consist of two twinned particles, were known to be pathogens, but their range seemed to be limited, and they looked like promising vehicles to carry genes for desirable new traits into plants. Gemini viruses lack of the necessary enzymes, they are emerging as serious enemies that are devastating crops worldwide. The geminivirus family derives its name from the unusual twin icosahedral (geminate) capsid structure of the members. Each paired particle encapsidates a single molecule of covalently closed, circular, single stranded DNA (ssDNA) varying in size from 2.5 kb to 3.0 kb, depending on the virus. The genomic ssDNA is replicated in the nucleus of the host cell by a rolling-circle mechanism utilizing double-stranded DNA (dsDNA) intermediates in a manner similar to the ssDNA-containing bacteriophages. Viral dsDNAs, which include open and covalently closed (supercoiled) circular forms, are organized as minichromosomes apparently complexed with histone proteins (Pilartz and Jeske, 1992). The nucleus is also the site of virus assembly, and nuclear inclusions consisting of large numbers of virus particles have been observed in infected cells by transmission electron microscopy (Esau 1977; Kim et. al., 1978; Rushing et al., 1987).

"In the past, not much attention was given to geminiviruses because they hunkered down in remote places," says Claude Fauquet, Director of the International Laboratory for Tropical Agricultural Biotechnology at the Donald Danforth Plant Science Center in St. Louis. But global transportation networks have aided the spread of the virus and its carrier, the white fly,

Robert Goodman, University of Wisconsin, Madison, and Brian Harrison and colleagues of the Scottish Crop Research Institute in Dundee first described geminiviruses in the mid-1970s, but records from Africa suggest they probably caused diseases of maize and cotton there in the 1930s or earlier. They have also been attacking bean fields in the tropics of the Western Hemisphere for decades. But geminiviruses didn't draw much scientific attention until early in this decade, when serious outbreaks flared up in Africa, India, Pakistan, southern Europe, South and Central America, and elsewhere.

The viruses also struck in the Caribbean in the early 1990s, and the first serious geminivirus infections appeared in Florida and Texas in the United States at about the same time. Since then, they have been spotted, in 1997, as far north as Virginia, South Carolina, and Tennessee.

Geminiviruses are versatile and can infect everything from monocots such as maize to dicots such as cassava and tomato. The infections can produce leaf mottling that interferes with photosynthesis, decreasing yields of starchy foods such as cassava, and they also disrupt flower and fruit formation in crops such as tomato, pepper, and cotton.

The transportation networks that are spreading the virus from tropical to temperate regions have also brought different species. There are more than 66 of them into close contact with virus. This may have set the stage for the viruses to swap DNA, a phenomenon first noted in 1997 by Fauquet and Roger Beachy. So far, the researchers have identified more than 1000 recombinant strains, in some cases involving species whose DNA sequences indicate that they are not closely related. "It is like a recombinant between a human and a chimpanzee." And while the virus generates new, virulent strains, measures to control it are falling short. Plowing under plant debris after harvest limits the food supply for the white flies and the virus, but the measure has limited effectiveness, as do insecticides. White flies are developing resistance to commonly used chemicals.

So researchers are searching for resistant cultivars, created either by classical crossbreeding or through genetic engineering. Crossbreeding has produced more resistant cassava, beans, and tomatoes, although the new resistant varieties have such drawbacks as small fruit or poor taste. And although modern gene transfer techniques are faster and more precise than classical plant breeding, the diverse, changeable viruses have already demonstrated that they can outwit the genetic engineers.

To create virus-resistant varieties, researchers equip the plants with the genes for certain viral proteins--a modification that, for unknown reasons, protects the plant cells from the same virus. In 1992, in the wake of the devastation of the Florida tomato crop, virologist Jane, Bradenton, and Ernest Hiebert, University of Florida, put a gene for geminivirus replicase the enzyme that copies the viral DNA into tomatoes. The resulting plants had good horticultural qualities and were resistant to the virus, but before the new tomato variety could be commercialized, a new gemini variant moved in on Florida tomatoes that had no problems infecting the plants.

So researchers are also attempting to endow plants with broad-scale resistance to the rapidly expanding family of gemini pathogens. One strategy Fauquet and his colleagues are exploring is to equip plants with a protein from a bacterial virus that binds viral DNA and prevents movement of geminiviruses within plants, thereby blocking infection. They report good lab results and plan to test their transgenic plants in greenhouses in the spring of 2000. Researchers are also hoping to arm susceptible plants with a protein that attacks a structural element to the replicase proteins of many geminis.

FAMILY GEMINIVIRIDAE

General Features

The geminivirus family derives its name from the unusual twin icosahedral (geminate) capsid structure of its members. Each paired particle encapsidates a single molecule of covalently closed, circular, single stranded DNA (ssDNA) varying in size from 2.5 kb to 3.0 kb, depending on the virus. They have a circular DNA with a molecular weight of 0.7 – 0.8 x

10^6 (about 2, 500 base pairs). In the case of some gemini viruses, it has been proven, in the case of others, it is assumed, that the genome consists of two molecules of DNA of almost equal size, but different sequence. The nucleotide sequences of some species are known.

The genomic ssDNA is replicated in the nucleus of the host cell by a rolling-circle mechanism utilizing double-stranded DNA (dsDNA) intermediates in a manner similar to the ssDNA-containing bacteriophages. Viral dsDNAs, which include open and covalently closed (supercoiled) circular forms, are organized as minichromosomes apparently complexed with histone proteins (Pilartz and Jeske 1992). The nucleus is also the site of virus assembly, and nuclear inclusions consisting of large numbers of virus particles have been observed in infected cells by transmission electron microscopy (see, e.g., Esau 1977; Kim et al., 1978; Rushing et al., 1987).

Viral Genome Organization

The geminivirus family is diverse and can be divided into at least three subgroups based on genome organization, host range, and type of insect vector (Matthews 1991). Subgroup I comprise viruses that have a single component genome, are transmitted by leafhoppers, and usually infect monocotyledonous plants. Well-characterized subgroup I pathogens include maize streak virus (MSV) and wheat dwarf virus (WDV).

The second subgroup until recently contained only a single member, beet curly top virus (BCTV). BCTV also has a monopartite genome and is transmitted by a leafhopper vector, but it infects only dicotyledonous plants. The genome organization of BCTV also differs in significant ways from MSV and its relatives (Figure 1). Subgroup III contains viruses with monopartite genomes (-2.8 kb) as well as bipartite viruses that have two genome components, each of similar size (-2.6 kb). In the latter, the two genomic DNAs, designated A and B, differ in sequence except for a common region of 200-250 bp that is nearly identical in the genome components of any given virus, but differs between viruses. Both genome components are required for infectivity (Hamilton et al., 1983; Stanley 1983), although DNA A contains all viral information necessary for replication and encapsidation (Rogers et al. 1986; Townsend et al., 1986; Sunter et al., 1987). The B component provides genes whose functions are required for cell-to-cell and systemic spread of the virus from the inoculation site.

Genome Structure

Two families of plant viruses, the *Geminiviridae* and *Circoviridae*, have ssDNA genomes. In some Gemini viruses the genome is a single covalently closed circle of about 2.5 – 3.0 kb. The genome of others consists of two such circles of about the same size. Generally known as DNAs 1 and DNAs 2 or A and B. Goodman (1977) showed that isolated Gemini virus DNA was infectious.

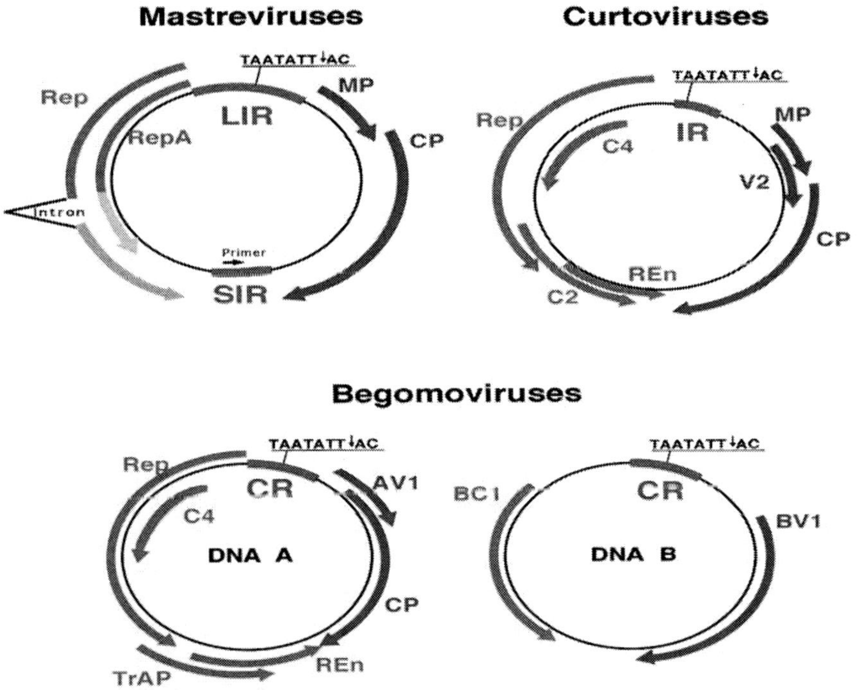

Figure 1. Genome organization of the three geminivirus genera. The genetic organization of the dsDNA forms are represented. The sequences regulating DNA replication and transcriptional activity are located in the intergenic regions. Mastreviruses contain a large (LIR) and a small (SIR) intergenic region, to which a small DNA molecule (primer) is associated. The invariant TAATATT_AC sequence located in the LIR (mastreviruses), IR (curtoviruses) and CR (begomoviruses) containing the initiation site (_) of rolling-circle DNA replication is shown. Arrows indicate the viral proteins which have been named according to either their function, if known, or their genetic location and direction of transcription: MP, movement protein; CP, capsid protein; Rep, replication protein; TrAP, transcriptional activator; REn, replication enhancer. The RepA protein is also shown in mastreviruses.

There are four genera have been described within the family *Geminiviridae* that are distinguished from one another by their genome organization and their vectors. The Mastrevirus, the Curtovirus, the Begomovirus and the Topocuvirus. The genomes of all four genera comprise circular ssDNA and either as one species or divided between two species. The main feature of the genome that these genomes are contained in geminate virus particles consisting of two incomplete icosahedra. The genera are distinguished on their genome organizations. However all Gemini viruses have a conserved genome sequence. There is a large non-coding intergenic region, termed the common region, in the two component Gemini viruses, features of which are found in the similar genomic position in the single component Gemini viruses. This region has sequences capable of forming a hairpin loop. Within this loop is a conserved sequence TAATATTAC found in all Gemini viruses.

1. **Genus Mastrevirus** (type species: Maize streak virus)(Palmer and Rybicki, 1998) This genus contains the economically important MSV. All members have narrow host ranges and, with the exception of TYDV and BeYDV which infect dicotyledenous species, their host ranges are limited to species in the Poaceae. Mastrevirus are transmitted by leafhoppers in a circulative, non-propagative manner.

They have genomes comprising a single component of ssDNA. The nucleotide sequence of the MSV genome revealed a single DNA circle of 2687 nucleotides that had no detectable overall sequence similarity with genomes of geminivirus with two DNA components (Mullineaux et. al., 1984). Analysis of ORFs with potential for polypeptides of 10kDA or larger revealed four potential coding regions arranged as shown in figure. There are two intergenic regions, the long intergenic region (LIR) and the short intergenic region (SIR). This is notable that not all but most of the transcription starts in the LIR and terminates in the SIR. As can be seen from figure two of the ORFs ar on the viral strand (given the designation V) and two are on the complementary strand (given the designation C). ORF V1 encodes the cell to cell movement protein, ORF V2 the virion coat protein and ORFs C1 and C2 encode the two replication associated proteins.

2. **Genus Curtovirus** (type species: Beet curly top virus) the type member of this genus used to be economically important virus of sugar beet and it has a wide host range. Cutoviruses are transmitted by leafhopper in the circulative non-propagative manner. They have a monopartite ssDNA genome. The sequence of 2993 nucleotides in an infectious clone of BCTV has been established (Stanley et. al. 1986). The organization of the single component genome revels that curto virus DNAs produce six or seven proteins. The Product of ORF V1 is the virion coat protein that of ORFs V2 is involved in regulation of the genomic ss and ds DNA levels, and that of ORF V3 facilitates the cell to cell movement of the virus.The four proteins encoded by the complementary sense ORFs are involved in virus replication. ORF C1 encodes the Rep protein, C3 a protein analogous to the replication enhancer protein of begomoviruses and C4 a protein tha can initiate cell division;the function of the product of ORF C2 is unknown.
3. **Genus Begomovirus** (type species: Bean gold mosaic virus) this is the largest Gemini virus family member of which have narrow host range among dicotyledonous species. They are transmitted by white flies. Most members of this genus have the genomes divided between two DNA molecules though the genomes of some members are monopartite. The full nucleotide sequence of ACMV DNA showed that the genome consists of two ssDNA circles of similar size (Stanley and Gay, 1983). Stanley demonstrated that both DNAs are needed for infectivity. The nucleotide sequences of the two DNAs are very different except for a 200 nucleotide non coding region, the common region which is almost identical on each DNA. The two DNAs have been arbitrarily labeled A and B. Most of the old world and all the new world begomo viruses have two component genomes. The two component begomo viruses encode five or six genes on DNA A, one or two on the virion-sense strand (termed AV) and four in the complementary sense strand (AC). DNA B encodes two proteins, one on the virion sense strand (BV) and one on the complementary strand (BC). The product of ORF AV1 is the coat protein that of AV2 the cell to cell movement protein. The proteins encoded by the complementary sense of DNA A are all involved with replication and expression.AC1 is the Rep protein, AC2 the transcriptional activator protein and AC3 the replication enhancer protein; the function of AC4is unknown.BV1 encoded on the virion sense of DNA B, is the nuclear shuttle protein and BC1 is involved in cell to cell movement.

4. **Genus Topocuvirus** (type species: Tomato pseudo-curly top virus) this is a very recently designated genus that has been split off from the Curtovirus genus. It has a similar genome organization to the curtoviruses but it transmitted by the treehopper, *Micruta malleifera*. The single component genome of TPCTV has properties of both mastreviruses and begomoviruses (Briddon et. al., 1996). The coat protein has features more akin to the leafhopper-transmitted mastreviruses than to those transmitted by whiteflies whereas the organization of the complementary-sense genes is similar to that of the single-component begomoviruses.

Replication of Gemini Virus

Many features of the Gemini virus replication cycle have recently been elucidated, though there are still several points that are poorly understood. The geminivirus replicative cycle relies entirely on DNA intermediates and occurs within the nucleus of the infected cell. Recent and rapid advances in the elucidation of the molecular and cellular biology of geminiviruses make them excellent model systems to study DNA replication and cell cycle events in plants. In addition, insights into their replication cycle and the molecular mechanisms of virus-cell interactions should help in the rational design of novel strategies for disease control. The elucidation of the replication cycle has revealed several aspects of the normal cell cycle as Gemini virus replication depends upon many host functions. (Hanley Bowdoin et. al. 1999, Gutierrez, 1999, 200a). Gemini viruses are circular single-stranded DNA viruses, which replicate through rolling circle replication (RCR) mechanism. A considerable body of evidence supports the idea that geminivirus replication occurs by a rolling-circle mechanism (Saunders et al. 1991; Stenger et al. 1991). Rolling-circle replication (RCR) is also employed by the ssDNA-containing coliphages (e.g., 4x174) (Kornberg and Baker 1992) and certain bacterial plasmids (Koepsel et al. 1985; Gros et al. 1987). The parvoviruses employ an analogous rolling-hairpin mechanism (Im and Muzyczka 1990). A characteristic feature of RCR is the involvement of a replication initiator protein (Rep) with a nicking-closing activity similar to that found in topoisomerases.

The initiation of RCR is by the replication initiator protein (Rep) encoded by the DNA replicons. The current analysis was carried out between replication initiator proteins of all groups of organisms undergoing RCR to understand the evolution of the geminivirus Rep. The geminivirus Rep is a multifunctional protein: i.e. it cleaves and ligates single-stranded DNA in the invariant sequence of the hairpin loop (Laufs *et al* 1995; Orozco and Hanley-Bowdoin 1996), hydrolyzes ATP (Desbiez *et al* 1995; Orozco *et al* 1997) and has a site-specifi c topoisomerase activity (Pant *et al* 2001). Rep interacts with itself (Orozco *et al* 1997), the viral replication enhancer protein AC3 (Settlage *et al* 1996), and a maize homologue of the cell cycle regulatory protein, retinoblastoma (Ach *et al* 1997).The DNA - DNA replication of Gemini viruses should result in less mutagenic variation than RNA-RNA replication. However there is evidence for significant genomic variation in at least two two of the geminivirus genera, the begomoviruses and mastreviruses. There iswidespread genomic and serological variation in the begomoviruses (Harrison and Robinson, 1999).

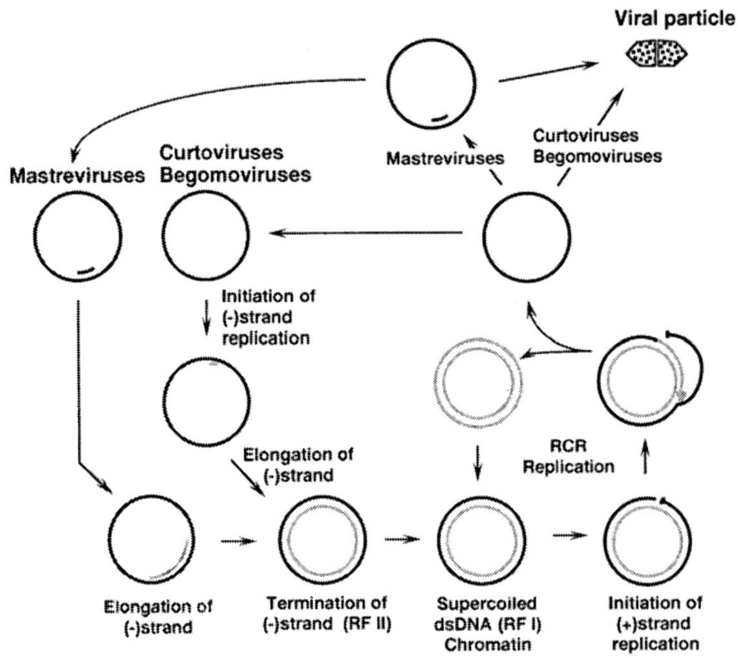

Figure 2. Summary of the geminivirus replicative cycle.

Transmission of Virus

The virus is not easily transmitted by mechanical ways. It seems to be confined to the phloem tissue, where it causes degeneration, other tissues being apparently toxic to it. Virus particles exhibit considerable resistance to a number of common disinfectants and can remain active for 10 months in dried phloem exudate, 4 months in dried sugarbeet tissue and 6 months in dried sugarbeet leafhoppers. The main vectors are phloem-feeding leafhoppers, such as Circulifer tenellus in North America and C. opacipennis in the Mediterranean area. There seem to be two types of transmission of the virus by the insect: a purely mechanical transmission by contaminated mouthparts, and circulative transmission in which there is a delay of 21-24 h in the development of infective power of the insect. It is not thought that the virus multiplies in the insect vector. The leafhoppers can acquire the virus in a matter of minutes during sucking on infested material and retain it for a month or even more (Duffus, 1986). For more information, see Bennett (1935; 1971), Kheyri & Alimoradi (1969), Smith (1972).

Insects (greenhouse whitefly, grasshoppers, and others) help usually in spreading Gemini-viruses in nature. These viruses can cause considerable damage to agriculture. Diseases caused by Gemini viruses have long been recognized as a limitation to the cultivation of several important crops, including maize, cassava, bean, squash, cucurbits, and tomato, in tropical and subtropical regions of the world. More recently, Gemini-virus diseases, particularly those transmitted by whiteflies, have become an even greater threat to agriculture due to the appearance of a new and more aggressive whitefly biotype (Brown et al. 1995).

Whiteflies (Aleyrodidae)

These genera of viruses are transmitted by whiteflies; the Begomovirus of the Geminiviridae family is one of the suitable examples. Most of these viruses are found in the tropics and subtropics where they can be substantial importance.

Begomoviruses are transmitted exclusively by *Bemisia tabaci* whereas the whitefly transmitted closterovirus and the crini-viruses are transmitted by glass house whitefly, *Trialeuroides vaporariorum,* (BPYV, TICV &CCSV), *T.abulilonca* (DVCV and AYV), *B.tabaci* (CYSDV and SPSVV) *and B.argentifolia*(LCV).

The most studied vector is *B.tabaci* which is important in the epidemiology of several major diseases. Only the first instar of the larva is mobile, and it does not move far. Adults are winged, and many generations may be produce in a year. The nymphs of B.tabaci are phloem feeders. The B type of *B.tabaci*, some times name *B. argentifolia* is of increasing importance as a vector.

The general morphology of whiteflies is reviewed by Gill (1990) and details of the ultra structure of the mouthparts of *B.tabaci* have been studied by Rosell et. al. (1995). Adult whitefly mouthparts are similar to those of other homopterans, especially aphids and comprise the labrum, the lamium and the stylets. (Jansen et. al., 1989) Most bipartite begomoviruses are transmitted in the persistent circulative manner. Virus particles have been observed in the gut epithelial cells and associated with salivary glands of whitefly vectors and thought to follow a similar circulative route to those of luteoviruses (Cohen and Antignus, 1994;Hunter et. al. 1998).

Unique about Gemini Virus

Much attention has been focused on the Gemini viruses as potential gene vectors because of their DNA genomes and the fact that the small size of the genomes makes them convenient for in vitro manipulations. (Mullineaux et. Al 1992; Stanley 1993; Timmermans et. Al. 1994; Palmer and Rybicki, 1997) Nevertheless this small size may restrict the amount of viral DNA that can be deleted (Davies et. Al. 1987) this point is counterbalanced by the fact that for some Gemini viruses a viable coat protein and encapsidation are not necessary for successful inoculation by mechanical means or for systemic movement through the plant. (Gardiner et. al. 1988). Isolated circular DNA alone is not infectious. In infected host cells, the nucleus holds the chief amount of viral DNA. It is therefore assumed that the nucleus is also the place of its replication. It looks as if – analogous to the TMV – double-stranded intermediates would exist (= replication state: RF), since double-stranded DNA has been isolated from infected cells. Gemini viruses are of interest to genetic engineers, since they can infect monocots where their DNA enters the nucleus, too, and since the interest in monocot vectors is still large. The bean golden mosaic-virus (BGMV), the cassava latend-virus (CLV), the tomato golden-mosaic-virus (TGMV), the maize-streak virus (MSV), and the abutilon mosaic-virus belong all to the Gemini-virus family.

There are some other potential difficulties (Davies et. Al 1987).Recombination can occur to give parent type molecules. Most Gemini-viruses are restricted mainly to the phloem and associated cells. However the wide host range of Gemini-viruses makes them of considerable interest. The ability of some members to infect cereal crops would be particularly useful, except for the fact that they are not seed transmitted and are mechanically transmitted only with difficulty.

Some model experiments have shown that a Gemini-virus can be used to introduce and express foreign genes in plants. Various workers have shown that an intact coat protein gene is not essential for TGMV replication and movement.

ACKNOWLEDGMENT

R.K.Gaur is thankful to Department of Biotechnology, and Department of Science and Technology, New Delhi, India for providing financial support for the present study.

REFERENCES

Ach R A, Durfee T, Miller A B, Taranto P, Hanley-Bowdoin L, Zambriski P C and Gruissem W (1997). An alternatively spliced, multigene family in maize encodes retinoblastoma-related proteins which can interact with a plant D type cyclin and a geminivirus replication protein; *Mol. Cell. Biol.* **17** 5077–5086 Bennett, C.W. (1935) Studies on properties of the curly-top virus. *Journal of Agricultural Research 50*, 211-241.

Bennett, C.W. (1971) The curly top disease of sugarbeet and other plants. *Monograph, American Phytopathological Society No. 7*, 81 pp.

Briddon, R.W., Bedford, I.D., Tsai, J.H., Markham, P.G. (1996) Analysis of the nucleotide sequence of the treehopper transmitted Gemini virus, tomato pseudocurly top virus, suggests are combinant origin. *Virology 219*, 387-395.

Brown, J.K., D.R. Frohlich, and R. Rosell. (1995). The sweet potato / silver leaf whiteflies: Biotypes of *Bemisiu tabaci* Genn., or a species complex? *Annu. Rev. Entomol. 40*:511-534

Cohen and Antignus, (1994). Tomato yellow leaf curl virus, A whitefly borne Gemini virus of tomatoes. *Adv. Dis. Vector Res.10*, 259-288.

Crisanto Gutierrez, Elena Ramirez-Parra, M. Mar Castellano, Andres P. Sanz-Burgos, Alejandro Luque, Riccardo Missich (2004) Gemini virus DNA replication and cell cycle interactions *Veterinary Microbiology, Volume 98,* Issue 2, 4 February 2004, Pages 111-119.

Davies, J.W., Townsend, R., and Stanley, J. (1987) The Structure expression functions & possible exploitation of Gemini-virus genomes. *Adv, Plant Sci. 4*:31-52.

Desbiez, C., C. David, A. Mettouchi, J. Laufs, and B. Gronenborn. (1995). Rep protein of tomato yellow leaf curl Gemini virus (TYLCV) has an ATPase activity required for viral DNA replication. *Proc. Nutl. Acad. Sci. 92*: 5640-5644.

Donson, J., G.P. Accotto, M.I. Boulton, P.M. Mullineaux, and J.W. Davies. (1987). The nucleotide sequence of a Gemini virus from *Digitaria sunguinalis. Virology 161*:160-169

Duffus, J.E. (1986). Beet curly top. In: *Compendium of beet diseases and insects*, pp. 31-32. American Phytopathological Society Press, St Paul, Minnesota, USA.

Esau, K. (1977). Virus-like particles in nuclei of phloem cells in spinach leaves infected with the curly top virus. J. *Ultrustruct Res. 61*: 78-88.

Gardiner, W.E., Sunter, *G,* Rushing, A.E. Rogers, S.G. and Bisaro. D.M. (1987) Independent encapsidation of tomato golden mosaic virus A component DNA in transgenic plants. *Plant Mol. Biol. 8*: 477-484.

Gardiner, W.E., Sunter, G., Brand, L., Elmer, J.S., Rogers, S.G., & Bisaro, D.M., (1988) Genetic analysis of tomato golden mosaic virus;The coat protein is not required for systemic spread or symptom development. *EMBO J. 7*, 899-904.

Gill, R.J., (1990). The morphology of whiteflies: In D.Gerling (ed.) *Whitflies: Their Bionomics*, Pest status & management.P.P.13-46, Intercept Andover. U.K.

Goodman, R.M. (1977). Infectious DNA from a whitefly-transmitted virus of *Phaseolus vulgaris. Nature 266*: 54-55.

Gros, M.F., H. te Riele, and S.D. Ehrlich. (1987). Rolling circle replication of single stranded DNA plasmid pC194. *EMBOJ. 6*: 3863-3869.

Gutierrez (1999) Gemini virus DNA replication. *Cell. Mol.Life Sci.56*, 313-329.

Gutierrez (2000a) DNA replication and cell cycle in plants: learning from Gemini viruses. *EMBO. J. 19*, 792-799.

Gutierrez (2000b) Geminivirus and the plant cell cycle. *Plant Mol. Biol. 43*, 763-772.

Hamilton, W.D.O., D.M. Bisaro, R.H.A. Coutts, and K.W. Buck. (1983). Demonstration of the bipartite nature of the genome of a single-stranded DNA plant virus by infection with the cloned DNA components. *Nucleic Acids Res. 11*: 7387-7396.

Hamilton, W.D.O., R.C. Sanders, R.H.A. Coutts, and K.W. Buck. (1981). Characterization of tomato golden mosaic virus as a Gemini virus. *FEMS Microbiol. Lett. 11*: 263-267.

Hanley Bowdoin, L., Settlage, S.B., Orozco, B.M., Nagar, S. and Robertson, D. (1999) Gemini viruses: Models for plant DNA replication, transcription and cell cycle regulation. *Crit. Rev. Plant. Sci.18*, 71-106.

Hanley-Bowdoin, L., Elmer, J.S. and Rogers, S.G. (1988) Transsient Expression of heterologous RNAs using Tomato Golden Mosaic Virus.Nucl.Acids.Res.*16*, 10511-10528.

Hanley-Bowdoin, L., J.S. Elmer, and S.G. Rogers. (1990). Expression of functional replication protein from tomato golden mosaic virus in transgenic tobacco. *Proc. Natl. Acad. Sci. 87*: 1446-1450.

Harrison, B. D. (1985) Advances in geminivirus research Annu. Rev. Phytopathol. 23, 55-82.

Harrison, B.D. and Robinson, D.J. (1999) Natural Genomic and antigenic variation in whitefly transmitted geminiviruses (Begomoviruses) *Anu.Rev. Phytopathol.37*, 369-398.

Hatta, T. (1976). Recognition and measurement of small isometric virus particles in thin sections. *Virology 69*, 237-245.

Hunter, W. B., Hiebert, E., Webb, E., Tsai, J. H. and Polston, J. E. (1998) Location of Gemini viruses in the whitefly *Bemisia tabaci* (Homoptera: Aleyroididae), *Plant Dis. 82*, 1147-1151.

Im, D.S. and N. Muzyczka. (1990). The AAV origin binding protein Rep68 is an ATP dependent site-specific endo-nuclease with DNA helicase activity. *Cell 61*: 447-457.

Janssen, J.A.M., Tjallingii, W.F. and Van Lenteren, J.C. (1989) Electrical recording and Ultrastructure of stylet penetration by the green house whitefly, *Entomol. Exp. Appl. 52*, 69-81.

Kheyri, M.; Alimoradi, I. (1969) *The leafhoppers of sugarbeet in Iran and their role in curly-top virus disease*. Sugarbeet Seed Institute Karaj, Entomological Research Division, Tehran, Iran.

Kim, K.S., T.L. Shock, and R.M. Goodman. (1978). Infection of *Phaseolus vulgaris* by bean golden mosaic virus: Ultra-structural aspects. *Virology 89*: 22-33.

Koepsel, R.R., R.W. Murray, W.D. Rosenblum, and S.A. Khan. (1985). The replication initiator protein of plasmid pT181 has sequence-specific endo-nuclease and topoisomerase activity. *Proc. Natl. Acad. Sci. 82*: 6845-6849.

Kornberg, A. and T.A. Baker. (1992). *DNA replication.* W.H. Freeman, San Francisco.

Laufs, J., W. Traut, F. Heyraud, V. Matzeit, S.G. Rogers, J. Schell, and B. Gronenborn. (1995). In *vitro* cleavage and ligation at the viral origin of replication by the replication protein of tomato yellow leaf curl virus. *Proc. Natl. Acad. Sci. 92*: 3879-3883.

Mullineaux, P.M., J.W. Davies, and C.J. Woolston. (1992). Gemini viruses as gene vectors. In *Genetic engineering with plant viruses* (ed. T.M.A. Wilson and J.W. Davies), pp. 187-215. CRC Press, Boca Raton, Florida.

Mullineaux, P.M., Donson, J., Marris-krsinich, B.A.M., Boulton, M.I. and Davis, J.W. (1984).The nucleotide sequence of Maize streak virus DNA. *EMBO.J.3*, 3063-3068.

Orozco, B. M., Miller, A. R., Settlage, S. B. and Hanley-Bowdoin, L. *(*1997) Functional domains of a geminivirus replication protein. *J. Biol.Chem. 272*, 9840-9846.

Orozco, B.M. and. Hanley-Bowdoin, L. (1994) A DNA sequence required for Gemini virus replication also mediates transcriptional regulation. *Plant Cell 6*:1157-1170

Orozco, B.M., Gladfelter, H.J., Settlage, S. B., Eagle, P. A. Gentry, R. and Hanley-Bowdoin, L. (1998) Multiple cis elements contribute to geminivirus origin functions. *Virology, 242*, 346-356.

Palmer, K.E. and Rybicki, E.P. (1997) The use of geminiviruses in biotechnology and plant molecular biology with particular focus on master viruses. *Planr Sci., 129*, 115-130.

Palmer, K.E. and Rybicki, E.P.(1998) The Molecular biology of master viruses. *Adv. virus Rec. 50*, 183-234.

Pant V, Gupta D, Choudhury N R, Malathi V G, Varma A and Mukherjee S K (2001) Molecular characterization of the Rep protein of the black gram isolate of Indian mungbean yellow mosaic virus; *J. Gen. Virol. 82* 2559–2567

Pilartz, M. and H. Jeske. (1992). Abutilon mosaic virus double-stranded DNA is packed into mini-chromosomes. *Virology 189*: 800-802.

R. Rosell. (1995). The sweet potato / silver leaf whiteflies: Biotypes of *Bemisiu tabaci* Genn., or a species complex? *Annu. Rev. Entomol. 40*:511-534

Rogers, S.G., D.M. Bisaro, R.B. Horsch, R.T. Fraley, N.L. Hoffmann, L. Brand, J.S. Elmer, and A.M. Lloyd. (1986). Tomato golden mosaic virus A component DNA replicates autonomously in transgenic plants. *Cell 45*: 593-600.

Rushing, A.E., G. Sunter, W.E. Gardiner, R.R. Dute, and D.M. Bisaro. (1987). Ultra-structural aspects of tomato golden mosaic virus infection in tobacco. *Phytopathology 77*: 1231-1236

Saunders K., Lucy A. and Stanley J. (1991) DNA forms of the Gemini virus African cassava mosaic virus consistent with a rolling-circle mechanism of replication. *Nucleic Acids Res. 19*: 2325–2330

Saunders, K., A. Lucy, and J. Stanley. (1991). DNA forms of the Gemini virus African cassava mosaic virus consistent with a rolling circle mechanism of replication. *Nucleic Acids Res. 19*: 2325-2330.

Saunders, K., (1992). RNA-primed complementary-sense DNA synthesis of the Gemini virus African cassava mosaic virus. *Nucleic Acids Res. 20*: 6311-6315.

Science, 26 November, p. 1662.

Settlage S. B., Miller A. B. and Hanley-Bowdoin L. (1996) Interactions between Gemini virus replication proteins. *J. Virol. 70*: 6790–6795

Smith, K.M. (1972) *A textbook of plant virus diseases* (3rd edition), pp. 76-87. Longman, London, UK.

Stanley J. (1985), the molecular biology of Gemini viruses. *Adv. Virus Res. 30*: 139–177

Stanley J. (1993) Gemini viruses: plant viral vectors. *Curr. Opin. Genet. Dev. 3*: 91–96

Stanley J. (1995) Analysis of African cassava mosaic virus recombinant suggests strand nicking occurs within the conserved nonanucleotide motif during the initiation of rolling circle DNA replication. *Virology 206*: 707–712

Stanley J. and Latham J. R. (1992) A symptom variant of beet curly top Gemini virus produced by mutation of open reading frame C4. *Virology 190*: 506–509

Stanley J. and Townsend R. (1985) Characterization of DNA forms associated with cassava latent virus infection. *Nucleic Acids Res. 13*: 2189–2206.

Stanley J., Latham J. R., Pinner M. S., Bedford I. and Markham P. G. (1992) Mutational analysis of the mono-partite Gemini virus beet curly top virus. *Virology 191*: 396–405

Stanley, J., Markham, P.G., Callis, R. J. and Pinner, M.S. (1986) The nucleotide sequence of an infectious clone of the Gemini virus beet curly top virus. *EMBO.J. 5*, 1761-1767.

Stanley, J. (1983) Infectivity of the cloned geminivirus genome requires sequences from both DNAs. *Nature (London) 305*, 643-645.

Stanley, J. and Gay, M.R. (1983) Nucleotide sequence of Cassara latent virus DNA. *Nature (London), 301* 260-262.

Stenger D. C., Revington G. N., Stevenson M. C. and Bisaro D. M. (1991) Replicational release of Gemini virus genomes from tandemly repeated copies: evidence for rolling-circle replication of a plant viral DNA. *Proc. Natl. Acad. Sci. USA 88*: 8029–8033

Sunter, G. and Bisaro, D. N. (1991) Transactivation in a Gemini virus: AL2 gene product is needed for coat protein expression. *Virology, 180*, 416-419.

Sunter, G., D.C. Stenger, and D.M. Bisaro. (1994) Heterologous complementation by Gemini virus AL2 and AL3 genes. *Virology 203*: 203-210.

Sunter, G., K.W. Buck, and R.H.A. Coutts. (1985) S1-sensitive sites in the super coiled double-stranded form of tomato golden mosaic virus DNA component B: Identification of regions of potential alternative secondary structure and regulatory function. *Nucleic Acids Res. 13*: 4645-4659.

Sunter, G., M.D. Hartitz, and D.M. Bisaro. (1993) Tomato golden mosaic virus leftward gene expression: Auto regulation of Gemini virus replication protein. *Virology 195*:275-280

Sunter, G., M.D. Hartitz, S.G. Hormuzdi, C.L. Brough, and D.M. Bisaro. (1990) Genetic analysis of tomato golden mosaic virus. ORF AL2 is required for coat protein accumulation while ORF AL3 is necessary for efficient DNA replication. *Virology 179*:69-77

Sunter, G., W.E. Gardiner, A.E. Rushing, S.G. Rogers, and D.M. Bisaro. (1987) Independent encapsidation of tomato golden mosaic virus a component DNA in transgenic plants. *Plant Mol. Biol. 8*: 477-484.

Timmermans et. Al. 1994; Timmermans M. C. P., Das O. P. and Messing J. (1994) Gemini viruses and their uses as extrachromosomal replicons. *Annu. Rev. Plant. Physiol. 45*: 79–112

Townsend, R., J. Watts, and J. Stanley. (1986) Synthesis of viral DNA forms in *Nicotiana plumbaginafolia* protoplasts inoculated with cassava latent virus (CLV); evidence for independent replication of one component of the CLV genome. *Nucleic Acids Res.* *14*:1253-1265.

In: Emerging Geminiviral Diseases and their Management
Editors: P. Sharma, R. K. Gaur et al. pp. 209-234
ISBN 978-1-61668-620-8

Chapter 11

ROLE OF DNA SATELLITES IN GEMINIVIRUS DISEASE COMPLEXES

Imran Amin, Muhammad Ilyas, Shahid Mansoor, Rob W. Briddon and Muhammad Saeed*

National Institute for Biotechnology and Genetic Engineering, Faisalabad, Pakistan.

ABSTRACT

Viruses of the family *Geminiviridae* are responsible for many economically important crop diseases worldwide. The majority of these diseases are caused by geminiviruses that are whitefly transmitted and are placed in the genus *Begomovirus*. Within the last ten years there has been a fundamental shift in our understanding of the nature of begomoviruses. It has become evident that, in contrast to the situation in the New World where all native viruses have genomes consisting of two components, in the Old World the majority of begomoviruses are monopartite and most of these associate with single-stranded DNA satellites. Two classes of such molecules were identified at the turn of the Century. The alphasatellites are satellite-like and have not been extensively investigated. Their precise function in begomovirus infections remains unclear but they are nevertheless of evolutionary interest. The betasatellites are true satellites and are, in many cases, an essential factor for begomoviruses to infect their hosts. These, at first glance, "simple" molecules have proven to be complex with respect to their interactions with both their helper begomoviruses and host plants. This chapter will outline the advances made in studying the alpha- and betasatellites over the 10 years since their discovery. This greater understanding of how they interact with both the host and the helper virus may provide useful information for their control as well as an insight into plant developmental processes.

* Correspondence concerning this article should be addressed to: Muhammad Saeed, E-mail: msaeed@nibge.org; saeed_hafeez@yahoo.com.

THE GEMINIVIRUSES

Viruses belonging to the family *Geminiviridae* have an almost worldwide impact on agricultural productivity. Historically the first record of geminivirus infection may have been made as early as 725 AD in the Man'yoshu, a classical anthology of Japanese poetry. This record was in a poem attributed to the Empress Koken that described the autumnal appearance of *Eupatorium* plants in the summer, an observation which has been linked with yellow vein disease of this perennial shrub (Inouye & Osaki, 1980). The striking yellow symptoms have since been shown to be caused by a geminivirus/satellite disease complex (Saunders et al., 2003). Although the diseases caused by geminiviruses represent serious constraints to agriculture, little was known about the causal agents of the diseases until the isolation of virus particles with a unique twinned quasi-isometric morphology associated with maize streak and beet curly top diseases (Bock et al., 1974, Mumford, 1974). This attribute provided the name geminivirus, from Gemini, the sign of the zodiac symbolized by twins (Harrison et al., 1977), and has remained a unifying feature of this family of viruses.

In groundbreaking research it was demonstrated that the so-called geminate particles associated with Cassava latent virus, *Maize streak virus* (MSV) and *Bean golden mosaic virus* (BGMV) contained circular single-stranded (ss)DNA (Goodman, 1977b, Harrison et al., 1977), and that this genomic DNA was infectious when re-introduced to plants by mechanical inoculation (Goodman, 1977a). Evidence was provided for BGMV and *Tomato golden mosaic virus* (TGMV) to suggest that at least some geminiviruses had divided genomes (Bisaro et al., 1982, Haber et al., 1983, Hamilton et al., 1982). Shortly thereafter, the nucleotide sequence of Cassava latent virus [subsequently renamed *African cassava mosaic virus* (ACMV)] was established, and infectious clones were used to demonstrate the bipartite genome structure of this virus (Stanley, 1983, Stanley & Gay, 1983).

Since this first demonstration of an unusual group of phytopathogenic viruses with, at the time, the unusual property of having ssDNA genomes, numerous disease of crops, weeds and ornamental plants have been shown to be caused by similar viruses. In 1977 the four confirmed ssDNA viruses were grouped together under the name "geminiviruses" (Harrison, 1977) which was upgraded to the status of family (*Geminiviridae*) in 1995; at which time the family consisted of 11 confirmed members (Murphy et al., 1995).Geminiviruses are at present classified into four genera, *Mastrevirus*, *Topocuvirus*, *Curtovirus* and *Begomovirus* based on their genome organization, host ranges, and insect vectors (Stanley et al., 2005).

The genus *Topocuvirus* contains only a single species, *Tomato pseudo curly top virus,* which occurs in the southern United States of America. This virus is the only known treehopper-transmitted geminivirus and has a monopartite genome that resembles the genomes of the other monopartite dicot-infecting geminiviruses [Figure 1; (Simons & Coe, 1958), (Briddon et al., 1996)]. Viruses of the genus *Curtovirus* (type species *Beet curly top virus*) are monopartite, transmitted by leafhoppers and infect only dicotyledonous plants and have been identified in both the New World (NW) and the Old World (OW) (Stanley et al., 2005). Viruses of the genus *Mastrevirus* (type species *Maize streak virus*) are transmitted by leafhoppers, have a single component genome, occur only in the OW and infect either monocots or dicots (Howell, 1984, Liu et al., 1997, Morris et al., 1992, Mullineaux et al., 1984, Nahid et al., 2008). The genus *Begomovirus* contains only viruses transmitted by the whitefly *Bemisia tabaci* to dicotyledonous hosts. Begomoviruses native to the NW have genomes consisting of two ssDNA circles (known as components DNA-A and DNA-B). The begomoviruses originating from the OW, in contrast, have either monopartite or bipartite genomes. The genomes of monpartite begomoviruses are homologous to, and genetically indistinguishable from, the DNA-A components of bipartite begomoviruses. Recently it has become evident that the monopartite outnumber the bipartite begomoviruses in the OW and that the majority of the monopartite begomoviruses associate with satellites.

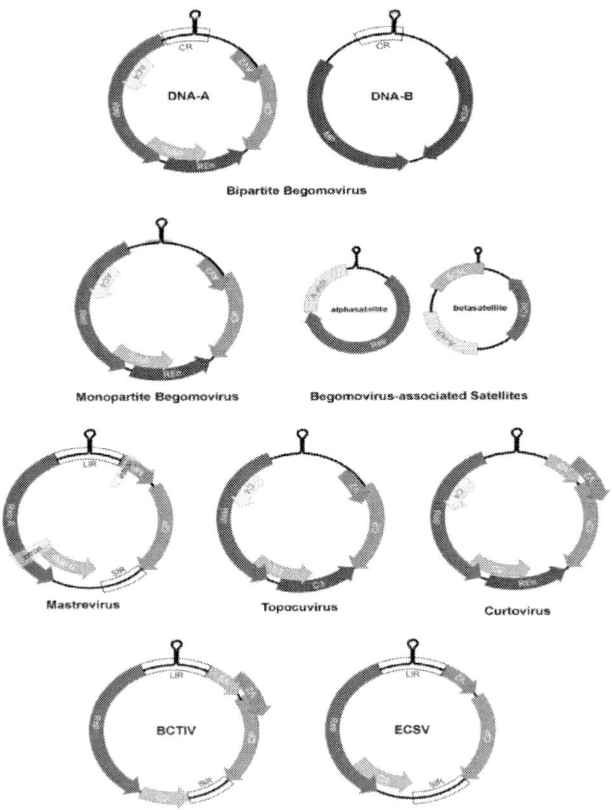

Figure 1. Typical genome organization of members of the four *Geminiviridae* genera and structures of the begomovirus-associated satellite (betasatellite) and satellite-like (alpghasatellite) molecules. The structure of both a bipartite and a monopartite begomovirus are shown. In addition the genome structures of the unusual *Beet curly top Iran virus* and *Eragrostis curvula streak virus* are shown. The position and orientation of virion-sense (V) and complementary-sense (C) genes are shown in relation to the initiation site of virion-sense DNA replication (by convention designated as nucleotide one), located within the conserved nonanucleotide TAATATTAC at the apex of a conserved hairpin structure. Mastreviruses contains two noncoding regions referred to as the large and small intergenic regions (LIR and SIR, respectively) and two introns (boxes). The DNA-A and –B components of bipartite begomoviruses contain an almost identical sequence of approximately 200 nucleotides, referred to as the common region (CR), that is usually located within the intergenic region. The begomoviruses originating from the New World, all of which are bipartite do not encode a V2 gene. The alpha- and betasatellites contain a region of sequence rich in adenine (A-rich), and the betasatellites contain a region of highly conserved sequence known as the satellite conserved region (SCR). The gene products (where their function is known) are indicated as coat protein (CP), replication-associated protein (Rep), transcriptional activator protein (TrAP), replication enhancer protein (REn), MP and nuclear shuttle protein (NSP). The only product encoded by betasatellites is known as βC1 whereas the alphasatellites encode a Rep.

The genomes (or genomic components) of geminiviruses consist of ssDNA circles that replicate in the nuclei of infected cells as double-stranded (ds), replicative-form intermediates by a rolling-circle mechanism. The encapsidated form of the virus DNA is referred to as the virion-strand. Transcription occurs from the replicative form with genes diverging from a non-coding region that habours cis-acting control elements and a predicted hairpin structure containing, within the loop, the sequence TAATATTAC (known as the nonanucleotide sequence). Virion-strand rolling-circle DNA replication initiates at the 3' adenine residue in the nonanucleotide sequence and this is, by convention, designated as position 1 for numbering the nucleotide positions on the sequence of the genome (genomic component). A further convention is to name open reading frames (ORFs; genes) as either encoded on the virion-sense strand (V) or complementary-sense strand (C) and, for bipartite begomoviruses, to prefix them with A or B when they are encoded on the DNA-A or DNA-B component, respectively.

The virion-sense strand of all geminiviruses encodes the coat protein [CP; ORF (A)V1] that is the only structural protein of these viruses, forming the distinctive geminiate particles, interacts with the insect vectors to facilitate transmission and is involved in movement in host plants (Hanley-Bowdoin et al., 1999). For NW begomoviruses this is the only virion-sense encoded gene. However, the OW begomoviruses encode an additional gene, (A)V2, the product of which is involved in movement in host plants. The V2 of the mastreviruses encodes the movement protein (MP) that mediates cell-to-cell movement in host plants. In addition to the CP, the curtoviruses encode two additional genes. The V2 encodes a protein involved in regulating ssDNA/dsDNA levels, whereas the V3 is involved in virus movement in hosts. The genes encoded by the topocuvirus TPCTV have not been investigated. However, sequence similarities suggest that they mirror the functions of genes encoded by begomoviruses.

The dicot-infecting geminiviruses (with the exception of the dicot-infecting mastreviruses) encode four genes on the complementary-sense strand (Figure 1). The replication- associated protein [Rep, encoded by ORF (A)C1] is a rolling-circle replication initiator protein with DNA nicking and joining activity. It is a sequence specific DNA binding protein that nicks the virion-strand ahead of the 3' adenine to initiate rolling circle replication and resolves new ssDNA circles by cutting and ligating. The protein recognizes small repeated sequences, known as "iterons", just upstream of the hairpin structure which are, for the most part, species specific. The specificity of a particular Rep for its cognate iterons is such that, in most instances, the Rep from one virus species will not recognize the iterons of a second species to initiate replication. The second complementary-sense gene [ORF (A)C2] encodes the a protein that may act as a transcription factor that up-regulate host-encoded genes and that may also be suppressor of RNA interference (RNAi – a host defense mechanism that recognizes and degrades foreign and aberrant RNAs (Baulcombe, 2004, Baulcombe, 2005, Hannon, 2002). For many begomoviruses the (A)C2 protein is also involved in up-regulating the late (virion-sense) genes and is known as the transcriptional activator protein [TrAP; (Sunter & Bisaro, 1992, Sunter et al., 1994)]. The replication enhancer protein [REn, ORF (A)C3] interacts with host factors, to create a cellular environment suitable for viral replication, and with Rep, to up-regulate its activity (Settlage et al., 2005). The (A)C4 protein (ORF C4) is for many viruses a pathogenicity determinant and in some cases a suppressor of RNAi (Vanitharani et al., 2004). The complementary-sense coding strategy of the mastreviruses is unusual amongst geminiviruses. The Rep is expressed from a splice transcript of the *RepA* and *RepB* (ORFs C1 and C2, respectively) *genes* (Wright et al., 1997). The RepA protein, translated from a non spliced transcript, interacts with host-encoded factors to create an environment suitable for virus replication and also is involved in up-regulating the expression of late, virion-sense genes. The DNA B component of bipartite begomoviruses encodes two additional genes; in the virion-sense the nuclear shuttle protein (NSP, ORF BV1) and in the complementary-sense the MP (ORF BC1). These act in a concerted manner to mediate cell-to-cell movement of the virus in plants (Hanley-Bowdoin et al., 1999).

Recently two viruses have been identified with characteristics that do not readily fit into the present classification system for geminiviruses. *Beet curly top Iran virus* has a virion-sense genome arrangement typical of that of curtoviruses but an unusual complementary-sense arrangement (Yazdi et al., 2008). It lacks the C3 gene (encoding REn) and has a C2 gene that could possibly be expressed from a spliced transcript spanning the C1 and C2 genes (a feature typical of mastreviruses). In addition the virus is reported to be transmitted by a vector not normally associated with curtoviruses, although the two species are closely related. The second virus, *Eragrostis curvula streak virus*, was identified in a wild grass (thus a monocot; so far only geminiviruses of the genus *Mastrevirus* have been identified in monocots) (Varsani et al., 2009). Its C2 gene appears to encode a homologue of the transcription factor (TrAP for bipartite begomoviruses) of the dicot infecting geminiviruses, rather than part of the Rep (RepB for mastreviruses). Both these viruses also have an unusual nonanucleotide sequence (TAAGATTC) and new genera may need to be established to accommodate them within the present taxonomic structure. However, a final decision on this will need to await infectivity studies with the cloned viruses to satisfy Koch's postulates.

Satellites Associated with Geminiviruses

In the field of virology a satellite is defined as a virus or nucleic acid that lacks extensive nucleotide sequence homology to the helper virus and is entirely dependent on helper virus for its replication and movement (Mayo et al., 2005). Satellite viruses encode a structural protein that encapsidates its own nucleic acid while satellite nucleic acids rely on the helper virus structural protein for encapsidation and do not necessarily encode additional non-structural proteins (Mayo et al., 2005). Another type of molecule, referred to as satellite-like virus or nucleic acid, depend on the helper virus for their replication but also provide a function that is necessary for the biological success of the helper virus. Thus such molecules are considered to be the part of helper virus genome.

The first satellite virus was identified in 1969 associated with the nepovirus *Tobacco ringspot virus* (Schneider, 1969) and since that time a large number of satellite RNAs, associated with several groups of plant viruses, have been reported (Mayo et al., 2005). The vast majority of satellite RNAs do not encode functional proteins but nevertheless can have a profound effect on the symptoms induced by their helper viruses, ranging from symptom amelioration to an increase in symptom severity (Simon et al., 2004).

The first DNA satellite discovered, referred to as the Tomato leaf curl virus-satellite (ToLCV-sat), was isolated from tomato plants originating from Australia infected with the monopartite begomovirus *Tomato leaf curl virus* (ToLCV) (Dry et al., 1997). This circular ssDNA molecule is small (682 nucleotides) with no ORF and little sequence similarity to its helper virus with the exception of sequences within the apex of two stem-loop structures, one containing the conserved geminivirus nonanucleotide (TAATATTAC) motif and the other containing a putative ToLCV Rep binding "iteron" motif (Behjatnia et al., 1998). ToLCV-sat is not required for ToLCV infectivity and has no effect on the symptoms induced by the helper virus but it is dependent on the helper begomovirus for its replication, movement in and between plants and thus it represents a true satellite. However, it was not until the start of the Millennium that the full significance of ToLCV-sat would become apparent.

The Discovery of Alpha- and Betasatellites

The first begomoviruses to be fully characterized were bipartite –ACMV (Stanley & Gay, 1983) and TGMV (Hamilton et al., 1984) – and the original clones of these viruses have since become the model systems for studying many aspects of the biology of geminiviruses. With the subsequent publication of the complete sequences of the type members of the other three genera, MSV (Mullineaux et al., 1984), BCTV (Stanley et al., 1986) and TPCTV (Briddon et al., 1996), the geminivirus community was quite settled with the idea that whitefly-transmitted geminiviruses (begomoviruses) possessed two component genomes, whereas geminiviruses transmitted by other vectors had genomes consisting of only a single component. However, in the early 1990s it became apparent that begomoviruses consisting of only a single component (lacking the DNA-B) exist. Koch's postulates were satisfied independently for two tomato-infecting viruses originating from the Mediterranean region – *Tomato yellow leaf curl virus* (TYLCV) and *Tomato yellow leaf curl Sardinia virus* (TYLCSV) (Kheyr-Pour et al., 1991, Navot et al., 1991). One of these, TYLCV, has subsequently become of almost global importance to tomato production (Mugiira et al., 2008, Tesoriero & Azzopardi, 2006).

During the mid to late 1990s an increasing number of apparently monopartite begomoviruses were identified for which Koch's postulates could not be satisfied. For two such viruses, *Ageratum yellow vein virus* [AYVV; associated with a disease of *Ageratum conyzoides* in Singapore; (Wong et al., 1993)] and *Cotton leaf curl Multan virus* [CLCuMV; associated with cotton leaf curl disease [CLCuD] in Pakistan; (Briddon & Markham, 2000)], clones were either not infectious, or poorly infectious and inducing atypical symptoms, upon reintroduction to the plant species they were isolated from (Briddon et al., 2000, Tan et al., 1995). This strongly suggested that the viruses lacked some factor required for infectivity of these hosts and initiated a search for additional ssDNA components – only ssDNA components were sought since the diseases were transmitted by the whitefly *B. tabaci* and begomoviruses were the only whitefly transmissible agents identified in diseased plants.

The first breakthrough came with pathogen causing CLCuD in Pakistan. Shotgun cloning from supercoiled DNA isolated from CLCuD affected material purified by cesium chloride/ethidium bromide density gradient centrifugation yielded an unusual molecule approximately half the size (~1400) of a begomovirus component (Mansoor et al., 1999). Although this component, which at the time was named DNA1, was of great scientific interest (as will be described below) it did not resolve the etiology of CLCuD or of AYVD, for which a very similar molecule was subsequently cloned (Saunders & Stanley, 1999) Inoculation of either virus with their respective DNA1 molecules has no effect on the infectivity of the virus to the natural host or major effects on the symptoms induced in alternate hosts (Mansoor et al., 1999, Saunders & Stanley, 1999). Nevertheless, subsequent studies have shown these molecules, which we now refer to collectively as alphasatellites, are very widespread in the OW and associated with many, but by no means all, monopartite begomoviruses (Briddon et al., 2004).

Driven on by this unexpected turn of events, efforts were redoubled to identify further ssDNA molecules, which were soon rewarded. It was noted that AYVD affected plants were associated with a number of apparently recombinant molecules of approximately half begomovirus components size consisting of known (begomovirus) sequence and unknown sequence. PCR amplifications using primers designed to this "unknown" sequence led to the isolation of a novel new component, at the time referred to as DNA β (Briddon et al., 2001, Saunders et al., 2000). Comparisons of the sequence of this molecule to sequences available in the databases showed it to have similarity to recombinant molecules earlier isolated from CLCuD affected cotton (Briddon et al., 2001, Briddon & Markham, 2000). Subsequently, by PCR, a homolog of the AYVD DNA β was isolated from CLCuD affected cotton. Inoculation of these two DNA β molecules, with their respective begomoviruses, resolved the etiology of AYVD and CLCuD, with the induction of the full range of disease symptoms in each host, *A. conyzoides* and cotton, respectively (Briddon et al., 2001, Saunders et al., 2000). These DNA β molecules have since then been shown to be representative of a large class of satellites, which are associated with the majority of monopartite begomoviruses (and thus occur only in the OW; (Briddon et al., 2003, Zhou et al., 2003) which we now refer to as betasatellites (Briddon & Mansoor, 2008).

Since their first identification, over 260 full-length betasatellite sequences have been deposited with the databases, an indication of their widespread nature, importance and also the ease with which they can be isolated by PCR and, more recently, by rolling circle amplification. They have proven to be extremely diverse and the sheer number and conflicting/confusing nomenclature used by researchers in the field to describe/define them necessitated a logical and uniform system of classification and nomenclature. Such a system was proposed by Briddon et al. (2008). It is this system of naming and classifying of betasatellites that will be used throughout the rest of this chapter.

Properties of Alpha- and Betasatellites

Although alphasatellite did not resolve the etiology of diseases such as AYVD and CLCuD, they were nevertheless interesting. The first alphasatellite clone identified by Mansoor et al. (1999) would likely have been discarded had it not been for a surprising similarity to the Rep encoding components of a second family of ssDNA viruses, the nanoviruses (family *Nanoviridae*) (Gronenborn, 2004). The nanoviruses are a family of multicomponent ssDNA viruses, with each component being approximately 1 to 1.1 kb in length, that are transmitted by aphids in a persistent, circulative manner. In common with geminiviruses, nanovirus genomes replicate by a rolling-circle mechanism. Recently Grigoras et al. (2009) have shown, using cloned components reintroduced into plants, that the genome of legume-infecting nanovirus consists of eight circular components. To maintain genome integrity, each virus encodes a rolling circle replication initiator protein, known as the masterRep (mRep) and encoded by the component known as DNA-R, with the specificity to replicate all *bona fide* virus components (Timchenko et al., 2000). This is thus analogous to the situation with bipartite begomoviruses, with the *bona fide* virus components sharing sequences that are recognized by Rep (iterons) to initiate rolling circle replication at the conserved hairpin. Such "iterons" have been identified, and confirmed experimentally, for one nanovirus (Herrera-Valencia et al., 2006). However, in addition to the DNA-R, some nanovirus isolates are additionally associated with further Rep-encoding components (Hu et al., 2007, Vetten et al., 2004). These are distinct from the *bona fide* components and are not replicated by the mRep. Rather the Rep that the additional components encode is capable of only initiating replication of the component that encodes it. These components can thus been seen as parasites of the nanovirus and are best described as satellite-like; remembering that satellites are defined as not capable of self-replication and having no sequence similarity to their helper viruses.

The alphasatellites associated with begomoviruses have a conserved structure and are typically slightly larger (~1380) than the betasatellites. These molecules have been identified, in the field, only with monopartite begomoviruses that are associated with betasatellites, suggesting that they provide some function useful to begomovirus-betasatellite complexes. However, not all begomovirus-betasatellite complexes are associated with alphasatellites (Briddon et al., 2004). Specifically no alphasatellites have been identified with begomovirus-betasatellite complexes originating from the Far East. For the CLCuD begomovirus-betasatellite complex affecting cotton in southern Asia, all cotton plants affected by the disease that were examined during the 1990s, were shown to harbour an alphasatellite. However, following the appearance of a resistance breaking strain of the virus complex, no alphasatellites have been detected in cotton. This is just one of several changes that have occurred to the CLCuD complex in recent years (Amrao, 2009).

The structure of full-length alphasatellites consists of one large gene in the virion-sense, encoding a rolling circle replication initiator protein (Rep), a region of sequence rich in adenine (A-rich) and a predicted hairpin structure containing, within the loop, the nonanucleotide sequence TAGTATTAC (Briddon et al., 2004, Mansoor et al., 1999, Saunders & Stanley, 1999). This nonanucleotide sequence is common to the nanoviruses. Typically for begomovirus-associated alphasatellites the TATA box of the presumed Rep promoter lies 5' of the stem-loop structure. This is a feature in common with many of the satellite-like Rep-encoding components associated with nanoviruses. In contrast, for most nanoviruses this feature lies 3' of the stem-loop structure for mRep-encoding components. Beyond determining the sequence, little analysis of alphasatellites has been conducted, although recently Huang et al. (2009) have adapted an alphasatellite for use in the expression of foreign sequences in plants.

Due to the ease with which they can be cloned, and their relatively small size, greater than 260 full-length betasatellite sequences are now available in the databases. Betasatellites are described as *bona fide* satellites since, in contrast to alphasatellites, they are not capable of autonomous replication and bear little sequence similarity to their helper begomoviruses. Comparisons of the nucleotide sequences of all available full-length betasatellites has identified three absolutely conserved features (Figure 1):

(1) A region of sequence that is rich in adenine (A-rich).

(2) A sequence which is highly conserved between distinct betasatellites. This encompasses a conserved hairpin structure containing, within the loop sequence, the nonanucleotide sequence TAA/GTATTAC, with similarity to the "*ori*" of virion-strand DNA replication of geminiviruses.
(3) A single ORF, conserved in both sequence and position.

The A-rich region is typically between 160 and 280 bases in length and has between 57 to 65% A content (the overall A content of betasatellitc is between 28 to 38%). The A-rich region is maintained for all DNA molecules, including defective molecules as well as the ToLCV-sat, and consists of blocks of up to 11 consecutive A residues. By phylogenetic analysis the sequences of the A-rich regions of betasatellites segregate with and form the same groups as the full-length sequences, suggesting that the sequence is conserved (Briddon et al., 2002). The functions of, and/or requirement for, the A-rich region in betasatellites remains unclear. As for the begomovirus-associated alphasatellites, as discussed earlier, it has been suggested that the A-rich sequence may represent a "stuffer" to fulfill the size limitations for movement by, and encapsidation in the CP of, begomoviruses. This in turn implies that betasatellites evolved from a smaller molecule (possibly by component capture, as is likely the case for begomovirus-associated alphasatellites), although there is no evidence to support this contention at this time. Tao & Zhou (2004) experimentally deleted the A-rich sequence of a Tomato yellow leaf curl China betasatellite (TYLCCNB) clone and showed that the resulting molecule, when inoculated to plants with TYLCCNV, remained competent for transreplication and encapsidation by the virus. This showed that the A-rich sequence is not absolutely required for replication and encapsidation and suggests that the mechanism for the maintenance of the A-rich sequence may be very subtle. A hypothesis for the possible role of the A-rich sequence, at least for betasatellites, has been proposed by Amin et al. (2006). This will be elaborated upon later.

In addition to the A-rich region, betasatellites contain a highly conserved sequence of approximately 80 nucleotides, referred to as the satellite-conserved region [SCR; (Briddon et al., 2003)]. This is the region to which a pair of universal primers were designed that proved to be a simple and robust method for the isolation of betasatellites (Briddon et al., 2002). The SCR encompasses a putative stem-loop structure that contains the nonanucleotide motif (for betasatellites TAA/GTATTAC) which, by analogy to geminiviruses, is the site where Rep introduces a nick during the initiation of virion-sense DNA replication. The fact that naturally occurring mutants lacking the βC1 coding region (Briddon et al., 2003) and mutants in which the βC1 coding region and A-rich region have been deleted *in vitro* (Qian & Zhou, 2005, Tao et al., 2004) are maintained by the helper begomovirus is consistent with the involvement of the SCR and stem-loop in replication. However, the identification of precise betasatellite sequences that contribute to replication awaits fine mapping studies, as discussed later.

Interactions between Betasatellites and their Helper Begomoviruses

The interaction with betasatellites is a feature of monopartite begomoviruses. Even the archetypal monopartite begomovirus TYLCV has not remained immune from the influence of betasatellites. A field isolate of TYLCV originating from Oman has been shown to associate with a betasatellite [Tomato yellow leaf curl betasatellite; (Khan et al., 2008)] and clones of TYLCV from isolates not associated with betasatellites have nevertheless been shown to be able to interact experimentally with betasatellites [specifically Tobacco leaf curl Japan betasatellite and Honeysuckle yellow vein mosaic betasatellite; (Ito et al., 2009)]. Betsatellites have only been identified in the OW, where monopartite begomoviruses are now known to outnumber the bipartite begomoviruses. Betasatellites have been identified in plants infected with bipartite begomoviruses, specifically with *Mungbean yellow mosaic India virus* (MYMIV) in India and Pakistan (Rouhibakhsh & Malathi, 2005, M. Ilyas, manuscript in preparation) and *Tomato leaf curl New Delhi virus* (ToLCNDV) in India (Raj et al., 2007). However, it has yet to be demonstrated that these are productive interactions (an interaction where the betasatellite provides a useful function and is selectively maintained) and not just transient interactions due, for example, to a co-infection between a bipartite begomovirus and a monopartite, betasatellite associated begomovirus. Surprisingly, Saunders et al. (2002) were able to show that AYVB could interact with *Sri Lanka cassava mosaic virus* DNA-A (in the absence of its cognate DNA-B) to induce symptoms typical of AYVD (vein yellowing and leaf curling) in *A. conyzoides*.

All begomoviruses native to the NW thus far characterized have a bipartite genome arrangement, although the monopartite begomovirus TYLCV has been inadvertently introduced to the NW, possibly on more than one occasion (Duffy & Holmes, 2007). This is in and of itself a catastrophe for tomato growers in the Americas and becomes even more significant with the possibility that the virus could interact with a betasatellite, were one to be introduced. The absence of betasatellites in the NW, however, cannot be taken to mean that NW begomoviruses are unable to interact with these satellites. In *N. benthamiana*, the bipartite NW begomovirus *Cabbage leaf curl virus* (CbLCuV) can maintain CLCuMB, albeit at a low level (Nawaz-Ul-Rehman et al., 2009). However, the same authors reported that CbLCuV was unable to maintain a Chili leaf curl betasatellite, indicating that there is specificity/selectivity in the interaction.

Maintenance of CLCuMB by CbLCuV required the presence of the DNA-B; in the absence of the component encoding MP and NSP no infection ensued (Nawaz-Ul-Rehman et al., 2009). This contrasts with the results of Saeed et al. (2007), again using CLCuMB, which showed that the betasatellite could functionally replace the DNA-B component of the bipartite OW begomovirus ToLCNDV (the implications of this are discussed in more detail below). The betasatellite in this case was maintained at high level. Taken together these findings suggest that CLCuMB is better adapted for maintenance by ToLCNDV than CbLCuV and that, possibly, the ability to dispense with the DNA-B, when interacting with the betasatellites could be due to the presence in the OW virus of the V2 gene, the product of which has been implicated in virus movement (Saeed et al., 2007). Coincidentally, it has been suggested that the apparent absence of monopartite begomoviruses in the NW (or rather the reason as to why NW begomoviruses did not evolve monopartite types) may be due to the lack of the V2 gene in NW begomoviruses; the implication being that the additional movement functions provided by the V2 gene product are required for this process of "monopartisation" (Saeed et al., 2008).

The begomoviruses that associate with betasatellites show a range of dependence upon the satellite, from full dependence to facultative. For example, the range of begomoviruses that interact with CLCuMB to cause CLCuD in southern Asia have an absolute requirement for the betasatellite (all cotton plants showing CLCuD symptoms contain CLCuMB), suggesting that this provides some essential function for the infection of cotton. In contrast, for *Tobacco curly shoot virus*, some field isolates of the virus are associated with a betasatellite [Tobacco curly shoot betasatellite (TbCSB)] whereas others are not (Li et al., 2005). Even more surprising is that for *Tomato yellow leaf curl Thailand virus* (TYLCTHV) there are field isolates that associate with a DNA-B component (Rochester et al., 1994) and other isolates that associate with a betasatellite [Tomato yellow leaf curl Thailand betasatellite; (Guo et al., 2009)]. TYLCTHV is anyway an unusual begomovirus since the DNA-A is experimentally infectious to plants in the absence of its cognate DNA-B component (Rochester et al., 1994). Taken together these findings show that the range of interactions of begomoviruses with secondary components (DNA-B or betasatellite) covers the whole spectrum. This may indicate that begomoviruses are evolving, although it is unclear in which direction. However, it is probably more likely that this represents an evolutionary adaptation which allows begomoviruses to more rapidly fill niches as they appear. It is interesting to note that many of the truly monopartite begomoviruses (including TYLCV, TYLCSV and ToLCV) occur, for the most part, in tomato and it is tempting to suggest that this introduced crop species has some part to play in the evolution of monopartite begomoviruses. It certainly has played a part in the global spread of at least one of these viruses.

In contrast to the alphasatellites, which encode their own Rep, betasatellites do not encode a Rep and rely entirely for their replication on the helper begomovirus. Geminivirus Rep is a sequence specific DNA binding protein (Hanley-Bowdoin et al., 1999). Rep recognizes and binds to sequences, known as iterons, within the "ori" to initiate rolling circle replication. This interaction is very specific and, for the most part, the Rep for a particular geminivirus species will only recognize the iterons of that species. This specificity also maintains the genome integrity of bipartite begomoviruses since the two components share sequences (known as the "common region") that encompass the "ori" (Figure 1). Early on it was realized that there was no such specificity for the interaction between begomoviruses and betasatellites. For example, AYVV was shown to be able to transreplicate many distinct betasatellites and change the symptom phenotype in *N. benthamiana* from an upward to a downward leaf curl [Figure 2; (Briddon et al., 2003, Bull et al., 2004, Saunders et al., 2000)]. For most begomovirus-betasatellite combinations the satellite does not contain the specific iteron sequences predicted to be required for by the Rep of the helper virus, suggesting that some unusual/atypical interaction was occurring. In light of the AYVV (one virus able to interact with multiple betasatellites) and the CLCuD [one betasatellite interacting with numerous begomoviruses, either singly or as multiple infections, to cause the disease; (Briddon et al., 2003, Mansoor et al., 2003b) situations, two opposing, though not necessary mutually exclusive, hypotheses have been proposed that since have been named the "universal Rep" and "universal iteron" hypotheses (Nawaz-ul-Rehman & Fauquet, 2009). The universal Rep hypothesis contends that begomoviruses that interact with betasatellites encode Rep proteins with relaxed recognition specificity, whereas the universal iteron hypothesis suggests that betasatellites contain iteron(-like) sequences that can interact with the Rep proteins of diverse begomoviruses. The available evidence (summarised above) is leaning in favour of the "universal iteron" hypothesis and was recently given a boost by the results of (Nawaz-Ul-Rehman et al., 2009). They showed that progeny CLCuMB, cloned after infections of *N. benthamiana* with the NW bipartite begomovirus CbLCuV, showed significant sequences changes and higher levels of transreplication by the virus when inoculated back to *N. benthamiana* plants with CbLCuV. This suggests that the betasatellite is able to rapidly evolve to be more efficiently replicated by a begomovirus with which the interaction is poor.

Figure 2. Distinct symptoms types induced by *Agerum yellow vein virus* (AYVV) in the presence and absence of betasatellites. (A) A healthy *Nicotiana benthamiana*. (B) A *N. benthamiana* infected with only AYVV. (C) *N. benthamiana* plants infected with AYVV and Ageratum yellow vein betasatellite, and (D) with AYVV and Cotton leaf curl Multan betasatellite.

Experiments in which plants were inoculated with TYLCCNV in the presence of both TYLCCNB and TbCSB showed that this virus was able to transreplicate both satellites initially (Qing & Zhou, 2009).However, over time the non-cognate betasatellite (TbCSB) decreased in titre leading, ultimately, to only the cognate (TYLCCNB) being detectable. This shows that there is specificity in the interaction between a begomovirus and a betasatellite and suggests that, at least with respect to transreplication, there is adaptation of one to the other and might suggest that the Rep protein of TYLCCNV interacts more efficiently with TYLCCNB than with TbCSB.

To identify betasatellite sequences involved in the transreplication of AYVB by the Rep encoded by the helper virus AYVV, Saunders et al. (2008) conducted a deletion mutagenesis analysis of this satellite. They showed that sequences spanning the βC1 gene, the A-rich region and the sequences between these two features are not required for transreplication of AYVB in *N. benthamiana*; mutants with these sequences were trans-replicated by AYVV, confirming the earlier conclusion based on naturally occurring betasatellite mutants and other studies that deleted the βC1 coding sequence (Briddon et al., 2003, Qazi et al., 2007, Zhou et al., 2003) They then sequentially deleted sequences that lie between the A-rich region and the SCR. This showed that a particular sequence containing a predicted hairpin structure, with a sequence similar to the predicted iteron sequence of AYVV in the loop, is essential for transreplication. This sequence aligns with the sequence of the ToLCV-sat shown to contain a predicted hairpin structure containing, within the loop, a sequence with identity to the predicted iteron of ToLCV. However, for both ToLCV and ToLCV-sat this sequence was shown not to be absolutely essential for replication (Dry et al., 1997). The study of Saunders et al. (2008) did not address the requirement for the SCR, or the stem-loop structure contained therein, for transreplication of AYVB by AYVV Rep. However, the highly conserved nature of the SCR between distinct betasatellites and the similarity of the hairpin structure to the origin of replication of geminiviruses strongly suggests that they play an important part.

The sequences identified by Saunders et al. (2008) as important for transreplication of AYVB fall within an area of sequence identified for CLCuMB as highly variable (Amin et al., 2006). The proximity of the A-rich region to this variable sequence led them to suggest that the variability may be generated by mis-incorporation during rolling circle replication due to a local deletion in adenine. The A-rich sequence may thus represent an adaptation of the betasatellites to generate sequence changes in the region of the component that interacts with helper begomovirus-encoded Rep. Certainly the sequence changes in CLCuMB following maintenance by CbLCuV support this argument, although the specific contribution of each mutation was not investigated (Nawaz-Ul-Rehman et al., 2009). Nevertheless, the situation with respect to the molecular basis for recognition of helper virus Rep of betasatellites is far from resolved and will not be so until the precise DNA sequences mediating the interaction with Rep are identified.

FUNCTIONS OF BETASATELLITES

A variety of host-specific symptoms are induced by begomovirus-betasatellite disease complexes ranging from a vein yellowing phenotype in *Ageratum*, *Eupatorium* and honeysuckle, that appears to have no significant adverse effect on plant growth and survival (Saunders et al., 2003, Were et al., 2005, Wong et al., 1993), to severe leaf curl, chlorosis, enations and stunting in crops such as cotton, tobacco and tomato, that can have a significant effect on productivity [Figure 3; (Briddon & Markham, 2000, Cui et al., 2004, Harrison et al., 1997)]. The early studies that resolved the aetiology of AYVD and CLCuD clearly demonstrated that the newly identified betasatellites made an important contribution to the disease phenotype (Briddon et al., 2001, Saunders et al., 2000). In the absence of the betasatellite the helper begomoviruses were poorly infectious to the hosts from which they were isolated, inducing atypical and mild symptoms (Briddon et al., 2000, Saunders et al., 2000). Stable integration of a dimeric construct of the AYVB into the *N. benthamiana* genome resulted in severe developmental abnormalities in transgenic plants, indicating that the component encodes a pathogenicity determinant that is functional in the absence of the helper begomovirus (Saunders et al., 2004). Transcript mapping of the betasatellite for such AYVB transgenic plants identified a single transcript, in the complementary-sense, spanning the βC1 gene (Saunders et al., 2004), which was later confirmed for CLCuMB, again with plants transformed with a dimeric construct (Saeed et al., 2005).

Figure 3. Examples of symptoms induced by various begomovirus-betasatellite complexes in their natural hosts. (A) Cotton (*Gossypium hirsutum*) from Faisalabad, Pakistan showing typical symptoms of cotton leaf curl disease caused by one or more of seven distinct begomoviruses in association with Cotton leaf curl Multan betasatellite (CLCuMB). (B) Chili (*Capsicum annum*) from Multan, Pakistan showing typical symptoms of chilli leaf curl disease caused by an as yet unidentified begomovirus in association with Chili leaf curl betasatellite. (C) Yellow vein symptoms of *Croton glandulosus* from Faisalabad, Pakistan caused by Croton yellow vein mosaic virus in association with Croton yellow vein mosaic betasatellite. (D) Yellow vein symptoms of *Lonicera japonica* (honeysuckle originating from Norwich, UK caused by *Honeysuckle yellow vein virus* in association with Honeysuckle yellow vein betasatellite. (E) Yellow vein and leaf curling symptoms of *A.* conyzoides originating from Faisalabad, Pakistan caused by an as yet unidentified begomovirus in association with Ageratum yellow leaf curl betasatellite. (F). CLCuD-like symptoms of *Hibiscus rosa-sinnensis* originating from Faislabad, Pakistan caused by Cotton leaf curl Multan virus and CLCuMB.

Initial, circumstantial evidence that betasatellite pathogenicity is attributable to βC1 activity was provided by the analysis of naturally occurring betasatellite mutants with the βC1 coding sequences missing. The findings were confirmed for studies in which the βC1 coding sequences were either deleted or mutated (Qian & Zhou, 2005, Saeed et al., 2005, Saunders et al., 2004). For the AYVB studies involving the mutation of the βC1 gene showed that, although the mutant betasatellite lacking βC1was maintained by AYVV in *N. benthamiana*, it did not induce the altered phenotype (from upward leaf curling typical of AYVV to downward leaf curling typical of the interaction of AYVV with intact AYVB) (Saunders et al., 2004). However, in the presence of the mutated AYVB, AYVV was not infectious to Ageratum. Similar studies mutating the βC1 of TYLCCNVB showed that βC1 mutation ameliorated symptoms (leading to symptoms typical of the helper virus (TYLCCNV) in the absence of TYLCCNB) and did not allow TYLCCNV to infect either *N. tabacum* or tomato.

In contrast to the AYVV/AYVB studies, mutation of TYLCCNB βC1 reduced the accumulation of both the helper virus and the betasatellite in plants. Similar mutation studies with CLCuMB, but using ToLCV as the helper virus, showed that mutation of the βC1 abolished the typical CLCuD-like symptoms in *N. tabacum* (Saeed et al., 2005). Taken together these findings indicate that the βC1 protein encoded by betasatellites is a pathogenicity determinant and plays a major part in determining host range of the begomovirus-betasatellite complex.

Interestingly, for AYVD betasatellite mutants in which the first in-frame methionine codon of βC1, located nine nucleotides downstream of the major transcript start, had been replaced with a nonsense codon, were still able to produce attenuated symptoms in *Ageratum* (Saunders et al., 2004), suggesting that alternative initiation codons could be used for βC1 expression, as was likely the case for naturally occurring betasatellite mutants (Briddon et al., 2003). This also showed that the truncated βC1 protein maintained most of its activity, which is consistent with the finding that the C-terminal βC1 sequences are more conserved (and thus likely more important for the function of the protein) than those at the N-terminus.

The cell cycle is altered upon infection of geminiviruses in order to establish a cellular environment for virus replication. The major player in this case is the Rep that binds to plant homologues of the retinoblastoma tumor suppressor protein, thereby removing the block that otherwise prevents terminally differentiated cells reentering S-phase (Ach et al., 1997, Hanley-Bowdoin et al., 2004, Xie et al., 1995). This activity may result in endoreduplication rather than cell division (Grafi et al., 1996, Nagar et al., 1995). However, the C4 proteins encoded by the monopartite begomovirus ToLCV and the curtovirus BCTV are important symptom determinants that induce abnormal phloem cell division (hyperplasia) and a characteristic vein swelling phenotype in infected *N. benthamiana, N. tabacum*, tomato and sugar beet plants (Krake et al., 1998, Latham et al., 1997, Stanley & Latham, 1992). Since the phenotype is induced in the absence of other viral genes, it is possible that C4 induces ectopic division in undifferentiated cells, for example, cambium cells that normally remain competent for mitosis. Although structurally unrelated, the βC1 of AYVB has been shown to induce a remarkably similar phenotype when expressed in *N. benthamiana* from either a transgene or a *Potato virus X* (PVX) based vector (Saunders et al., 2004). Furthermore, βC1 encoded by TYLCCNB has a profound effect on the development of spongy and palisade parenchyma cells when expressed as a transgene in *N. tabacum* (Cui et al., 2005a). The CLCuMB, in association with a number of different begomoviruses, causes cellular re-differentiation, manifested by vein swelling and greening, and the production of leaf-like growths in cotton (Briddon & Markham, 2000, Mansoor et al., 2003a). Surprisingly, expression of the βC1 of CLCuMB from a PVX vector has been shown to induce the whole range of CLCuD symptoms in *N. tabacum*, (Qazi et al., 2007). The βC1 in this was shown to induce proliferation of the spongy parenchyma as well as a significant enlargement of these cells. Hyperplasia and hypertrophy may thus be common effects of βC1. However, it remains to be seen if these striking effects on plant development play an essential role in maintenance of the disease or if they are an unnecessary side effect resulting from a possible recent adaptation of these begomovirus/betasatellites to new hosts. It is possible that the hyperplasia induced by βC1 may be the reason for, in many cases, the apparent increase in virus DNA levels seen in begomovirus-betasatellite infected plants in comparison to plants only infected with the virus (Briddon et al., 2001, Saunders et al., 2000); thus an increase in the number of cells containing virus. However, only studies with protoplast systems will provide an answer to this question.

The first response of plants against invading viruses is the host defense response RNAi (also known as RNA silencing or post-transciptional gene silencing (Baulcombe, 2005). To counter this viruses encode gene products that can suppress the RNAi pathway and these genes are referred to as suppressors of gene silencing (Dunoyer et al., 2004). The βC1 proteins encoded by TYLCCNB (Cui et al., 2005b), Bhendi yellow vein mosaic betasatellite (Gopal et al., 2006), Tomato leaf curl Java betasatellite (Kon et al., 2007) and CLCuMB (I. Amin, manuscript in preparation) have been shown to have suppressor of gene silencing activity. These findings are thus in agreement with the general contention that viral (in this case satellite associated with virus) pathogenicity determinants are suppressors (Brigneti et al., 1998).

Several recent studies have provided an insight to the possible mechanism of symptom induction by βC1. Yeast-two-hybrid studies have shown that the βC1 encoded by CLCuMB interacts with host ubiquitin-conjugating (UBC) enzyme *SlUBC3;* a functional UBC protein of the E2 class (Eini et al., 2009). UBC plays a major role in the ubiquitin proteosome pathway which is responsible for the selective degradation of abnormal and short lived regulatory proteins, including transcription factors, cell cycle proteins and signal transducers, and thus plays a major role in the turnover of eukaryotic cellular products. Mutation of a predicted mysitoylation motif in CLCuMB βC1 abolished the interaction with *SlUBC3* and also abolished the ability to induce the symptoms when the betasatellite with the mutated βC1 was inoculated to *N. tabacum* with ToLCV. These result showed that the interaction of βC1 with *SlUBC3* is essential for symptom induction. It is unclear how this interaction results in symptoms. Possibly the interaction acts to extend the half-life of the protein by interfering with the proteasome pathway which might otherwise be rapidly degraded. This is consistent with the finding that *N. tabacum* mutants in which the ubiquitin pathway has been suppressed exhibit a virus-like leaf curling phenotype (Bachmair et al., 1990) and Yang et al. (2008) have shown βC1 to be unstable and susceptible to degradation by 26S proteasome.

Transgenic expression of the TYLCCNB *βC1 gene* in *Arabidopsis thaliana* has been shown to virus-like symptoms, reminiscent of infections of *N. benthamiana* with TYLCCNB and its helper virus TYLCCNV, consisting of upward curled leaves, radialized leaves with the tumorous outgrowth of tissues from abaxial surfaces, and sterile flowers (Yang et al., 2008). Transgenic expression of the TYLCCNB βC1 in *A. thaliana* results in plants exhibiting virus-like symptoms which are associated with a reduction in miR165/166 levels and an increase in the levels of transcripts of *PHABULOSA (PHB)* and *PHAVOLUTA (PHV)*; two genes encoding transcription factors involved in controlling adaxial-abaxial polarity (McConnell & Barton, 1998, McConnell et al., 2001). ASYMMETRIC LEAVES 1 (AS1) and ASYMMETRIC LEAVES 2 (AS2) regulate leaf development as an AS1/AS2 complex. Strikingly, the βC1 phenotype was similar to plants overexpressing AS2 at the morphological and molecular level and βC1 was shown to be able to partially complement an *as2* mutation and to bind directly to AS1. Additionally, βC1 was shown to selectively suppress selective jasmonic acid-responsive genes that are implicated in plant defense against insects. Thus as well as manipulating the plant to be suitable for virus multiplication, βC1 may be additionally ensuring onward transmission of the virus-satellite complex by countering host defenses to the vector, *B. tabaci*.

In plants auxin is a key hormone responsible for modulating many aspects of plant growth, including leaf/shoot/root architecture, organ patterning, vascular development and tropic response to light and gravity (Millner, 1995). Auxin response factors (ARFs) are the major class of transcription factor that control auxin signaling by binding to cis-elements in the promoter regions of auxin-inducible genes (Ulmasov et al., 1999). Expression of the CLCuMB βC1 gene from a PVX vector in *N. benthamiana* was shown to up-regulate the levels of miR160 (I. Amin, manuscript in preparation). miR160 negatively regulates ARF 17, which is involved in controlling leaf and flower morphology (Teotia et al., 2008). This finding provides a link between the major pathogenicity determinant of begomovirus-betasatellite complexes and the phytohormones controlling growth and development.

Recently evidence has been provided that βC1 may provide cell-to-cell movement activity to helper begomoviruses. CLCuMB was shown to complement the movement of the DNA-A of the bipartite begomovirus TLCNDV in the absence of DNA-B in tomato. A mutation of the βC1 gene abolished this activity, showing the phenomenon to be mediated by this gene (Saeed et al., 2007, Saeed 2008). While this suggests the involvement of βC1 in virus movement, nuclear localization of TYLCCNVB βC1 protein (Cui et al., 2005a, Cui et al., 2005b) argues against a direct role in facilitating movement across the plasma membrane. It is possible that situation here is analogous to that reported for the potexvirus PVX. The p25 protein is one of three 'triple gene block' (TGB) proteins of potexviruses that are required for cell-to-cell movement of the virus (Angell et al., 1996, Beck et al., 1991). It is an RNA helicase (Kalinina et al., 2002) that moves cell-to-cell (Yang et al., 2000), modifies plasmodesmata (Angell et al., 1996) and is a suppressor of RNA silencing (Voinnet et al., 2000). Bayne et al (2005) showed by random mutagenesis of PVX p25 that suppression was required for cell-to-cell movement. The authors suggested that possibly p25 suppressor activity could be required to prevent the spread of a mobile silencing signal, which might prime adjacent cells to counter the incoming virus. Thus it is possible that βC1 complementation of movement of the DNA-A component of a bipartite begomovirus could similarly be countering a silencing phenomenon that counters virus cell-to-cell spread, rather than acting as a classical virus movement protein. Similar results have recently been obtained with CLCuMB and the DNA-A component of bipartite begomovirus *Mungbean yellow mosaic India virus* (M. Ilyas, manuscript in preparation). This work additionally showed that the betasatellite is able to extend the host range of MYMIV to *N. benthamiana*, a species which is not otherwise a host for this virus.

Kumar et al. (2006) reported that the βC1 protein of Bhendi yellow vein mosaic betasatellite interacts with the CP of its helper virus, *Bhendi yellow vein mosaic virus*, in a yeast-two-hybrid assay. This is consistent with βC1 providing movement functions for the virus. Although bipartite begomoviruses do not require the CP for movement in plants (Pooma et al., 1996), it is absolutely required for the systemic infections of monopartite begomoviruses; implying that the viruses move either as intact virions or as nucleoprotein complexes involving the CP (Boulton et al., 1989, Briddon et al., 1989, Hormuzdi & Bisaro, 1993, Rigden et al., 1993). Thus, although the results of Cui et al. (2005), showing βC1 interacts with DNA, might have suggested that βC1 could directly move viral/satellite DNA, this is unlikely in light of the results of Saeed et al. (2008). They showed that βC1 is unable to complement a mutation of the CP of ToLCV (a monopartite begomovirus). The available evidences suggests that geminivirus CP is required for long distance spread of the virus in the phloem; bipartite begomoviruses with the CP gene mutated take longer to establish a systemic infection than the wild type virus, since the cell-to-cell movement mediated by the DNA-B encoded genes is slower than spread via phloem (Pooma et al., 1996). Also, most monopartite geminiviruses which have been investigated are phloem limited (Hanley-Bowdoin et al., 1999), whereas many bipartite begomoviruses infect cells in addition to those associated with phloem (Jeske, 2009), presumably mediated by the DNA-B genes.

Conclusion

It is evident that, in common with the proteins encoded by their helper begomoviruses, the single gene product of betasatellites is multifunctional. For many of the viruses that associate with betasatellites, it encodes the dominant pathogenicity determinant. The protein acts to up-regulate viral and satellite DNA levels in plants, possibly by inducing cell division, and may enhance virus movement in plants. The central theme in all the functions appears to be its suppressor of RNAi activity. Many suppressors act by binding small RNAs (Lakatos et al., 2006) and some, including the TrAP of the bipartite begomovirus *East African cassava mosaic Cameroon virus* (Chellappan et al., 2005), also bind miRNAs. These are endogenous small non-coding RNAs which are involved in gene regulation at the posttransiptional level in eukaryotes (Carrington & Ambros, 2003). It is thus possible that the effects of βC1 are due to its suppressor activity, possibly by binding small RNAs, including miRNAs. This property has yet to be shown for βC1, but undoubtedly investigations into this possibility will be forthcoming soon.

The interaction of the helper begomoviruses with betasatellites is less clear. The available evidence suggests that betasatellites encode iteron-like sequences, more than likely in a highly variable region of sequence immediately upstream of the A-rich region. It is possibly the variability, or rather its ability to rapidly change, that is allowing betasatellites to interact with such a large range of begomoviruses. Some work to conclusively identify the sequences involved is urgently required.

The recent identification of betasatellites associated with bipartite begomoviruses, the ability of betasatellites to rapidly adapt to new helper begomoviruses (including those present in the NW) and the global spread of a virus that can interact with betasatellite (TYLCV) are causes for concern. In areas where betasatellites already occur, these complexes cause major losses to agricultural productivity – CLCuD in southern Asia being a prime example.

References

Ach, R. A., Durfee, T., Miller, A. B., Taranto, P., Hanley-Bowdoin, L., Zambryski, P. C. & Gruissem, W. (1997). RRB1 and RRB2 encode maize retinoblastoma-related proteins that interact with a plant D-type cyclin and geminivirus replication protein. *Molecular and Cellular Biology 17*, 5077-5086.

Amin, I., Mansoor, S., Amrao, L., Hussain, M., Irum, S., Zafar, Y., Bull, S. E. & Briddon, R. W. (2006). Mobilisation into cotton and spread of a recombinant cotton leaf curl disease satellite. *Archives of Virology 151*, 2055-2065.

Amrao, L. (2009). Characterization and infectivity analysis of begomovirus components found in cotton-based agro-ecosystem. *Ph.D.Thesis, Quaid-i-Azam University, Islamabad Pakistan.*

Angell, S. M., Davies, C. & Baulcombe, D. C. (1996). Cell-to-cell movement of Potato virus X is associated with a change in the size-exclusion limit of plasmodesmata in trichome cells of *Nicotiana clevelandii*. *Virology 216*, 197-201.

Bachmair, A., Becker, F., Masterson, R. V. & Schell, J. (1990). Perturbation of the ubiquitin system causes leaf curling, vascular tissue alterations and necrotic lesions in a higher plant. *EMBO Journal 9*, 4543-9.

Baulcombe, D. (2004). RNA silencing in plants. *Nature 431*, 356-363.

Baulcombe, D. (2005). RNA silencing. *Trends in Biochemical Sciences 30*, 290-293.

Bayne, E. H., 1, D. V. R., 2, S. Y. M., 2 & Baulcombe, a. D. C. (2005). Cell-to-cell movement of Potato Potexvirus X is dependent on suppression of RNA silencing. *Plant Journal 44*, 471–482.

Beck, D. L., Guilford, P. J., Voot, D. M., Andersen, M. T. & Forster, R. L. S. (1991). Triple gene block proteins of white clover mosaic potexvirus are required for transport. *Virology 183*, 695–702.

Behjatnia, S. A., Dry, I. B. & Rezaian, M. A. (1998). Identification of the replication-associated protein binding domain within the intergenic region of tomato leaf curl geminivirus. *Nucleic Acids Research 26*, 925-931.

Bisaro, D. M., Hamilton, W. D. O., Coutts, R. H. A. & Buck, K. W. (1982). Molecular cloning and characterisation of the two components of tomato golden mosaic virus. *Nucleic Acids Research 10*, 4913- 4922.

Bock, K. R., Guthrie, E. J. & Woods, R. D. (1974). Purification of maize streak virus and its relationship to viruses associated with streak diseases of sugar cane and *Panicum maximum*. *Annals of Applied Biology 77*, 289-296.

Boulton, M. I., Steinkellner, H., Donson, J., Markham, P. G., King, D. I. & Davies, J. W. (1989). Mutational analysis of the virion-sense genes of maize streak virus. *Journal of General Virology 70*, 2309-2323.

Briddon, R. W., Bedford, I. D., Tsai, J. H. & Markham, P. G. (1996). Analysis of the nucleotide sequence of the treehopper-transmitted geminivirus, tomato pseudo-curly top virus, suggests a recombinant origin. *Virology 219*, 387-394.

Briddon, R. W., Bull, S., Mansoor, S., Amin, I. & Markham, P. G. (2002). Universal primers for the PCR-mediated amplification of DNA β; a satellite associated with some monopartite begomoviruses. *Molecular Biotechnology 20*, 315-318.

Briddon, R. W., Bull, S. E., Amin, I., Mansoor, S., Bedford, I. D., Dhawan, P., Rishi, N., Siwatch, S. S., Abdel-Salam, A. M. & Markham, P. G. (2003). Diversity of DNA β; a satellite molecule associated with some monopartite begomoviruses. *Virology 312*, 106-121.

Briddon, R. W., Bull, S. E., Amin, I., Mansoor, S., Bedford, I. D., Rishi, N., S., S., Zafar, Y., Abdel-Salam, A. M. & Markham, P. G. (2004). Diversity of DNA 1: a satellite-like molecule associated with monopartite begomovirus-DNA β complexes. *Virology 324*, 462 - 474.

Briddon, R. W. & Mansoor, S. Beta DNA satellite. Cotton Leaf Curl Disease. In: Mahy, B. W. J. & van Regenmortel M. H. V. *Encyclopedia of Virology,* Volume 5 pp *0-0* Elsevier Oxford; 2008, pp.

Briddon, R. W., Mansoor, S., Bedford, I. D., Pinner, M. S. & Markham, P. G. (2000). Clones of cotton leaf curl geminivirus induce symptoms atypical of cotton leaf curl disease. *Virus Genes 20*, 17-24.

Briddon, R. W., Mansoor, S., Bedford, I. D., Pinner, M. S., Saunders, K., Stanley, J., Zafar, Y., Malik, K. A. & Markham, P. G. (2001). Identification of DNA components required for induction of cotton leaf curl disease. *Virology 285*, 234-243.

Briddon, R. W. & Markham, P. G. (2000). Cotton leaf curl virus disease. *Virus Research 71*, 151-159.

Briddon, R. W., Watts, J., Markham, P. G. & Stanley, J. (1989). The coat protein of beet curly top virus is essential for infectivity. *Virology 172*, 628-633.

Brigneti, B., Voinnet, O., Li, W.-X., Ji, L.-H., Ding, S.-W. & Baulcombe, D. C. (1998). Viral pathogenicity determinants are suppressors of transgene silencing in *Nicotiana benthamiana*. *EMBO Journal 17*, 6739-6746.

Bull, S. E., Tsai, W.-S., Briddon, R. W., Markham, P. G., Stanley, J. & Green, S. K. (2004). Diversity of begomovirus DNA β satellites of non-malvaceous plants in east and south east Asia. *Archives of Virology 149*, 1193 - 1200.

Carrington, J. C. & Ambros, V. (2003). Role of microRNAs in plant and animal development. *Science 301*, 336-338.

Chellappan, P., Masona, M. V., Vanitharani, R., Taylor, N. J. & Fauquet, C. M. (2005). Broad spectrum resistance to ssDNA viruses associated with transgene-induced gene silencing in cassava. *Plant Molecular Biology 56*, 601-611.

Cui, X.-F., Li, Y.-Q., Hu, D.-W. & Zhou, X.-P. (2005a). Expression of a begomoviral DNA β gene in transgenic *Nicotiana plants* induced abnormal cell division. *Journal of Zhejiang University Science 6B*, 83-86.

Cui, X., Li, G., Wang, D., Hu, D. & Zhou, X. (2005b). A begomovirus DNA β-encoded protein binds DNA, functions as a suppressor of RNA silencing, and targets the cell nucleus. *Journal of Virology 79*, 10764-10775.

Cui, X., Tao, X., Xie, Y., Fauquet, C. M. & Zhou, X. (2004). A DNA β associated with Tomato yellow leaf curl China virus is required for symptom induction. *Journal of Virology 78,* 13966-13974.

Dry, I., Krake, L. R., Rigden, J. E. & Rezaian, M. A. (1997). A novel subviral agent associated with a geminivirus: the first report of a DNA satellite. *Proceedings of the National Academy of Sciences, USA 94*, 7088--7093.

Duffy, S. & Holmes, E. C. (2007). Multiple introductions of the Old World begomovirus, Tomato yellow leaf curl virus (TYLCV), into the New World. *Applied and Environmental Microbiology 73*, 7114-7117.

Dunoyer, P., Lecellier, C.-H., Parizotto, E. A., Himber, C. & Voinnet, O. (2004). Probing the microRNA and small interfering RNA pathways with virus-encoded suppressors of RNA silencing. *Plant Cell 16*, 1235-1250.

Eini, O., Dogra, S., Selth, L. A., Dry, I. B., Randles, J. W. & Rezaian, M. A. (2009). Interaction with a host ubiquitin-conjugating enzyme is required for the pathogenicity of a geminiviral DNA beta satellite. *Molecular Plant Microbe Interactions 22*, 737-46.

Goodman, R. M. (1977a). Infectious DNA from a whitefly-transmitted virus of *Phaseolus vulgaris*. *Virology 266*, 54-55.

Goodman, R. M. (1977b). Single-stranded DNA genome in a whitefly-transmitted plant virus. *Virology 83*, 171-179.

Gopal, P., Kumara, P. P., Sinilal, B., Jose, J., Yadunandama, A. K. & Ushaa, R. (2006). Differential roles of C4 and C1 in mediating suppression of post-transcriptional gene silencing: evidence for transactivation by the C2 of Bhendi yellow vein mosaic virus, a monopartite begomovirus. *Virus Research 123*, 9-18.

Grafi, G., Burnett, R. J., Helentjaris, T., Larkins, B. A., DeCaprio, J. A., Sellers, W. R. & Kaelin, W. G. (1996). A maize cDNA encoding a member of the retinoblastoma protein family: involvement in endoreduplication. *Proceedings of the National Academy of Sciences USA 93*, 8962-8967.

Grigoras, I., Timchenko, T., Katul, L., Grande-Perez, A., Vetten, H. J. & Gronenborn, B. (2009). Reconstitution of authentic nanovirus from multiple cloned DNAs. *Journal of Virology 83*, 10778-10787.

Gronenborn, G. (2004). Nanoviruses: genome organisation and protein function. *Veterinary Microbiology 98*, 103-110.

Guo, W., Yang, X., Xie, Y., Cui, X. & Zhou, X. (2009). Tomato yellow leaf curl Thailand virus-[Y72] from Yunnan is a monopartite begomovirus associated with DNA β. *Virus Genes 38*, 328-33.

Haber, S., Howarth, A. J. & Goodman, R. M. (1983). Restriction map and southern analysis of the bean golden mosaic virus genome. *Virology 129*, 469.

Hamilton, W. D. O., Bisaro, D. M. & Buck, K. W. (1982). Identification of novel DNA forms in tomato golden mosaic virus infected tissue. evidence for a two component genome. *Nucleic Acids Research 19*, 4901-4912.

Hamilton, W. D. O., Stein, V. E., Coutts, R. H. A. & Buck, K. W. (1984). Complete nucleotide sequence of the infectious cloned DNA components of tomato golden mosaic virus: potential coding regions and regulatory sequences. *EMBO Journal 3*, 2197-2205.

Hanley-Bowdoin, L., Settlage, S. B., Orozco, B. M., Nagar, S. & Robertson, D. (1999). Geminviruses: models for plant DNA replication, transcription, and cell cycle regulation. *Critical Reviews in Plant Sciences 18*, 71-106.

Hanley-Bowdoin, l., Settlage, S. B. & Robertson, D. (2004). Reprogramming plant gene expression: a prerequisite to geminivirus DNA replication. *Molecular Plant Pathology 5*, 149-156.

Hannon, G. J. (2002). RNA interference. *Nature 418*, 244-251.

Harrison, B. D., Barker, H., Bock, K. R., Guthrie, E. J., Meredith, G. & Atkinson, M. (1977). Plant viruses with circular single-stranded DNA. *Nature 270*, 760-762.

Harrison, B. D., Zhou, X., Otim-Nape, G. W., Liu, Y. & Robinson, D. J. (1997). Role of a novel type of double infection in the geminivirus-induced epidemic of severe cassava mosaic in Uganda. *Annals of Applied Biology 131*, 437-448.

Herrera-Valencia, V. A., Dugdale, B., Harding, R. M. & Dale, J. L. (2007). Mapping the 5' ends of banana bunchy top virus gene transcripts. *Archives of Virology 152*, 615-620.

Hormuzdi, S. G. & Bisaro, D. M. (1993). Genetic analysis of beet curly top virus: evidence for three virion sense genes involved in movement and regulation of single- and double-stranded DNA levels. *Virology 193*, 900-909.

Howell, S. H. (1984). Physical structure and genetic organisation of the genome of maize streak virus (Kenyan isolate). *Nucleic Acids Research 12*, 7359-7375.

Hu, J.-M., Fu, H.-C., Lin, C.-H., Su, H.-J. & Yeh, H.-H. (2007). Reassortment and concerted evolution in Banana bunchy top virus genomes. *Journal of Virology 81*, 1746–1761.

Huang, C., Xie, Y. & Zhou, X. (2009). Efficient virus-induced gene silencing in plants using a modified geminivirus DNA1 component. *Plant Biotechnology Journal 7*, 254-65.

Inouye, T. & Osaki, T. (1980). The first record in the literature of the possible plant virus disease that appeared in Manyoshu, a Japanese classic anthology, as far back as the time of the 8th century. *Annals of the Phytopathological Society of Japan. 46*, 49-50.

Ito, T., Kimbara, J., Sharma, P. & Ikegami, M. (2009). Interaction of tomato yellow leaf curl virus with diverse betasatellites enhances symptom severity. *Archives of Virology 154*, 1233-1239.

Jeske, H. (2009). Geminiviruses. *Current Topics in Microbiology and Immunology 331*, 185-226.

Kalinina, N. O., Rakitina, D. V., Solovyev, A. G., Schiemann, J. & Morozov, S. Y. (2002). RNA helicase activity of the plant virus movement proteins encoded by the first gene of the triple gene block. *Virology 296*, 321–329.

Khan, A. J., Idris, A. M., Al-Saady, N. A., Al-Mahruki, M. S., Al-Subhi, A. M. & Brown, J. K. (2008). A divergent isolate of tomato yellow leaf curl virus from Oman with an associated DNA beta satellite: an evolutionary link between Asian and the Middle Eastern virus-satellite complexes. *Virus Genes 36*, 169-76.

Kheyr-Pour, A., Bendahmane, M., Matzeit, V., Accotto, G. P., Crespi, S. & Gronenborn, B. (1991). Tomato yellow leaf curl virus from Sardinia is a whitefly-transmitted monopartite geminivirus. *Nucleic Acids Research 19*, 6763-6769.

Kon, T., Sharma, P. & Ikegami, M. (2007). Suppressor of RNA silencing encoded by the monopartite tomato leaf curl Java begomovirus. *Archives of Virology 152*, 1273-1282.

Krake, L. R., Rezaian, M. A. & Dry, I. B. (1998). Expression of the tomato leaf curl geminivirus C4 gene produces viruslike symptoms in transgenic plants. *Molecular Plant-Microbe Interactions 11*, 413-417.

Kumar, P., Ushaa, R., Zrachya, A., Levyb, Y., Spanov, H. & Gafni, Y. (2006). Protein–protein interactions and nuclear trafficking of coat protein and beta C1 protein associated with Bhendi yellow vein mosaic disease. *Virus Research 122*, 127–136.

Lakatos, P. L., Fischer, S., Lakatos, L., Gal, I. & Papp, J. (2006). Current concept on the pathogenesis of inflammatory bowel disease-crosstalk between genetic and microbial factors: pathogenic bacteria and altered bacterial sensing or changes in mucosal integrity take "toll"? *World Journal of Gastroenterology 12*, 1829-41.

Latham, J. R., Saunders, K., Pinner, M. S. & Stanley, J. (1997). Induction of cell division by beet curly top virus gene *C4*. *Plant Journal 11*, 1273-1283.

Li, Z. H., Xie, Y. & Zhou, X. P. (2005). Tobacco curly shoot virus DNAP is not necessary for infection but intensifies symptoms in a host-dependent manner. *Phytopathology 95*, 902-908.

Liu, H., Boulton, M. I. & Davies, J. W. (1997). Maize streak virus coat protein binds single- and double-stranded DNA *in vitro*. *Journal of General Virology 78*, 1265-1270.

Mansoor, S., Amin, I., Iram, S., Hussain, M., Zafar, Y., Malik, K. A. & Briddon, R. W. (2003a). The breakdown of resistance in cotton to cotton leaf curl disease in Pakistan. *New Disease Reports*, [http://www.bspp.org.uk/ndr/].

Mansoor, S., Briddon, R. W., Bull, S. E., Bedford, I. D., Bashir, A., Hussain, M., Saeed, Zafar, M. Y., Malik, K. A., Fauquet, C. & Markham, P. G. (2003b). Cotton leaf curl disease is associated with multiple monopartite begomoviruses supported by single DNA β. *Archives of Virology 148*, 1969-1986.

Mansoor, S., Khan, S. H., Bashir, A., Saeed, M., Zafar, Y., Malik, K. A., Briddon, R. W., Stanley, J. & Markham, P. G. (1999). Identification of a novel circular single-stranded DNA associated with cotton leaf curl disease in Pakistan. *Virology 259*, 190-199.

Mayo, M. A., Leibowitz, M. J., Palukaitis, P., Scholthof, K.-B. G., Simon, A. E., Stanley, J., Taliansky, M., 2005. Satellites. In: Fauquet, C. M., Mayo, M.A., Maniloff, J., Desselberger, U., Ball, L.A. (Eds.), *VIIIth report of the international taxonomy of viruses*. Elsevier/Academic Press, London; pp. 1163-1169.

McConnell, J. R. & Barton, M. K. (1998). Leaf polarity and meristem formation in Arabidopsis. *Development 125*, 2935–2942.

McConnell, J. R., Emery, J., Eshed, Y., Bao, N., Bowman, J. & Barton, M. K. (2001). Role of PHABULOSA and PHAVOLUTA in determining radial patterning in shoots. *Nature 411*, 709–713.

Millner, P. A. (1995). The auxin signal. *Current Opinion in Cell Biology 7*, 224-223.

Morris, B. A. M., Richardson, K. A., Haley, A., Zhan, X. & Thomas, J. E. (1992). The nucleotide sequence of the infectious cloned DNA component of tobacco yellow dwarf virus reveals features of geminiviruses infecting monocotyledonous plants. *Virology 187*, 633-642.

Mugiira, R. B., Liu, S. S. & Zhou, X. (2008). Tomato yellow leaf curl virus and Tomato leaf curl Taiwan virus invade south-east coast of China. *Journal of Phytopathology 156*, 217-221.

Mullineaux, P. M., Donson, J., Morris-Krsinich, B. A. M., Boulton, M. I. & Davies, J. W. (1984). The nucleotide sequence of maize streak virus DNA. *EMBO Journal 3*, 3063-3068.

Mumford, D. L. (1974). Purification of curly top virus. *Phytopathology 64*, 136-142.

Murphy, F. A., Fauquet, C. M., Bishop, D. H. L., Ghabrial, S. A., Jarvis, A. W., Martelli, G. P., Mayo, M. A. & Summers, M. D. (1995). *Virus Taxonomy: Sixth Report of the International Committee on Taxonomy of Viruses*. Wien, New York: Springer-Verlag. pp 585.

Nagar, S., Pedersen, T. J., Carrick, K. M., Hanley-Bowdoin, L. & Robertson, D. (1995). A geminivirus induces expression of a host DNA synthesis protein in terminally differentiated plant cells. *Plant Cell 7*, 705-719.

Nahid, N., Amin, I., Mansoor, S., Rybicki, E. P., van der Walt, E. & Briddon, R. W. (2008). Two dicot-infecting mastreviruses (family Geminiviridae) occur in Pakistan. *Archives of Virology 153*, 1441-1451.

Navot, N., Pichersky, E., Zeidan, M., Zamir, D. & Czosnek, H. (1991). Tomato yellow leaf curl virus: a whitefly-transmitted geminivirus with a single genomic component. *Virology 185*, 151-161.

Nawaz-ul-Rehman, M. S. & Fauquet, C. M. (2009). Evolution of geminiviruses and their satellites. *FEBS Letters 583*, 1825-32.

Nawaz-Ul-Rehman, M. S., Mansoor, S., Briddon, R. W. & Fauquet, C. M. (2009). Maintenance of an Old World betasatellite by a New World helper begomovirus and possible rapid adaptation of the betasatellite. *Journal of Virology 83*, 9347-9355.

Pooma, W., Gillette, W. K., Jeffrey, J. L. & Petty, I. T. D. (1996). Host and viral factors determine the dispensability of coat protein for bipartite geminivirus systemic movement. *Virology 218*, 264-268.

Qazi, J., Amin, I., Mansoor, S., Iqbal, M. J. & Briddon, R. W.. (2007). Contribution of the satellite encoded gene beta C1 to cotton leaf curl disease symptoms. *Virus Research 128*, 135-139.

Qian, Y. & Zhou, X. (2005). Pathogenicity and stability of a truncated DNAβ associated with *Tomato yellow leaf curl China virus*. *Virus Research 109*, 159-163.

Qing, L. & Zhou, X. (2009). Trans-replication of, and competition between, DNA β satellites in plants inoculated with Tomato yellow leaf curl China virus and Tobacco curly shoot virus. *Phytopathology 99*, 716-720.

Raj, S. K., Khan, M. S., Snehi, S. K. & Roy, R. K. (2007). Yellow vein netting of Bimili jute (Hibiscus cannabinus L.) in India caused by a strain of Tomato leaf curl New Delhi virus containing DNA b. *Australasian Plant Disease Notes 2*, 45-47.

Rigden, J. E., Dry, I. B., Mullineaux, P. M. & Rezaian, M. A. (1993). Mutagensis of the virion-sense open reading frames of tomato leaf curl virus. *Virology 193*, 1001-1005.

Rochester, D. E., DePaulo, J. J., Fauquet, C. M. & Beachy, R. N. (1994). Complete nucleotide sequence of the geminivirus tomato yellow leaf curl virus, Thailand isolate. *Journal of General Virology 75*, 477-485.

Rouhibakhsh, A. & Malathi, V. G. (2005). Severe leaf curl disease of cowpea – a new disease of cowpea in northern India caused by *Mungbean yellow mosaic India virus* and a satellite DNA β. *New Disease Reports*, 7 [http://www.bspp.org.uk/ndr/].

Saeed, M. (2008). Limitations observed in the use of agroinoculation for geminivirus research. *Virus Genes 37*, 434-435

Saeed, M., Behjatania, S. A. A., Mansoor, S., Zafar, Y., Hasnain, S. & Rezaian, M. A. (2005). A Single complementrary-sense transcript 0f a geminiviral DNA β satellite is determinant of pathogenicity. *Molecular Plant Microbe Interaction 18*, 7-14.

Saeed, M., Mansoor, S., Rezaian, M. A., Briddon, R. W. & Randles, J. W. (2008). Satellite DNA beta overrides the pathogenicity phenotype of the C4 gene of tomato leaf curl virus but does not compensate for loss of function of the coat protein and V2 genes. *Archives of Virology 153*, 1367-72.

Saeed, M., Zafar, Y., Randles, J. W. & Rezaian, M. A. (2007). A monopartite begomovirus-associated DNA beta satellite substitutes for the DNA B of a bipartite begomovirus to permit systemic infection. *Journal of General Virology 88*, 2881-9.

Saunders, K., Bedford, I. D., Briddon, R. W., Markham, P. G., Wong, S. M. & Stanley, J. (2000). A novel virus complex causes *Ageratum* yellow vein disease. *Proceedings of the National Academy of Sciences, USA 97*, 6890-6895.

Saunders, K., Bedford, I. D., Yahara, T. & Stanley, J. (2003). The earliest recorded plant virus disease. *Nature 422*, 831.

Saunders, K., Briddon, R. W. & Stanley, J. (2008). Replication promiscuity of DNA-β satellites associated with monopartite begomoviruses; deletion mutagenesis of the Ageratum yellow vein virus DNA-{beta} satellite localizes sequences involved in replication. *Journal of General Virology 89*, 3165-72.

Saunders, K., Norman, A., Gucciardo, S. & Stanley, J. (2004). The DNA β satellite component associated with ageratum yellow vein disease encodes an essential pathogenicity protein (βC1). *Virology 324*, 37-47.

Saunders, K., Salim, N., Mali, V. R., Malathi, V. G., Briddon, R. W., Markham, P. G. & Stanley, J. (2002). Characterisation of Sri Lankan cassava mosaic virus and Indian cassava mosaic virus: evidence for acquisition of a DNA B component by a monopartite begomovirus. *Virology 293*, 63-74.

Saunders, K. & Stanley, J. (1999). A nanovirus-like component associated with yellow vein disease of *Ageratum conyzoides*: evidence for interfamilial recombination between plant DNA viruses. *Virology 264*, 142-152.

Schneider, L. R. (1969). Satellite-like particle of tobacco ringspot virus that resembles tobacco ringspot virus. *Science 166*, 1627-1629.

Settlage, S. B., See, R. G. & Hanley-Bowdoin, L. (2005). Geminivirus C3 protein: replication enhancement and protein interactions. *Journal of Virology 79*, 9885-9895.

Simon, A. E., Roossinck, M. J. & Havelda, Z. (2004). Plant virus satellite and defective interfering RNAs: new paradigms for a new century. *Annual Review of Phytopathology 42*, 415-437.

Simons, J. N. & Coe, D. M. (1958). Transmission of pseudo-curly top virus in Florida by a treehopper. *Virology 6*, 43-48.

Stanley, J. (1983). Infectivity of the cloned geminivirus genome requires sequences from both DNAs. *Nature 305*, 643-645.

Stanley, J., Bisaro, D. M., Briddon, R. W., Brown, J. K., Fauquet, C. M., Harrison, B. D., Rybicki, E. P. & Stenger, D. C. (2005). Geminiviridae. In: *Virus Taxonomy, VIIIth Report of the International Committee on Taxonomy of Viruses*. Edited by Fauquet, C. M., Mayo, M. A., Maniloff,. J., Desselberger U. & Ball L. A. London: Elsevier/Academic Press; pp. 301-326.

Stanley, J. & Gay, M. G. (1983). Nucleotide sequence of cassava latent virus DNA. *Nature 301*, 260-262.

Stanley, J. & Latham, J. R. (1992). A symptom variant of beet curly top geminivirus produced by mutation of open reading frame C4. *Virology 190*, 506-509.

Stanley, J., Markham, P. G., Callis, R. J. & Pinner, M. S. (1986). The nucleotide sequence of an infectious clone of the geminivirus beet curly top virus. *EMBO Journal 5*, 1761-1767.

Sunter, G. & Bisaro, D. M. (1992). Transactivation of geminivirus AR1 and BR1 gene expression by the viral AL2 gene product occurs at the level of transcription. *Plant Cell 4*, 1321-1331.

Sunter, G., Stenger, D. C. & Bisaro, D. M. (1994). Heterologous complementation by geminivirus AL2 and AL3 genes. *Virology 203*, 203-210.

Tan, P. H. N., Wong, S. M., Wu, M., Bedford, I. D., Saunders, K. & Stanley, J. (1995). Genome organization of ageratum yellow vein virus, a monopartite whitefly-transmitted geminivirus isolated from a common weed. *Journal of General Virology 76*, 2915-2922.

Tao, O. & Zhou, X. (2004). A modified viral satellite DNA that suppresses gene expression in plants. *Plant Journal 38*, 850-860.

Tao, X., Qing, L. & Zhou, X. (2004). Function of A-rich region in DNA β associated with *Tomato yellow leaf curl China virus*. *Chinese Science Bulletin 49*, 1490-1493.

Teotia, P. S., Mukherjee, M. S. & Mishra, N. S. (2008). Fine tuning of auxin signaling by miRNAs. *Physiology and Molecular Biology of Plants 14*, 81-90.

Tesoriero, L. & Azzopardi, S. (2006). Tomato yellow leaf curl virus in Australia. *Primefacts, June 2006*.

Timchenko, T., Katul, L., Sano, Y., de Kouchkovsky, F., Vetten, H. J. & Gronenborn, B. (2000). The master Rep concept in nanovirus replication: identification of missing genome components and potential for natural genetic reassortment. *Virology 274*, 189-195.

Ulmasov, T., Hagen, G. & Guilfoyle, T. J. (1999). Activation and repression of transcription by auxin-response factors. *Proceedings of the National Academy of Sciences, USA 96*, 5844–5849.

Vanitharani, R., Chellappan, P., Pita, J. S. & Fauquet, C. M. (2004). Differential roles of AC2 and AC4 of cassava geminiviruses in mediating synergism and suppression of posttranscriptional gene silencing. *Journal of Virology 78*, 9487-9498.

Varsani, A., Shepherd, D. N., Dent, K., Monjane, A. L., Rybicki, E. P. & Martin, D. P. (2009). A highly divergent South African geminivirus species illuminates the ancient evolutionary history of this family. *Virology Journal 6*, 36.

Vetten, H. J., Chu, P. W. G., Dale, J. L., Harding, R., Hu, J., Katul, L., Kojima, M., Randles, J. W., Sano, Y. & Thomas, J. E. (2004). *Nanoviridae*. In: *Virus Taxonomy, VIIIth Report of the International Committee on Taxonomy of Viruses*. Edited by Fauquet, C. M., Mayo, M. A., Maniloff,. J., Desselberger U. & Ball L. A. London: Elsevier/Academic Press; pp. 343-352

Voinnet, O., Lederer, C. & Baulcombe, D. C. (2000). A viral movement protein prevents spread of the gene silencing signal in Nicotiana benthamiana. *Cell. 103*, 157-167.

Were, H. K., Takeshita, M., Furuya, N. & Takanami, Y. (2005). Molecular characterization of a new begomovirus infecting honeysuckle in Kobe, Japan. *Journal of Faculty of Agriculture., Kyushu University. 50*, 61– 71.

Wong, S. M., Swanson, M. M. & Harrison, B. D. (1993). A geminivirus causing vein yellowing of Ageratum conyzoides in Singapore. *Plant Patholology. 42*, 137- 139.

Wright, E. A., Heckel, T., Groenendijk, J., Davies, J. W. & Boulton, M. I. (1997). Splicing features in maize streak virus virion- and complementary-sense gene expression. *Plant Journal 12*, 1285-1297.

Xie, Q., Suárez-López, P. & Gutiérrez, C. (1995). Identification and analysis of a retinoblastoma binding motif in the replication protein of a plant DNA virus: requirement for efficient viral DNA replication. *EMBO Journal 14*, 4073-4082.

Yang, J.-Y., Iwasaki, M., Machida, C., Machida, Y., Zhou, X. & Chua, N.-H. (2008). βC1, the pathogenicity factor of TYLCCNV, interacts with AS1 to alter leaf development and suppress selective jasmonic acid responses. *Genes & Developments 22*, 2564–2577.

Yang, Y., Ding, B., Baulcombe, D. C. & Verchot, J. (2000). Cell-to-Cell Movement of the 25K Protein of Potato virus X Is Regulated by Three Other Viral Proteins. *Molecular Plant-Microbe Interactions 13*, 599–605.

Yazdi, H. R., Heydarnejad, J. & Massumi, H. (2008). Genome characterization and genetic diversity of beet curly top Iran virus: a geminivirus with a novel nonanucleotide. *Virus Genes 36*, 539-45.

Zhou, X., Xie, Y., Tao, X., Zhang, Z., Li, Z. & Fauquet, C. M. (2003). Characterization of DNA β associated with begomoviruses in China and evidence for co-evolution with their cognate viral DNA-A. *Journal of General Virology 84*, 237-247.

INDEX

D

E

J

K

L

M

R

S

W

X

Y

Z